P9-DIY-404

Study Guide
for Stewart's

SINGLE VARIABLE
CALCULUS
Early Transcendentals

FOURTH EDITION

RICHARD ST. ANDRE
Central Michigan University

 **Brooks/Cole**
Thomson Learning.

Pacific Grove • Albany • Belmont • Boston • Cincinnati • Johannesburg • London • Madrid
Melbourne • Mexico City • New York • Scottsdale • Singapore • Tokyo • Toronto

Assistant Editor: *Carol Ann Benedict*
Marketing Team: *Caroline Croley, Debra Johnston*
Production Coordinator: *Dorothy Bell*

Cover Illustration: *dan clegg*
Printing and Binding: *Webcom Limited*

COPYRIGHT © 1999 by Brooks/Cole
A division of Thomson Learning
The Thomson Learning logo is a trademark used herein under license.

For more information, contact:
BROOKS/COLE
511 Forest Lodge Road
Pacific Grove, CA 93950 USA
www.brookscole.com

All rights reserved. No part of this work may be reproduced, transcribed or used in any form or by any means—graphic, electronic, or mechanical, including photocopying, recording, taping, Web distribution, or information storage and/or retrieval systems—without the prior written permission of the publisher.

For permission to use material from this work, contact us by
web: www.thomsonrights.com
fax: 1-800-730-2215
phone: 1-800-730-2214

Printed in Canada

10 9 8 7 6 5 4 3 2 1

ISBN 0-534-36820-4

Preface

This Study Guide is designed to supplement the first eleven chapters of Calculus Early Transcendentals, 4th edition, by James Stewart. It may also be used with Single Variable Calculus Early Transcendentals, 4th edition. If you later go on to multivariable calculus, you will want to obtain the multivariable volume of this Study Guide.

Your text is well written in a very complete and patient style. This Study Guide is not intended to replace it. You should read the relevant sections of the text and work problems, lots of problems. Calculus is learned by doing; it is not a spectator sport.

This Study Guide captures the main points and formulas of each section and provides short, concise questions that will help you understand the essential concepts. Every question has an explained answer. The pages are perforated so that you can detach them. The two-column format allows you to cover the answer portion of a question while working on it and then uncover the given answer to check you solution. Working in this fashion leads to greater success than simply perusing the solutions. Students have found this Study Guide helpful for reviewing for examinations.

Technology can be a tool to help understand calculus concepts by drawing intricate graphs, solving or approximating difficult equations, doing numerically intense work, and performing symbolic manipulations. We have included additional question in sections called "Technology Plus" at the end of each chapter. For these questions you should use a graphing calculator or computer with a computer algebra system software program.

As a quick check of your understanding of a section work a page of On Your Own questions is located toward the back of the Study Guide. These are all multiple choice type questions — the kind you might see on an exam in a large-sized calculus class. You are "on your own" in the sense that an answer, but no solution, is provided for each question.

I hope that you find this Study Guide helpful in understanding the concepts and solving the exercises in Calculus, 4th edition.

Richard St. Andre

Table of Contents

11 Infinite Sequences and Series

On Your Own

Answers to On Your Own

Please cut page down center line.
Use the half page to cover the right
column while you work in the left.

Chapter 1 — Functions and Models

"WE HAVE REASON TO BELIEVE BINGLEMAN IS AN IRRATIONAL NUMBER HIMSELF."

© 1999 by Sidney Harris.

Section 1.1 Four Ways to Represent a Function

The notion of a function is central to all of calculus. This section defines what a function is as well as several concepts that go along with functions.

Concepts to Master

A. Definition of a function; Function value; Domain; Range; Independent variable; Dependent variable

B. Four ways to describe a function; Tables of values; graphs

C. Implied domain

D. Piecewise Defined Function

E. Symmetry (even and odd functions)

F. Increasing; Decreasing

Summary and Focus Questions

Page 12

A. A function f is a rule that associates pairs of numbers: to each number x in a set (the *domain*) there is associated another real number, denoted $f(x)$, the *value* of f at x.

The set of all images ($f(x)$ values) is the *range of f*. The variable x is the *independent* variable and $y = f(x)$ is the dependent variable.

Example: The function S, "to each nonnegative number, associate its square root," has domain $[0, \infty)$ and rule of association given by:

$$S(x) = \sqrt{x}$$

The value of S at $x = 25$ is 5 because $S(25) = \sqrt{25} = 5$. S has no value at -7 since $\sqrt{-7}$ is not a real number; -7 is not in the domain of S.

1) Sometimes, Always, Never:
Both of the points $(2, 5)$ and $(2, 7)$ could be pairs of a function f.

Never. $f(2)$ cannot be both 5 and 7.

2) True or False:
$x = y^2$ defines y as a function of x.

False, since $(4, 2)$ and $(4, -2)$ satisfy the equation. Note that x is a function of y.

3) A track is in the shape of two semicircular ends and straight sides as in the figure. Find the distance around the track as a function of the radius.

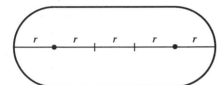

Each straight side is $3r$ and each semicircular end is πr. The distance d is $d = 2(3r + \pi r)$ or $d = (6 + 2\pi)r$.

Page
14

B. The four ways to describe the rule of association for a function are

   1. in words, a verbal description

   2. with a table of values which lists various domain values and corresponding range values

   3. with the *graph* of $y = f(x)$; that is, all points (x, y) in the Cartesian plane that make $y = f(x)$ true. The y coordinate of a point on the graph of function f is the value of f associated with the x coordinate.

   4. with an explicit algebraic formula or equation.

Example: Here are four ways to represent the square root function.

   1. "to each nonnegative number, associate its square root."

   2.

x	0	1	2	3	...
y	0	1	$\sqrt{2}$	$\sqrt{3}$	...

   (Note: For this function with infinite domain, we can only list a few x values with corresponding y values and hope that the meaning is clear.)

   3.

   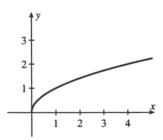

   (Here, again, with an infinite domain, we can only represent a portion of the function.)

   4. $S(x) = \sqrt{x}$, where $x \geq 0$.

4) Write 3 other ways to describe each function:

a) "to each real number, associate one more than its square"

i) By table of values

x	y
−3	10
−2	5
−1	2
0	1
1	2
2	5
3	10

(A partial list of values)

ii)

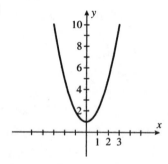

iii) $f(x) = x^2 + 1$

b)

x	1	2	3	4	5
y	5	4	3	2	1

i) "to each of $x = 1, 2, 3, 4, 5$ associate 6 minus the value of x"

ii)

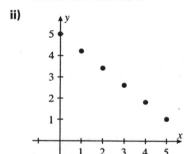

Note: Unless we know the domain is all real numbers, we should not "connect the dots" to form a line.

iii) For $x = 1, 2, 3, 4, 5$, let $f(x) = 6 - x$.

5) Given the graph of f below, what is the value of f for each of these?

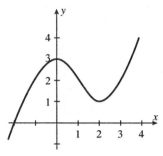

a) $f(2)$

1 (since the pair $(2, 1)$ is on the graph)

b) $f(0)$

3

c) $f(3)$

2

d) $f(f(0))$ (Hint: First determine $f(0)$.)

Since $f(0)$ is 3, $f(f(0)) = f(3) = 2$.

e) $f(7)$

7 is not pictured on the x-axis; we can only surmise from the trend of the graph that $f(7)$ will be a large positive number.

6) Sketch the graph of $f(x) = \sqrt{x^2 - 16}$.

Here are several computed values: $(4, 0)$, $(5, 3)$, $(6, \sqrt{20})$, $(-4, 0)$, $(-5, 3)$.
Squaring both sides of $\sqrt{x^2 - 16}$ gives
$y^2 = x^2 - 16$, or $x^2 - y^2 = 16$. The graph is the top half of a hyperbola.

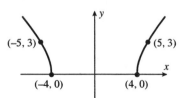

Note: About all we can do now to sketch graphs is plot points and recognize certain graphs from the form of the equation.

C. The domain of a function f, if unspecified, is understood to be the largest set of the real numbers x for which $f(x)$ exists. Thus $f(x) = \dfrac{1}{\sqrt{x-5}}$ has implied domain $(5, \infty)$. The range of $f(x)$ is $(0, \infty)$.

7) Find the domain of $f(x) = \sqrt{x^2 - 16}$.

For $f(x)$ to exist, $x^2 - 16 \geq 0$. Thus $(x - 4)(x + 4) \geq 0$. The solution set is $x \leq -4$ or $x \geq 4$. The domain is $(-\infty, -4] \cup [4, \infty)$.

8) Find the domain of $f(x) = \dfrac{1}{|x| + 1}$.

Since $|x| \geq 0$ for all x, $|x| + 1$ is never 0. The domain is all real numbers.

9) What is the domain and range of the function whose graph is given below?

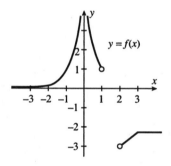

The domain is $(-\infty, 0) \cup (0, 1) \cup (2, \infty)$. This is all the x values for which there is a point (x, y) on the graph. The range is $(-3, -2) \cup (0, \infty)$—the set of all y-coordinates used in the graph.

Page
18

D. A piecewise defined function divides its domain into parts and uses a different formula on each part. For example:

$$f(x) = \begin{cases} \dfrac{1}{x} & \text{if } x \geq 1 \\ 3x & \text{if } x < 1 \end{cases}$$

has a domain of all real numbers split up into $(-\infty, 1)$ and $[1, \infty)$. Use the appropriate rule to draw the graph over each part.

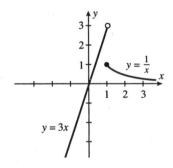

10) The phone company charges a flat rate for local calls of $12.60 per month with an additional charge of $0.05 per call after 55 calls. Express the monthly bill as a function of the number of local calls.

Let x = the number of local calls. Then

$$M(x) = \begin{cases} 12.60 & \text{if } x \in [0, 55] \\ 12.60 + 0.05(x - 55) & \text{if } x \in (55, \infty) \end{cases}$$

Page
20

E. Sometimes one part of a graph of a function will be a reflected image of another part:

$y = f(x)$ is *even* means $f(x) = f(-x)$ for all x in the domain.

An even function is symmetric about the y-axis.

$y = f(x)$ is *odd* means $f(x) = -f(-x)$ for all x in the domain.

An odd function is symmetric with respect to the origin.

11) If $(7, 3)$ is on the graph of an odd function, what other point must also be on the graph?

$(-7, -3)$. This is because
$3 = f(7) = -f(-7)$, so $f(-7) = -3$.

12) Is $f(x) = 10 - 2x$ even or odd?

$f(x) = 10 - 2x$
$f(-x) = 10 - 2(-x) = 10 + 2x$
$f(x)$ is *not even* since $f(x) \neq f(-x)$.
$-f(-x) = -(10 + 2x) = -10 - 2x$
$f(x)$ is *not odd* since $f(x) \neq -f(-x)$.

F. A function f is *increasing* on an interval I if for all $x_1, x_2, \in I$, $x_1 < x_2$ implies $f(x_1) < f(x_2)$. In other words, as x gets larger, so does $f(x)$. f is *decreasing* means that $x_1 < x_2$ implies $f(x_1) > f(x_2)$ for all $x_1, x_2, \in I$.

Page 21

13. Given the graph of f below, on what intervals is f increasing? decreasing?

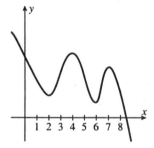

f is decreasing on each of these intervals: $(-\infty, 2], [4, 6], [7, \infty)$. f is increasing on $[2, 4]$ and on $[6, 7]$.

14) Is $f(x) = x^2 + 4x$ increasing on $[-1, 2]$?

Yes. The graph of f is given below. As x increases from -1 to 2, $f(x)$ increases from -3 to 12.

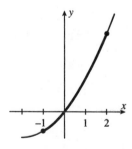

Section 1.2   **Mathematical Models**

This section describes several types of functions that are useful in modeling (describing) real-world phenomenon.

Concepts to Master

A. Characteristics of these modeling functions: linear, polynomial, power, root, rational, algebraic, trigonometric, exponential, logarithmic, and transcendental

B. Fit a math model to a real-world phenomenon; Interpolation; Extrapolation

Summary and Focus Questions

Page 25

A. A mathematical model is a mathematical description for a real-world phenomenon. For example, $C = \frac{5}{9}(F - 32)$ models the relationship between temperature measured in degrees Fahrenheit and degrees Celsius.

Example: If a hardware store makes a profit of \$1.38 for each latch hasp it sells, then $P(x) = 1.38x$ models the profit on selling x such hasps.

Many types of functions can serve as models; some include:

Type	Description	Examples
Linear	$f(x) = mx + b$	$f(x) = 5x + 6$ $f(x) = -\pi x + 3$
Polynomial	$f(x) = a_n x^n + \ldots + a_1 x + a_0$ (n is the degree)	$f(x) = 6x^2 + 3x + 1$ $f(x) = -5x^{12} + 7x$
Power	$f(x) = x^a$ $a = -1, 1, 2, 3, 4, \ldots$	$f(x) = x^7$ $f(x) = x^{-1}$
Root	$f(x) = x^a$ where $a = \frac{1}{2}, \frac{1}{3}, \frac{1}{4}, \ldots$	$f(x) = \sqrt{x},$ $f(x) = x^{1/10}$
Rational	$f(x) = \frac{P(x)}{Q(x)},$ where P, Q are polynomials	$f(x) = \frac{7x+6}{x^2 + 10}$ $f(x) = \frac{3}{2-x}$
Algebraic	$f(x)$ is obtained from polynomials, using algebra $(+, -, \cdot, /,$ roots, composition$)$	$f(x) = \sqrt{16 - x^2}$ $f(x) = \sqrt[3]{\frac{x^3}{x+1}}$

Trigonometric	sine, cosine, tangent, ...	$f(x) = \tan x$ $f(x) = \cos 3x$
Exponential	$f(x) = a^x$ (for $a > 0$)	$f(x) = 2^x$ $f(x) = (1/3)^x$
Logarithmic	$f(x) = \log_a(x)$ (for $a > 0$)	$f(x) = \ln x$ $f(x) = \log_4(x + 1)$

A transcendental function is a non-algebraic function. (Examples: trigonometric, exponential and logarithmic functions are transcendental functions.)

1) True or False:
$f(x) = x^3 + 6x + x^{-1}$ is a polynomial.

False, (because of the x^{-1} term).

2) True or False:
$f(x) = x + \dfrac{1}{x}$ is rational.

True. $\left(x + \dfrac{1}{x} = \dfrac{x^2 + 1}{x}\right)$.

3) True or False:
Every rational function is an algebraic function.

True.

4) $f(x) = 6x^{10} + 12x^4 + x^{11}$ has degree _____.

11.

5) True or False:
$f(x) = 2^\pi$ is an exponential function.

False. 2^π is constant (approximately 8.825).

Page
27

B. Success fitting models to given problems takes practice and some insight into how variables and data values are related. Sometimes the graph of a phenomenon given in tabular data will look rather like a certain type of model. For example, if the data appears to be periodic, a trigonometric function will likely be involved; something that seems to grow very rapidly may involve an exponential or a polynomial of a high degree.

Example: A tenant pays a non-refundable $150 deposit upon leasing an apartment and $450 per month. Rent is paid in advance. A linear model for the total amount paid to the landlord after x months is:

$$C(x) = 150 + 450x.$$

Interpolation for a model $y = f(x)$ is the estimation of a functional value for an x value between the observed values of x. *Extrapolation* is the prediction of a functional value outside the observed values.

Example: The amounts of sulfur dioxide particulates in the air around Rose City is given in the table at the right. A scatter plot of the data suggests that a model for the amount of particulates after x days is a quadratic (polynomial of degree 2).

day	parts per million
1	48
2	32
3	28
5	27
6	33
7	47

For the missing day, $x = 4$, we might estimate (interpolate) that the amount of sulfur dioxide is about $\frac{28 + 27}{2} = 27.5$ parts per million. Extrapolation much beyond the 7 days would involve a great deal of uncertainty.

6) Which type of function model is associated with each scatter diagram?

a)

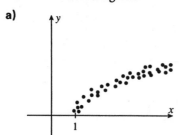

Logarithmic, such as $y = \log_2 x$.

b)

Rational, such as $y = 10 + \dfrac{1}{x} = \dfrac{10x + 1}{x}$.

c)

Linear, such as $y = \dfrac{1}{2}x + 5$.

d)

There does not appear to be a functional relationship between x and y.

7) To determine whether pennies change weight as they age (from factors such as wear from handling) the average weight of a sample of 100 pennies was calculated for each of the years 1985, 86, 88, 90, 91 and 92.

year	Average weight in sample
1985	1.912
1986	1.913
1988	1.914
1990	1.914
1991	1.915
1992	1.915

a) Draw a scatter diagram and determine whether the model seems linear.

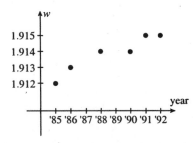

The model appears to be linear.

b) Assuming a linear model, find the model that passes through the 1985 and 1991 data points.

The slope of the line through (85, 1.912) and (91, 1.915) is

$$\frac{1.915 - 1.912}{91 - 85} = \frac{.003}{6} = .0005$$

The equation is

$$y - 1.915 = .0005(x - 91)$$

or

$$y = .0005x + 1.8695.$$

c) Estimate the average weight of a 1987 penny and a 1995 penny.

For 1987, $y = .0005(87) + 1.8695 = 1.913$
For 1995, $y = .0005(95) + 1.8695 = 1.917$

d) Does estimating the weight of a 2025 penny seem reasonable?

No. Pennies in the future will probably have the same weight as new ones have today.

8) Karen lives 10 miles from work in a busy city. Sketch a model of the time it takes her to drive home from work versus the times she leaves the office (4:00, 4:30, 5:00, 5:30, 6:00, 6:30). Is the model linear?

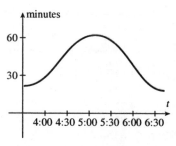

We have assumed that it takes her about 20 minutes in light traffic and nearly an hour in heavy rush hour traffic to drive from work to home. The model is not linear.

Section 1.3 New Functions from Old Functions

This section describes several ways to create "complicated" functions from simple ones. Knowing how to do this will come in handy when you learn formulas for derivatives of simple functions and need to calculate derivatives of more complex functions.

Concepts to Master

A. Translations and stretchings of functions; Reflecting

B. Combining functions (sum, difference, product, quotient, and composition)

Summary and Focus Questions

Page 38

A. Adding or subtracting a constant to either the dependent or the independent variable shifts the graph up, down, left, or right.

For $c > 0$:

Function	Effect on graph of $y = f(x)$
$y = f(x - c)$	shift *right* by c units
$y = f(x + c)$	shift *left* by c units
$y = f(x) + c$	shift *upward* by c units
$y = f(x) - c$	shift *downward* by c units

Multiplying either the dependent or the independent variable by a nonzero constant will stretch or compress a graph.

For $c > 1$:

Function	Effect on graph of $y = f(x)$
$y = cf(x)$	Stretch *vertically* by a factor of c.
$y = \frac{1}{c}f(x)$	Compress *vertically* by a factor of c.
$y = f(cx)$	Compress *horizontally* by a factor of c.
$y = f\left(\frac{x}{c}\right)$	Stretch *horizontally* by a factor of c.

Changing the sign of either the dependent or independent variable will reflect the graph about an axis.

Function	Effect on graph of $y = f(x)$
$y = -f(x)$	Reflect about x-axis.
$y = f(-x)$	Reflect about y-axis.

These operations may be combined to produce the graph of a "complex" function from the graph of a simpler function.

1) Given the graph below, sketch each:

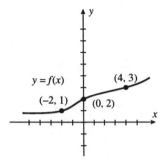

a) $y = 2f(x)$

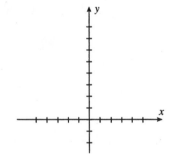

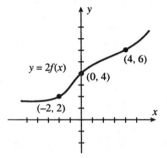

b) $y = f(x) - 2$

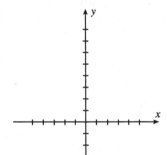

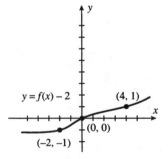

c) $y = f(-x)$

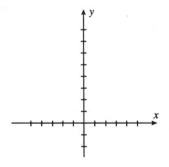

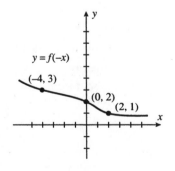

d) $y = \frac{1}{2}f(x) + 3$

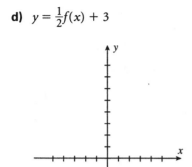

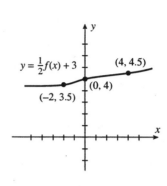

$y = \frac{1}{2}f(x) + 3$ (4, 4.5)

(0, 4)

(−2, 3.5)

Page 42

B. Given two functions $f(x)$ and $g(x)$:

1) The *sum* $f + g$ associates to each x the number $f(x) + g(x)$.

2) The *difference* $f - g$ associates to each x the number $f(x) - g(x)$.

3) The *product* fg associates to each x the number $f(x) \cdot g(x)$.

4) The *quotient* $\frac{f}{g}$ associates to each x the number $\frac{f(x)}{g(x)}$, provided that $g(x) \neq 0$.

For example, the sum, difference, product, and quotient of $f(x) = x^2$ and $g(x) = 2x - 8$ are:

$$(f + g)(x) = x^2 + 2x - 8$$
$$(f - g)(x) = x^2 - 2x + 8$$
$$fg(x) = x^2(2x - 8)$$
$$\frac{f}{g}(x) = \frac{x^2}{2x - 8}, \text{ for } x \neq 4.$$

The function $h(x) = \frac{x^2(x+1)}{2+x}$ may be written as $h = \frac{fg}{k}$ where $f(x) = x^2$, $g(x) = x + 1$, and $k(x) = 2 + x$.

For $f(x)$ and $g(x)$, the *composite* $f \circ g$ is the function that associates to each x the same number that f associates to $g(x)$, that is:

$$(f \circ g)(x) = f(g(x))$$

To compute $f \circ g(x)$, first compute $g(x)$, then compute f of that result.

For example, if $f(x) = x^2 + 1$ and $g(x) = 3x$, then for $x = 5$:

$$(f \circ g)(5) = f(g(5)) = f(3(5)) = f(15) = 15^2 + 1 = 226.$$

Finally, to determine the components f and g of a composite function, $h(x) = (f \circ g)(x)$, requires an examination of which function is applied first.

For example, for $h(x) = \sqrt{x + 5}$, we may use $g(x) = x + 5$ (it is applied first, before the square root) and $f(x) = \sqrt{x}$. Then

$$h(x) = \sqrt{x + 5} = \sqrt{g(x)} = f(g(x)).$$

2) Find $f + g$ and $\dfrac{f}{g}$ for $f(x) = 8 + x$, $g(x) = \sqrt{x}$. What is the domain of each?

$(f + g)(x) = 8 + x + \sqrt{x}$ with domain $= [0, \infty)$. $\dfrac{f}{g}(x) = \dfrac{8 + x}{\sqrt{x}}$ with domain $= (0, \infty]$.

3) Write $f(x) = \dfrac{2\sqrt{x + 1}}{x^2(x + 3)}$ as a combination of simpler functions.

There are many answers to this question. One is $f = \dfrac{hk}{mn}$ where $h(x) = 2$, $k(x) = \sqrt{x + 1}$, $m(x) = x^2$, $n(x) = x + 3$.

4) Find $f \circ g$ where:

a) $f(x) = x - 2x^2$ and $g(x) = x + 1$

$f \circ g(x) = f(g(x)) = f(x + 1)$
$= (x + 1) - 2(x + 1)^2.$

b) $f(x) = \sin x$ and $g(x) = \sin x$

$f \circ g(x) = f(\sin x) = \sin(\sin x)$
This is not the same as $\sin^2(x)$.

5) Rewrite $h(x) = (x^2 + 3x)^{2/3}$ as a composition, $f \circ g$.

Let $f(x) = x^{2/3}$ and $g(x) = x^2 + 3$.
$f \circ g(x) = f(g(x)) = f(x^2 + 3x)$
$= (x^2 + 3x)^{2/3}.$

Thus $h = f \circ g$.

Section 1.4 Graphing Calculators and Computers

Concepts to Master

Determining an appropriate viewing rectangle.

Summary and Focus Questions

A *viewing rectangle* for a function $y = f(x)$ is a rectangle:

$$\{(x, y) | a \le x \le b, c \le y \le d\}$$

corresponding to a calculator or computer screen within which a portion of the graph of $y = f(x)$ is displayed. Selecting the correct window coordinates can be tricky—choose a, b, c, and d so that the display brings out the most salient features of the graph.

Example: Choose an appropriate viewing window for $f(x) = 12 - x^2$.

Since the graph will have no y values greater than 12 ((0, 12) is a high point—a salient feature), $d = 13$ is a good choice. The graph is symmetric about the y-axis and crosses the x-axis (other salient features) between 3 and 4 and between -4 and -3. Good choices for a and b are $a = -5$ and $b = 5$. $f(5) = -13$, so $c = -15$ will work. The graph and viewing window are shown below:

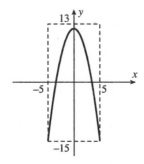

window: $[-5, 5]$ by $[-15, 13]$

1) Find an appropriate viewing rectangle for each:

a) $y = f(x)$ whose graph is below:

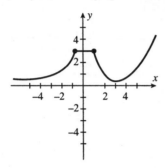

Your choices may vary. A good viewing window is $-4 \le x \le 5$, $-1 \le y \le 5$, or $[-4, 5]$ by $[-1, 5]$.

2) Is $[-50, 50]$ by $[-100, 100]$ an appropriate viewing rectangle for $y = \sin x$?

No. Since $-1 \le y \le 1$, the y scale is too large. Also, since sine is periodic, only a few periods need to be displayed. A window such as $[-12, 12]$ by $[-2, 2]$ would be more appropriate.

Section 1.5 Exponential Functions

An exponential function contains a variable in an exponent. This section defines exponential functions and gives some of their properties. Exponential functions may be used to model many types of phenomenon, including population growth and decay.

Concepts to Master

A. Definition, properties and graph of an exponential function
B. Model growth and decay with an exponential function
C. Definition of e

Summary and Focus Questions

Page
56

A. Let a be a positive constant. The *exponential function* with base a is $f(x) = a^x$ and is defined as follows:

Type of value of x	Definition of a^x
n, a positive integer	$a^n = a \cdot a \cdot a ... (n \text{ times})$
0	$a^0 = 1$
$-n$ (n, a positive integer)	$a^{-n} = \frac{1}{a^n}$
$\frac{p}{q}$, a rational number	$a^{p/q} = \sqrt[q]{a^p}$
x, irrational	a^x is approximately a^r, where r is a rational number near a.

The graph of $y = a^x$ is:

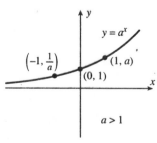

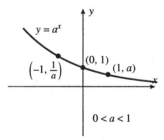

Properties of exponentials include:

$$a^{x+y} = a^x a^y \qquad a^{x-y} = \frac{a^x}{a^y} \qquad a^{xy} = (a^x)^y \qquad (ab)^x = a^x b^x$$

1) By definition, $2^{4/3} =$ _____.

$\sqrt[3]{2^4}$

2) If x is a rational number near $\sqrt{3}$, then 3^x is an approximation of _____.

$3^{\sqrt{3}}$

3) Graph $y = (0.7)^x$.

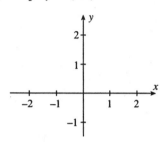

4) True or False:

   a) $(x + y)^a = x^a + y^a$

False.

   b) $x^{a-b+c} = \dfrac{x^a x^c}{x^b}$

True.

5) For large values of x, which is larger, $100x^4$ or $\dfrac{4^x}{100}$?

For $x \geq 15$, $\dfrac{4^x}{100} > 100x^4$.

Page
59

B. Exponential functions can be used to model growth and decay phenomenon.

Example: A population of tribles is known to double its size in 60 days. To what size has an initial population of 100 grown to after 1 year (360 days)?

If $P(t)$ is the population after t days,

$$P(0) = 100$$
$$P(60) = 200 = 100 \cdot 2$$
$$P(120) = 400 = 100 \cdot 2^2$$

and in general $P(t) = 100 \cdot 2^{t/60}$.

Thus $P(360) = 100 \cdot 2^{360/60} = 100 \cdot 2^6 = 6{,}400$.

6) The half life of a certain radioactive isotope is 100 years. To what size has a 12 mg sample disintegrated after 500 years?

Let $P(t)$ be the mass after t years. We are given $P(0) = 12$. $P(100) = 6 = 12 \cdot 2^{-1}$. $P(200) = 3 = 12 \cdot 2^{-2}$ and, in general, $P(t) = 12 \cdot 2^{-t/100}$.

$P(500) = 12 \cdot 2^{-500/100} = \dfrac{12}{2^5} = \dfrac{3}{8}$ mg.

7) The price of Macrosoft stock at the close of each day for a week is given in this table.

Mon	Tue	Wed	Thu	Fri
2.30	3.96	6.57	11.33	19.26

If x represents the number of days since Monday, which exponential model best describes the stock prices?

a) $y = 2.3^x$

b) $y = 2.3(1.7)^x$

c) $y = (2.3)^{-x}$

d) $y = (2.3)(1.7)^{-x}$

b) Here is a table of values for $y = 2.3(1.7)^x$:

x	0	1	2	3	4
y	2.30	3.91	6.65	11.30	19.21

**Page
61**

C. The number *e* is that real number so that the slope of the tangent line to $y = e^x$ at $(0, 1)$ is exactly one. In chapter 3 we will see that the value of *e*, correct to five decimal places is

$$e \approx 2.71828.$$

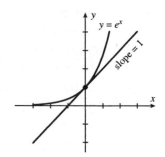

8) For the graphs below, which is the graph of $y = e^x$ and which is $y = 3^x$?

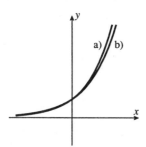

Since $e < 3$, **a)** is the graph of $y = 3^x$ and **b)** is graph of $y = e^x$.

9) Simplify to *e* to a power:

$$\frac{(e^x \cdot e^3)^2}{e}$$

$$\frac{(e^x \cdot e^3)^2}{e} = \frac{(e^{x+3})^2}{e} = \frac{e^{2x+6}}{e} = e^{2x+6-1}$$
$$= e^{2x+5}.$$

10) If the slope of the tangent line to $y = a^x$ at $(0, 1)$ is 1, then $a =$ _____.

e

Section 1.6   Inverse Functions and Logarithms

If y is a function of x described in any one of the several ways (such as by rule, by table of values or by graph), then there are times when the same description may be used with the roles of x and y reversed, so that x is a function of y. For example, if C is the circumference of a circle and r is the radius then $C = 2\pi r$ may be used to define C as a function of r; the same equation may also be used to define r as a function of C. The two functions are inverses of each other. This section defines inverse functions and sets conditions when the inverse exists. It also defines $y = \log_a x$, where $a > 0$, as the inverse of the exponential function $y = a^x$.

Concepts to Master

A. One-to-one functions; Horizontal line test; Inverse of a function

B. Definition, properties and graph of $y = \log_a x$

Summary and Focus Questions

A. A function f is *one-to-one* means that for all x_1, x_2 in the domain, if $x_1 \neq x_2$, then $f(x_1) \neq f(x_2)$.

For example, $f(x) = 2x + 5$ is one-to-one because if $x_1 \neq x_2$ then $2x_1 + 5 \neq 2x_2 + 5$. The function $g(x) = x^2$ is not one-to-one $(4 \neq -4, \text{ yet } g(4) = 4^2 = (-4)^2 = g(-4))$.

Increasing and decreasing functions are one-to-one.

Horizontal Line Test
A function f is one-to-one iff no horizontal line intersects the graph of f more than once.

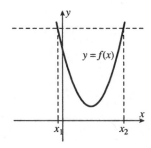

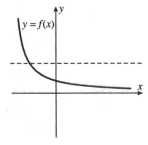

Horizontal Line Test fails
$x_1 \neq x_2$, but $f(x_1) = f(x_2)$

Horizontal Line Test succeeds

If f is one-to-one with domain A and range B then the *inverse function* f^{-1} has domain B and range A, and

$$f^{-1}(b) = a \text{ iff } f(a) = b.$$

Therefore:

a pair (a, b) satisfies $y = f(x)$ iff the pair (b, a) satisfies $y = f^{-1}(x)$.

a pair (a, b) is in the table that defines f iff the pair (b, a) is in the table that defines f^{-1}.

a point (a, b) is on the graph of f if and only if (b, a) is on the graph of f^{-1}.

These inverse properties hold

$f^{-1}(f(x)) = x$ for all x in the domain of f

$f(f^{-1}(x)) = x$ for all x in the domain of f^{-1}

To find the rule for $y = f^{-1}(x)$,

(1) solve the equation $y = f(x)$ for x.

(2) interchange x and y.

1) Is $f(x) = \sin x$ one-to-one?

No. For example, $0 \neq \pi$, yet $\sin 0 = 0 = \sin \pi$.

2) Is the function $y = g(x)$ given by this table one-to-one?

x	g(x)
1	10
2	9
3	8
4	7
5	8

No, $3 \neq 5$ but both $g(3)$ and $g(5)$ are 8.

3) Is the graph below the graph of a function with an inverse function?

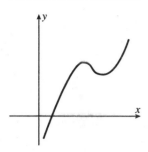

No. The function is not one-to-one.

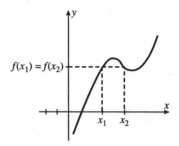

$x_1 \neq x_2$, but $f(x_1) = f(x_2)$.

4) Given the following graph of f, sketch the graph of f^{-1} on the same axis.

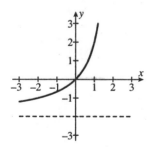

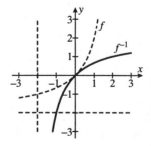

5) If f^{-1} is a function and $f(5) = 8$, then $f^{-1}(8) =$ _____.

5.

{}

28   Chapter 1   *Functions and Models*
{}

6) Write the inverse function of
$f(x) = 6x + 30$.

(1) $y = 6x + 30$

$y - 30 = 6x$

$x = \frac{1}{6}y - 5$

(2) $y = \frac{1}{6}x - 5$, so $f^{-1}(x) = \frac{1}{6}x - 5$.

7) True, False:
If f is one-to-one with domain A and range
B, then $f(f^{-1}(x)) = x$ for all $x \in A$.

False. $f(f^{-1}(x)) = x$ for all $x \in \underline{B}$.

Page
68

B. $\log_a x = y$ means $a^y = x$. This definition requires $x > 0$, since $a^y > 0$ for
all y. You should think of $\log_a x$ as the exponent that you put on a to get x.
The *logarithmic function* $y = \log_a x$ is defined as the inverse of the
exponential $y = a^x$. Thus, the function $y = \log_a x$ has these properties:

$\log_a(a^x) = x$

$a^{\log_a x} = x$, for $x > 0$

$\log_a(xy) = \log_a x + \log_a y$

$\log_a x^c = c \log_a x$

For $a > 1$, the graph of $y = \log_a x$
is given at the right.

In the special case where $a = e$, we use the
notation $\ln x$; therefore,
$\ln x$ means $\log_e x$.
$\ln e = \log_e e = 1$.
$\log_a x = \frac{\ln x}{\ln a}$.

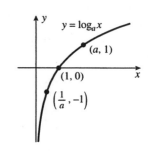

8) $\log_4 8 = $ _____.

Let $y = \log_4 8$.
Then $4^y = 8$
$2^{2y} = 2^3$
$2y = 3$
$y = \frac{3}{2}$.

9) True or False:
   a) $\log_a x^3 = 3 \log_a x$.

True.

b) $\log_2(10) = \log_2(-5) + \log_2(-2)$.

False. $\log_2(-5)$ and $\log_2(-2)$ are not defined.

c) $\ln(a - b) = \ln a - \ln b$.

False.

d) $\ln e = 1$.

True. In $\log_e$ notation this is $\log_e e = 1$.

e) $\ln\left(\frac{a}{b}\right) = -\ln\left(\frac{b}{a}\right)$.

True.

10) Write $2 \ln x + 3 \ln y$ as a single logarithm.

$2 \ln x + 3 \ln y = \ln x^2 + \ln y^3 = \ln(x^2 y^3)$.

11) For $x > 0$, simplify using properties of logarithms: $\log_2\left(\frac{4x^3}{\sqrt{2}}\right)$.

$\log_2\left(\frac{4x^3}{\sqrt{2}}\right) = \log_2 4x^3 - \log_2 \sqrt{2}$

$= \log_2 4 + \log_2 x^3 - \log_2 2^{1/2}$

$= \log_2 4 + 3 \log_2 x - \frac{1}{2} \log_2 2$

$= 2 + 3 \log_2 x - \frac{1}{2}(1)$

$= \frac{3}{2} + 3 \log_2 x.$

12) Solve for x:

a) $e^{x+1} = 10$.

By definition, this means $\ln 10 = x + 1$, so $x = \ln 10 - 1$.

b) $\ln(x + 5) = 2$.

By definition, this means $e^2 = x + 5$, so $x = e^2 - 5$.

c) $2^{3x} = 7$.

Take $\log_2$ of both sides:
$\log_2 2^{3x} = \log_2 7$
$3x = \log_2 7$
$x = \frac{1}{3} \log_2 7.$

13) $\log_2 10 = \dfrac{\ln \underline{\quad}}{\ln \underline{\quad}}$.

$\log_2 10 = \dfrac{\ln 10}{\ln 2}$

14) The magnitude of an earthquake on the Richter scale is $\log_{10}\left(\frac{I}{S}\right)$, where I is the intensity of the quake and S is the intensity of a "standard" quake. The Mexico City and San Francisco quakes were 6.4 and 7.1 on the Richter scale, respectively. How many times more powerful was the San Francisco quake than the Mexico City quake?

If the Mexico City quake intensity is I_M and the San Francisco quake is I_S then we need to find a value k for which $I_S = k \cdot I_M$.

$$7.1 = \log_{10}\left(\frac{I_S}{S}\right) = \log_{10}\left(\frac{kI_M}{S}\right)$$
$$= \log_{10} k + \log_{10}\left(\frac{I_M}{S}\right)$$
$$= \log_{10} k + 6.4.$$

Thus $7.1 = \log_{10} k + 6.4$

$\log_{10} k = 0.7$

$k = 10^{0.7} \approx 5$ times as powerful.

Technology Plus for Chapter 1

1) Use a graphing device to graph
$f(x) = ax - x^2$, for $a = 0, 1, 2, 3, 4, 5$.
Use $-2 \leq x \leq 7$.

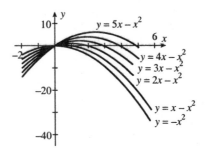

2) Use a graphing device to graph $f(x)$, $2f(x)$,
$4f(x)$, $-f(x)$, $-2f(x)$, and $-4f(x)$ where
$f(x) = 2x^2 - 4x$.
Use $-1 \leq x \leq 3$.

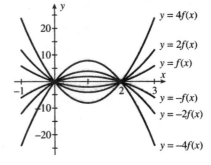

3) Solve graphically:
$3x = \tan x$ for $x \in \left[0, \frac{\pi}{2}\right]$.

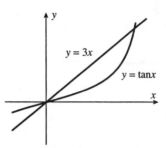

Graph both $y = 3x$ and $y = \tan x$ for $0 \le x \le \frac{\pi}{2}$. One solution is $x = 0$. Zooming in on the other, we see that $x \approx 1.32$.

4) Graph $f(x) = x^3 - x^2 + 0.335x + 2$ in the window $[-1, 3]$ by $[-1, 10]$. Is the function one-to-one?

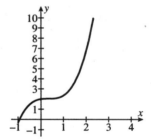

In this window it is difficult to tell. Zooming in near $x = 0.33$ we see that f is indeed always increasing and hence one-to-one.

Chapter 2 — Limits and Rates of Change

© 1999 by Sidney Harris.

Section 2.1   The Tangent and Velocity Problems

This section illustrates two types of problems to which calculus may be applied: finding the slope of a line tangent to a curve and finding the velocity of a moving object. These are two examples of instantaneous rate of change found as a "limit" of average rates of change.

Concepts to Master

A. Slope of secant line to the graph of a function; Slope of tangent line

B. Interpretation of slopes as velocities

Summary and Focus Questions

Page 85

A. A *secant line* at $P(a, b)$ for the graph of $y = f(x)$ is a line joining P and another point Q also on the graph. If Q has coordinates (c, d), then the slope of the secant line is

$$\frac{d - b}{c - a}$$

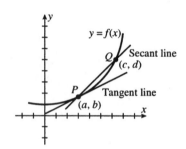

A *tangent line* to a graph of a function touches the graph once much like a tangent to a circle.

To find the slope of the tangent line at P, we select points Q closer and closer to P; the slopes of the secant lines become better and better estimates of the slope of the tangent line at P. Finally, our guess for the slope of the tangent line is the value that the slopes of the secant lines seem to be approaching as points Q get closer and closer to P.

Once we have determined the slope m of the tangent line to the curve, then using the point P as a point on the line, the equation of the tangent line is

$$y - b = m(x - a).$$

1) Let P be the point $(3, 10)$ on the graph of $y = f(x)$ and let Q be the differing points given in the table. Find the slopes of the secant lines PQ.

Q	Slope of PQ
(6, 18)	
(5, 15)	
(4, 12.3)	
(3.5, 11.1)	
(3.1, 10.21)	

Q	Slope of PQ
(6, 18)	$\frac{18 - 10}{6 - 3} = 2.6$
(5, 15)	$\frac{15 - 10}{5 - 3} = 2.5$
(4, 12.3)	$\frac{12.3 - 10}{4 - 3} = 2.3$
(3.5, 11.1)	$\frac{11.1 - 10}{3.5 - 3} = 2.2$
(3.1, 10.21)	$\frac{10.21 - 10}{3.1 - 3} = 2.1$

2) Let $P(1, 5)$ be a point on the graph of $f(x) = 6x - x^2$. Let $Q(x, 6x - x^2)$ be on the graph. Find the slope of the secant line PQ for each given x value for Q.

x	f(x)	Slope of PQ
3		
2		
1.5		
1.01		

x	f(x)	Slope of PQ
3	9	$\frac{9 - 5}{3 - 1} = 2$
2	8	$\frac{8 - 5}{2 - 1} = 3$
1.5	6.75	$\frac{6.75 - 5}{1.5 - 1} = 3.5$
1.01	5.0399	$\frac{5.0399 - 5}{1.01 - 1} = 3.99$

3) Use your answer to question 2 to guess the slope of the tangent to $f(x)$ at P.

The values appear to be approaching 4.

4) What is the equation of the tangent line at P in question 2?

$$y - 5 = 4(x - 1).$$

B. If $f(x)$ is interpreted as the distance an object is located from the origin along an x-axis at time x, then:

    a) the slope of the secant line from P to Q is the *average velocity* from P to Q.

    b) the slope of the tangent line at P is the *instantaneous velocity* at P.

Thus, using the distance function:
    calculating average velocity is performed in the same manner as
    calculating the slope of a secant line, and
    calculating instantaneous velocity is performed in the same manner as
    calculating the slope of a tangent line.

5) If a ball is $x^2 + 3x$ feet from the origin at any time x (in seconds), what is the instantaneous velocity when $x = 2$?

At $x = 2$, $f(2) = 2^2 + 3(2) = 10$.
Choose some x values approaching 2:

x	$f(x)$	Slope of Secant
3	18	$\dfrac{18 - 10}{3 - 2} = 8$
2.5	13.75	$\dfrac{13.75 - 10}{2.5 - 2} = 7.5$
2.1	10.71	$\dfrac{10.71 - 10}{2.1 - 2} = 7.1$

We guess the slope of the tangent, and thus the instantaneous velocity, is 7 ft/sec.

6) Let $f(x) = mx + b$, a linear position function. Sometimes, Always, or Never:

The average velocity for $f(x)$ from P to Q is the same as the instantaneous velocity at P.

Always.

7) The distance that a runner is from the starting line is given in this table:

t (seconds)	0	1	2	3	4
d (meters)	0	4	9	16	24

    a) Find the average velocity over the time intervals $[1, 4]$, $[1, 3]$, $[1, 2]$, and $[0, 1]$.

Interval	Average Velocity
$[1, 4]$	$\dfrac{24 - 4}{4 - 1} = \dfrac{20}{3} = 6.67$ m/s
$[1, 3]$	$\dfrac{16 - 4}{3 - 1} = \dfrac{12}{2} = 6$ m/s
$[1, 2]$	$\dfrac{9 - 4}{2 - 1} = \dfrac{5}{1} = 5$ m/s
$[0, 1]$	$\dfrac{4 - 0}{1 - 0} = \dfrac{4}{1} = 4$ m/s

    b) Estimate the instantaneous velocity at $t = 1$.

The intervals $[0, 1]$ and $[1, 2]$ give average velocities of 4 m/s and 5 m/s, respectively. We estimate the instantaneous velocity at $t = 1$ to be about 4.5 m/s, the average of the two average velocities.

Section 2.2 The Limit of a Function

This section introduces the concept of the limit of a function; a more exact definition will come in section 2.4. For now, we estimate limits by calculating functional values or observing trends from a graph. We also look at limits from just one side of the limit point (considering only lesser values of x or greater values of x) and limits that do not exist because the function values get very large.

Concepts to Master

A. Limit of a function; Numerical estimation of a limit; Estimates from a graph

B. Right- and left-hand limits; Relationships between limits and one-sided limits

C. Infinite limits; Vertical asymptotes

Summary and Focus Questions

Page 91

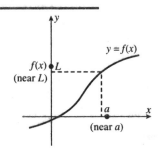

A. $\lim_{x \to a} f(x) = L$ means that as x gets closer and closer to the number a (but not equal to a), the corresponding $f(x)$ values get closer and closer to the number L. For now, ignore the value $f(a)$, if there is one, in determining a limit.

$\lim_{x \to a} f(x) = L$ is rather like a guarantee: if you choose x close enough to a on the x-axis, then $f(x)$ is guaranteed to be close to L on the y-axis.

To say that "$\lim_{x \to a} f(x)$ exists" means that there is some number L such that $\lim_{x \to a} f(x) = L$.

To evaluate $\lim_{x \to a} f(x)$ from the graph of $y = f(x)$, simply observe what value that corresponding $f(x)$ values approach on the y-axis as x values are chosen near a on the x-axis.

A somewhat risky method of calculating $\lim_{x \to a} f(x)$ is to select several values of x near a and observe the pattern of corresponding $f(x)$ values. Then guess the value $f(x)$ is approaching. Your answer may vary depending on the particular values of x used and whether you experience rounding errors from your calculator or computer.

Computer algebras and some calculators have "limit" commands that perform the calculations and make the guesses for you.

1) Answer each using the graph below:

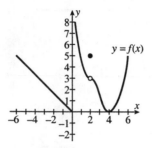

a) $\lim\limits_{x \to -5} f(x) =$

5.

b) $\lim\limits_{x \to 2} f(x) =$

3. (It does not matter that $f(2) = 5$.)

c) $\lim\limits_{x \to 0} f(x) =$

Does not exist.

d) $\lim\limits_{x \to 4} f(x) =$

0.

2) Find $\lim\limits_{x \to 2} |x - 5|$ by making a table of values for x near 2.

x	$f(x) = |x - 5|$
3	2
2.1	2.9
1.99	3.01
2.0003	2.9997

We guess that $\lim\limits_{x \to 2} |x - 5| = 3$.

3) Sketch three different functions, each of which has $\lim\limits_{x \to 4} f(x) = 2$.

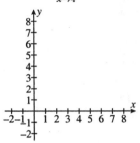

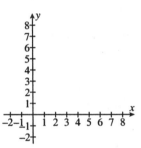

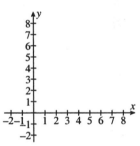

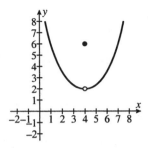

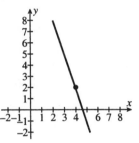

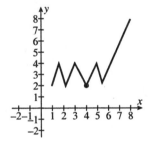

Your answers will probably not look much like the ones above except near the point $(4, 2)$—for x near 4, the $f(x)$ value should be near 2.

4) Estimate $\lim\limits_{x \to 1} \dfrac{1}{1 - x}$.

We make a table of values:

x	$f(x) = \dfrac{1}{1-x}$
1.5	-2
1.01	-100
0.99	100
1.0004	-2500

The values of $f(x)$ do not seem to congregate near a particular value. We guess than $\lim\limits_{x \to 1} \dfrac{1}{1 - x}$ does not exist.

Page 95

B. $\lim\limits_{x \to a^-} f(x) = L$, the *left-hand limit of f at a*, means that $f(x)$ gets closer and closer to L as x approaches a and $x < a$. It is called "left-hand" because it only concerns values of x less than a (to the left of a on the x-axis).

$\lim\limits_{x \to a^+} f(x) = L$, *the right-hand limit*, is a similar concept but involves only values of x greater than a.

If $\lim\limits_{x \to a^-} f(x)$ and $\lim\limits_{x \to a^+} f(x)$ both exist and are the same number L, then $\lim\limits_{x \to a} f(x)$ exists and is L.

If either one-sided limit does not exist, or if they both exist but are different numbers, then $\lim\limits_{x \to a} f(x)$ does not exist.

If $\lim\limits_{x \to a} f(x) = L$, then both one-sided limits exist and are equal L.

5) Answer each using the graph below:

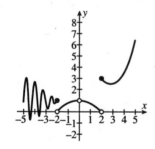

a) $\lim\limits_{x \to -2^-} f(x) = $ _____.

1.

b) $\lim\limits_{x \to -2^+} f(x) = $ _____.

0.

c) $\lim\limits_{x \to -2} f(x) = $ _____.

Does not exist.

d) $\lim\limits_{x \to 2^+} f(x) = $ _____.

3.

e) $\lim\limits_{x \to 2^-} f(x) = $ _____.

0.

f) $\lim_{x \to 2} f(x) = $ _____.

Does not exist.

g) $\lim_{x \to 0^-} f(x) = $ _____.

1.

h) $\lim_{x \to 0^+} f(x) = $ _____.

1.

i) $\lim_{x \to 0} f(x) = $ _____.

1.

6) $\lim_{x \to 3^-} \dfrac{3 - x}{|x - 3|} = $ _____.

1. For x near 3 with $x < 3$, $x - 3 < 0$.
Thus $|x - 3| = -(x - 3) = 3 - x$.
Therefore $\lim_{x \to 3^-} \dfrac{3 - x}{|x - 3|} = \lim_{x \to 3^-} \dfrac{3 - x}{3 - x} = 1$.

7) Find $\lim_{x \to 2} f(x)$, where

$$f(x) = \begin{cases} 3x + 1 & \text{if } x < 2 \\ 8 & \text{if } x = 2 \\ x^2 + 3 & \text{if } x > 2 \end{cases}$$

$\lim_{x \to 2^+} f(x) = \lim_{x \to 2^+} x^2 + 3 = 7$.
$\lim_{x \to 2^-} f(x) = \lim_{x \to 2^-} 3x + 7$.
Since both one-sided limits are 7,
$\lim_{x \to 2} f(x) = 7$.

8) Sometimes, Always, or Never:

If $\lim_{x \to 2} f(x)$ does not exist, then
$\lim_{x \to 2^+} f(x)$ does not exist.

Sometimes. True in the case $f(x) = \dfrac{1}{x - 2}$.
False in the case $f(x) = \dfrac{x - 2}{|x - 2|}$.

9) Sometimes, Always, or Never:

If $\lim_{x \to 2^+} f(x)$ does not exist, then $\lim_{x \to 2} f(x)$
does not exist.

Always.

10) The spherical helium balloons sold at the Four Ring Circus are known to burst when inflated if their diameters reach 80 cm. Describe the volume of a balloon, V, as a function of the diameter (x) and find $\lim\limits_{x \to 80^-} V(x)$.

The volume of a sphere is $\frac{4}{3}\pi(\text{radius})^3$, so

$$V(x) = \frac{4}{3}\pi\left(\frac{x}{2}\right)^3 = \frac{\pi}{6}x^3 \text{ for } 0 \leqslant x < 80.$$

$$\lim_{x \to 80^-} V(x) = \lim_{x \to 80^-} \frac{\pi}{6}x^3$$
$$= \frac{256,000}{3}\pi \approx 268,083 \text{ cm}^3.$$

Page 97

C. Sometimes $\lim\limits_{x \to a} f(x)$ does not exist because as x is assigned values that approach a, the corresponding $f(x)$ values grow larger without bound. In this case we say

$$\lim_{x \to a} f(x) = \infty.$$

Remember that $\lim\limits_{x \to a} f(x) = \infty$ still means $\lim\limits_{x \to a} f(x)$ does not exist, but it fails to exist in this special way.

$\lim\limits_{x \to a} f(x) = -\infty$ means $f(x)$ becomes smaller without bound as x approaches a. As with ordinary limits, for $\lim\limits_{x \to a^+} f(x) = \pm\infty$, consider only $x > a$, and for $\lim\limits_{x \to a^-} f(x) = \pm\infty$, consider only $x < a$.

A vertical asymptote for $y = f(x)$ is a vertical line $x = a$ for which at least one of these limits hold:

$$\lim_{x \to a^-} f(x) = \infty, \qquad \lim_{x \to a^-} f(x) = -\infty, \qquad \lim_{x \to a^+} f(x) = \infty, \qquad \lim_{x \to a^+} f(x) = -\infty.$$

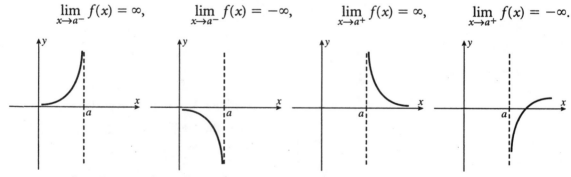

A function may have any number of vertical asymptotes or none at all. Some, like $y = \tan x$, have an infinite number.

11) Evaluate the limits given the graph of
$y = f(x)$ below:

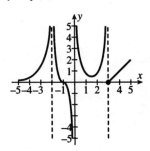

a) $\lim\limits_{x \to 2^+} f(x) =$ _____.

∞.

b) $\lim\limits_{x \to -2^-} f(x) =$ _____.

∞.

c) $\lim\limits_{x \to -2} f(x) =$ _____.

∞.

d) $\lim\limits_{x \to 3^-} f(x) =$ _____.

∞.

e) $\lim\limits_{x \to 3^+} f(x) =$ _____.

0.

f) $\lim\limits_{x \to 3} f(x) =$ _____.

Does not exist.

g) $\lim\limits_{x \to 0^+} f(x) =$ _____.

∞.

h) $\lim\limits_{x \to 0^-} f(x) =$ _____.

$-\infty$.

i) $\lim\limits_{x \to 0} f(x) =$ _____.

Does not exist.

12) $\lim\limits_{x \to 3} \dfrac{1}{|x - 3|} =$ _____.

∞. For x near 3, but not equal to 3, $|x - 3|$ is
a small positive number. Therefore, $\dfrac{1}{|x - 3|}$ is
a large positive number.

13) True or False: If $\lim\limits_{x \to a} f(x) = \infty$, and
$\lim\limits_{x \to a} g(x) = \infty$, then $\lim\limits_{x \to a} (f(x) - g(x)) = 0$.

False. For example, $f(x) = \dfrac{3}{x^2}$ and

$g(x) = \dfrac{2}{x^2}$. Then $\lim\limits_{x \to 0} \dfrac{3}{x^2} = \infty$, $\lim\limits_{x \to 0} \dfrac{2}{x^2} = \infty$,

but $\lim\limits_{x \to 0} \left(\dfrac{3}{x^2} - \dfrac{2}{x^2} \right) = \lim\limits_{x \to 0} \dfrac{1}{x^2} = \infty$ (not 0).

14) $\lim\limits_{x \to 3} \dfrac{x}{3 - x}$

For x near 3, $\dfrac{x}{3 - x}$ is the result of a number
near 3 divided by a number near 0.
Consider one-sided limits: $\lim\limits_{x \to 3^+} \dfrac{x}{3 - x} = -\infty$
because the denominator is negative, while
$\lim\limits_{x \to 3^-} \dfrac{x}{3 - x} = \infty$. Thus, $\lim\limits_{x \to 3} \dfrac{x}{3 - x}$ does not
exist.

15) What are the vertical asymptotes for the
function in question 11?

$x = -2$, $x = 0$ and $x = 3$.

16) Find the vertical asymptotes of
$f(x) = \dfrac{x}{|x| - 1}$ and sketch the graph.

Good places to check are where $f(x)$ is not
defined: at 1 and -1.

$\lim\limits_{x \to 1^+} \dfrac{x}{|x| - 1} = \infty$, and $\lim\limits_{x \to 1^-} \dfrac{x}{|x| - 1} = -\infty$,
so $x = 1$ is a vertical asymptote.

$\lim\limits_{x \to -1^+} \dfrac{x}{|x| - 1} = \infty$ and $\lim\limits_{x \to -1^-} \dfrac{x}{|x| - 1} = -\infty$,
so $x = -1$ is another vertical asymptote.

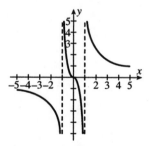

Later we will see that $f(x)$ has horizontal
asymptotes $y = 1$ and $y = -1$.

Section 2.3 Calculating Limits Using Limit Laws

You may perhaps look upon this section with some relief—in the sense that some useful tools are developed for evaluating some limits without resorting to estimations. For example if you know the limits of $f(x)$ and $g(x)$ at a point a then you can add them to get the limit of $(f + g)(x)$ at a. While this will take some of the guesswork out of finding limits, you may need to first use some algebraic manipulation to apply the results of this section.

Concepts to Master

A. Evaluation of limits of rational functions at points in their domain

B. Limit laws (theorems) for evaluating limits

C. Algebraic manipulations prior to using the limit laws

D. Squeeze Theorem

Summary and Focus Questions

Page 102

A. Limits of some functions at some points are easy to find:

> If $f(x)$ is rational and a is in the domain of f, then $\lim\limits_{x \to a} f(x) = f(a)$.

For example, $\lim\limits_{x \to 3} \dfrac{x}{x + 2} = \dfrac{3}{3 + 2} = \dfrac{3}{5}$ since 3 is in the domain of the rational function $f(x) = \dfrac{x}{x + 2}$.

Since polynomials are rational functions, the above applies to them as well, as in $\lim\limits_{x \to 2} (4x^3 - 6x^2 + 10x + 2) = 4(2)^3 - 6(2)^2 + 10(2) + 2 = 30$.

1) Find $\lim\limits_{x \to 6} (4x^2 - 10x + 3)$.

Since $4x^2 - 10x + 3$ is a polynomial, the limit is $4(6)^2 - 10(6) + 3 = 87$.

2) Find $\lim\limits_{x \to 1} \dfrac{x^2 - 6x + 5}{x - 1}$.

Since 1 is not in the domain of $f(x) = \dfrac{x^2 - 6x + 5}{x - 1}$, we cannot substitute $x = 1$ to find this limit. The substitution yields $\dfrac{0}{0}$ which will be a clue that the limit might still exist but must be found by a different method (see question 11). For $x = 1.001$,

$$f(x) = \frac{(1.001)^2 - 6(1.001) + 5}{1.001 - 1} \approx -3.999.$$

The limit appears to be -4.

3) Find $\lim\limits_{x \to 2} \dfrac{x^2 - 6x + 5}{x - 1}$.

Unlike question 2, this one is straightfoward.
$$\lim_{x \to 2} \frac{x^2 - 6x + 5}{x - 1} = \frac{2^2 - 6(2) + 5}{2 - 1} = -3.$$

Page 105

B. The eleven basic limit laws of this section, when applicable, allow you to evaluate complex limits by rewriting them as algebraic combinations of simpler limits as in:

$$\lim_{x \to 3} \sqrt{x^2 + x} = \sqrt{\lim_{x \to 3} (x^2 + x)} = \sqrt{\lim_{x \to 3} x^2 + \lim_{x \to 3} x} = \sqrt{9 + 3} = \sqrt{12}$$

The above was carried out in several steps to demonstrate the limit laws; since $x^2 + x$ is a polynomial, we could have concluded directly that $\lim\limits_{x \to 3} (x^2 + x) = 12$, and therefore that $\lim\limits_{x \to 3} \sqrt{x^2 + x} = \sqrt{12}$.
The limit laws fail when a zero denominator results (see part C for what to do in such cases).

4) True or False:
$$\lim_{x \to a} (f(x)g(x)) = \lim_{x \to a} f(x) \lim_{x \to a} g(x).$$

True, provided that both $\lim\limits_{x \to a} f(x)$ and $\lim\limits_{x \to a} g(x)$ exist.

5) True or False:
$$\lim_{x \to a} f(g(x)) = \lim_{x \to a} f(x) \lim_{x \to a} g(x).$$

False.

6) Evaluate $\lim\limits_{x \to 5} \sqrt[3]{\dfrac{4x + 44}{6x - 29}}$.

$$\lim_{x \to 5} \sqrt[3]{\frac{4x + 44}{6x - 29}} = \sqrt[3]{\lim_{x \to 5} \frac{4x + 44}{6x - 29}} = \sqrt[3]{\frac{64}{1}} = 4.$$

7) Evaluate $\lim\limits_{x \to 1} \dfrac{x^2(6x + 3)(2x - 7)}{(x^3 + 4)(x + 17)}$.

Do not multiply the numerator and denominator out. Applying the limit laws, this limit is
$$\frac{\left(\lim\limits_{x \to 1} x^2\right)\left(\lim\limits_{x \to 1} (6x + 3)\right)\left(\lim\limits_{x \to 1} (2x - 7)\right)}{\left(\lim\limits_{x \to 1} (x^3 + 4)\right)\left(\lim\limits_{x \to 1} (x + 17)\right)}$$
$$= \frac{(1)(9)(-5)}{(5)(18)} = -\frac{1}{2}.$$

8) Find $\lim\limits_{x\to 3} f(x)$ where

$$f(x) = \begin{cases} x^2 & \text{if } x \geq 3 \\ 6x - 4 & \text{if } x < 3 \end{cases}$$

Because $f(x)$ is defined piecewise, we need to consider one-sided limits.

$\lim\limits_{x\to 3^+} f(x) = \lim\limits_{x\to 3} x^2 = 9.$

$\lim\limits_{x\to 3^-} f(x) = \lim\limits_{x\to 3} 6x - 4 = 14.$

Since the one-sided limits do not agree, $\lim\limits_{x\to 3} f(x)$ does not exist.

9) Given the graphs below find $\lim\limits_{x\to 2} (f + g)(x)$.

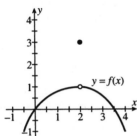

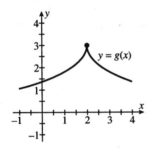

From the graphs, we have $\lim\limits_{x\to 2} f(x) = 1$ and $\lim\limits_{x\to 2} g(x) = 3$. Thus $\lim\limits_{x\to 2} (f + g)(x) = 4$.

10) $\lim\limits_{x\to\infty} 4^{1/x} = $ _____.

As $x \to \infty$, $\dfrac{1}{x} \to 0$. Thus $\lim\limits_{x\to\infty} 4^{1/x} = 4^0 = 1$.

C. To find a limit, first try the limit laws. If they fail, as in

$$\lim_{x \to 4} \frac{x^2 - 16}{x - 4} \text{ (gives } \tfrac{0}{0}, \text{ no information)},$$

you should next try algebraic techniques to rewrite the function in a form for which the limit theorems do work.

Division by zero is one common reason why the limit laws fail. Try to rewrite the function in a form that will not give a zero denominator. There are several techniques for this and often more than one technique is required in a given problem. The most common techniques are:

Cancellation:

When substitution of $x = a$ in $\lim_{x \to a} \frac{f(x)}{g(x)}$ results in $\frac{0}{0}$, $x - a$ may be a factor of both $f(x)$ and $g(x)$ and hence can be canceled.

Example: Find $\lim_{x \to 2} \frac{2x^2 - 4x}{x^2 - 4}$.

Here $x = 2$ gives $\frac{0}{0}$ but $\frac{2x^2 - 4x}{x^2 - 4} = \frac{2x(x - 2)}{(x + 2)(x - 2)} = \frac{2x}{x + 2}$, provided $x \neq 2$.

Thus, $\lim_{x \to 2} \frac{2x^2 - 4x}{x^2 - 4} = \lim_{x \to 2} \frac{2x}{x + 2} = \frac{2 \cdot 2}{2 + 2} = 1$.

Fraction Manipulation:

Some limits involve functions expressed as complex fractions. Rewriting the fraction in simpler terms may allow the limit laws to be used.

Example: Find $\lim_{x \to 1} \frac{\frac{1}{x} - x}{\frac{1}{x} - 1}$.

We rewrite the function by first finding a simpler numerator and denominator and then using cancellation:

$$\frac{\frac{1}{x} - x}{\frac{1}{x} - 1} = \frac{\frac{1}{x} - \frac{x^2}{x}}{\frac{1}{x} - \frac{x}{x}} = \frac{\frac{1 - x^2}{x}}{\frac{1 - x}{x}} = \frac{1 - x^2}{1 - x} = \frac{(1 - x)(1 + x)}{1 - x} = 1 + x, \text{ for } x \neq 1.$$

Thus, $\lim_{x \to 1} \frac{\frac{1}{x} - x}{\frac{1}{x} - 1} = \lim_{x \to 1} (1 + x) = 2$.

Rationalizing an Expression:

Sometimes if two radicals appear in the function, a process similar to using the conjugate of a complex number may be used to rewrite part of the function without radicals. For functions with $\sqrt{a} - \sqrt{b}$, use $\frac{\sqrt{a} + \sqrt{b}}{\sqrt{a} + \sqrt{b}}$. The result may be simplified using cancellation.

Find $\lim\limits_{x \to 0} \dfrac{\sqrt{2 + x} - \sqrt{2}}{x}$.

Here $\dfrac{0}{0}$ results, but if $\sqrt{2 + x} - \sqrt{2}$ is multiplied by $\sqrt{2 + x} + \sqrt{2}$, no radicals will remain in the numerator:

$$\dfrac{\sqrt{2 + x} - \sqrt{2}}{x} \cdot \dfrac{\sqrt{2 + x} + \sqrt{2}}{\sqrt{2 + x} + \sqrt{2}} = \dfrac{2 + x - 2}{x(\sqrt{2 + x} + \sqrt{2})} = \dfrac{1}{\sqrt{2 + x} + \sqrt{2}}.$$ We end up

with radicals in the denominator, but now we can use the limit laws.

$$\lim_{x \to 0} \dfrac{\sqrt{2 + x} - \sqrt{2}}{x} = \lim_{x \to 0} \dfrac{1}{\sqrt{2 + x} + \sqrt{2}} = \dfrac{1}{\sqrt{2} + \sqrt{2}} = \dfrac{1}{2\sqrt{2}}.$$

11) Evaluate $\lim\limits_{x \to 6} \dfrac{2x - 12}{x^2 - x - 30}$.

$$\dfrac{2x - 12}{x^2 - x - 30} = \dfrac{2(x - 6)}{(x - 6)(x + 5)}$$

$$= \dfrac{2}{x + 5}, \text{ for } x \neq 6.$$

$$\lim_{x \to 6} \dfrac{2}{x + 5} = \dfrac{2}{11}.$$

12) Evaluate $\lim\limits_{x \to 1} \dfrac{x^2 - 6x + 5}{x - 1}$.

$$\dfrac{x^2 - 6x + 5}{x - 1} = \dfrac{(x - 1)(x - 5)}{x - 1}$$

$$= x - 5, \text{ for } x \neq 1.$$

$$\lim_{x \to 1} (x - 5) = -4.$$

(Go back and look at how question 2 was answered).

13) Evaluate $\lim\limits_{x \to 3} \left[\dfrac{2x^2}{x - 3} + \dfrac{6x}{3 - x} \right]$.

$$\dfrac{2x^2}{x - 3} + \dfrac{6x}{3 - x} = \dfrac{2x^2}{x - 3} - \dfrac{6x}{x - 3}$$

$$= \dfrac{2x^2 - 6x}{x - 3} = \dfrac{2x(x - 3)}{x - 3} = 2x, \text{ for } x \neq 3.$$

$$\lim_{x \to 3} 2x = 6.$$

14) Evaluate $\lim\limits_{x \to 0} \dfrac{\dfrac{1}{4 + x} - \dfrac{1}{4}}{x}$.

$$\dfrac{\dfrac{1}{4 + x} - \dfrac{1}{4}}{x} = \dfrac{\dfrac{4}{(4 + x)4} - \dfrac{(4 + x)}{(4 + x)4}}{x}$$

$$= \dfrac{\dfrac{4 - (4 + x)}{(4 + x)4}}{x} = \dfrac{\dfrac{-x}{(4 + x)4}}{x} = \dfrac{-1}{(4 + x)4}.$$

$$\lim_{x \to 0} \dfrac{-1}{(4 + x)4} = -\dfrac{1}{16}.$$

15) Evaluate $\lim\limits_{x \to 3} \dfrac{\sqrt{x} - \sqrt{3}}{x - 3}$.

$$\frac{\sqrt{x} - \sqrt{3}}{x - 3} \cdot \frac{\sqrt{x} + \sqrt{3}}{\sqrt{x} + \sqrt{3}} = \frac{x - 3}{(x - 3)(\sqrt{x} + \sqrt{3})}$$

$$= \frac{1}{\sqrt{x} + \sqrt{3}}, \text{ for } x \neq 3.$$

$$\lim_{x \to 3} \frac{1}{\sqrt{x} + \sqrt{3}} = \frac{1}{2\sqrt{3}}.$$

16) Find $\lim\limits_{x \to 2} f(x)$, where

$$f(x) = \begin{cases} \dfrac{x^2 - 4}{x - 2} & \text{for } x < 2 \\ \dfrac{x^2 - x - 2}{x^2 - 4} & \text{for } x > 2 \end{cases}$$

We need to compare one sided limits:

$$\lim_{x \to 2^-} f(x) = \lim_{x \to 2^-} \frac{x^2 - 4}{x - 2}$$

$$= \lim_{x \to 2^-} \frac{(x - 2)(x + 2)}{x - 2} = \lim_{x \to 2^-} (x + 2) = 4.$$

$$\lim_{x \to 2^+} f(x) = \lim_{x \to 2^+} \frac{x^2 - x - 2}{x^2 - 4}$$

$$= \lim_{x \to 2^+} \frac{(x - 2)(x + 1)}{(x - 2)(x + 2)} = \lim_{x \to 2^+} \frac{x + 1}{x + 2} = \frac{3}{4}$$

Since the one-sided limits $\left(4 \text{ and } \frac{3}{4}\right)$ do not agree, $\lim\limits_{x \to 2} f(x)$ does not exist.

Page 108

D. If $g(x)$ is "trapped" between $f(x)$ and $h(x)$ (i.e., if $f(x) \leq g(x) \leq h(x)$ for all x near a, except $x = a$ then

$$\lim_{x \to a} f(x) \leq \lim_{x \to a} g(x) \leq \lim_{x \to a} h(x),$$

if each limit exists.

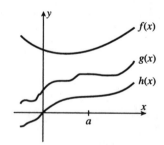

The Squeeze Theorem:

If $f(x) \leq g(x) \leq h(x)$ for all x

near a and $\lim\limits_{x \to a} f(x)$ and $\lim\limits_{x \to a} h(x)$

exist and are equal, then $\lim\limits_{x \to a} g(x)$

exists and is that same number.

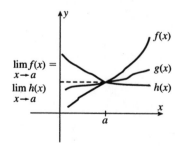

17) True or False: If $f(x) < g(x)$ for all x and $\lim\limits_{x \to a} f(x)$ and $\lim\limits_{x \to a} g(x)$ exist, then $\lim\limits_{x \to a} f(x) < \lim\limits_{x \to a} g(x)$.

False. Let $a = 0$, $f(x) = -x^2$ and

$$g(x) = \begin{cases} x^2 & \text{if } x \neq 0 \\ 1 & \text{if } x = 0 \end{cases}.$$

$f(x) < g(x)$, for all x, but

$$\lim_{x \to 0} f(x) = \lim_{x \to 0} -x^2 = 0 = \lim_{x \to 0} x^2$$
$$= \lim_{x \to 0} g(x).$$

18) For $0 \leq x \leq 1$, $x + 1 \leq 3^x \leq 2x + 1$. Use this to find $\lim\limits_{x \to 0^+} 3^x$.

Since $\lim\limits_{x \to 0^+} (x + 1) = 1$ and $\lim\limits_{x \to 0^+} (2x + 1) = 1$, by the Squeeze Theorem, $\lim\limits_{x \to 0^+} 3^x = 1$.

19) Sometimes, Always, or Never:

If $\lim\limits_{x \to a} f(x)$ and $\lim\limits_{x \to a} h(x)$ both exist and $f(x) < g(x) < h(x)$ for all x near a, then $\lim\limits_{x \to a} g(x)$ exists.

Sometimes. This is true for $f(x) = x^2 + 1$, $h(x) = -x^2 - 1$ and $g(x) = \sin x$ for $a = 0$: $\lim\limits_{x \to 0} \sin x = 0$. This is not true for $f(x) = x^2 + 1$, $h(x) = -x^2 - 1$ and $g(x) = \sin\left(\frac{1}{x}\right)$ for $a = 0$: $\lim\limits_{x \to 0} \sin \frac{1}{x}$ does not exist.

Section 2.4   The Precise Definition of a Limit

This section gives a precise definition of $\lim\limits_{x \to a} f(x) = L$. While phrased in ϵ and δ language, it helps to remember that the definition is consistent with the intuitive definition given earlier if you think of ϵ and δ as small numbers and note that $|f(x) - L|$ is the distance between $f(x)$ and L while $|x - a|$ is the distance between x and a.

Concepts to Master

A. Epsilon-delta (in ϵ-δ) definition of $\lim\limits_{x \to a} f(x) = L$

B. Epsilon-delta proof of $\lim\limits_{x \to a} f(x) = L$

C. Definitions of infinite limits

Summary and Focus Questions

Page 112

A. $\lim\limits_{x \to a} f(x) = L$ means for any given $\epsilon > 0$, there exists a $\delta > 0$ such that if $0 < |x - a| < \delta$, then $|f(x) - L| < \epsilon$.

This definition says that $f(x)$ will always be within ϵ units of L (that is $|f(x) - L| < \epsilon$) whenever x is within δ units of a (that is $|x - a| < \delta$). There are no universal techniques for determining δ in terms of ϵ but often this is accomplished by starting with $|f(x) - L| < \epsilon$ and replacing it with equivalent statements until finally the statement $|x - a| < \delta$ results. Reducing $|f(x) - L| < \epsilon$ down to $|x - a| < \delta$ may require several rewritings of $|f(x) - L| < \epsilon$ using factoring and cancellation. You may also need properties of absolute value, the most common being

$$|a \cdot b| = |a|\,|b| \text{ and}$$
$$|a| < b \text{ is equivalent to } -b < a < b.$$

Example: Given $\lim\limits_{x \to 3} (6x - 5) = 13$ and $\epsilon > 0$, find a suitable δ for the definition of the limit.

We start with $|f(x) - L| < \epsilon$ and simplify:
$$|f(x) - L| = |6x - 5 - 13| = |6x - 18| = |6(x - 3)| = |6|\,|x - 3| = 6|x - 3|.$$
Thus, $|6x - 5 - 13| < \epsilon$ is equivalent to $6|x - 3| < \epsilon$.

Divide by 6: $|x - 3| < \frac{\epsilon}{6}$. Our choice for δ is $\frac{\epsilon}{6}$.

When $f(x)$ is nonlinear, a simplified $|f(x) - L| < \epsilon$ may be in the form $|M(x)|\,|x - a| < \epsilon$ where $M(x)$ is some expression involving x. The procedure here is to find a constant K such that $|M(x)| < K$ for all x such that $|x - a| < p$ for some positive number p (often $p = 1$). The choice of δ is then the smaller of p and $\frac{\epsilon}{K}$, written $\delta = \min\left\{p, \frac{\epsilon}{K}\right\}$.

Example: Given $\lim\limits_{x \to 4} x^2 - 1 = 15$ and $\epsilon > 0$, find a corresponding $\delta > 0$.
$$|f(x) - L| = |x^2 - 1 - 15| = |x^2 - 16| = |(x + 4)(x - 4)| = |x + 4|\,|x - 4|.$$
For $p = 1$, if $|x - 4| < p$, then
$$|x - 4| < 1$$
$$-1 < x - 4 < 1$$
$$3 < x < 5$$
$$7 < x + 4 < 9$$

Thus, $|x + 4| < 9$. Therefore, using $K = 9$, when $|x - 4| < 1$, we have $|f(x) - L| = |x + 4|\,|x - 4| < 9|x - 4|$ and $9|x - 4| < \epsilon$ if $|x - 4| < \frac{\epsilon}{9}$. Thus choose $\delta = \min\left\{1, \frac{\epsilon}{9}\right\}$.

1) What part of the definition of $\lim\limits_{x \to a} f(x) = L$ guarantees that $x = a$ is not considered?

$0 < |x - a|.$

2) The choice of δ usually (does, does not) depend upon the size of ϵ.

Does. For most limits, the smaller the ϵ given, the smaller must be the corresponding choice of δ.

3) For the function in the figure below, what is an appropriate value of δ for the given ϵ in $\lim\limits_{x \to a} f(x) = L$?
  a) $\delta = p$
  b) $\delta = q$
  c) $\delta = p + q$
  d) None of these

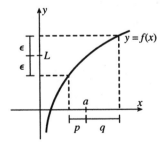

a) δ should be p or smaller, for then if x is within δ units of a, $f(x)$ will be within ϵ units of L.

4) For arbitrary $\epsilon > 0$, find a $\delta > 0$, in terms of ϵ such that the definition of limit is satisfied for $\lim\limits_{x \to 4} (2x + 1) = 9$.

$|f(x) - L| < \epsilon$ (Start)

$|2x + 1 - 9| < \epsilon$ (Replacement)

$|2x - 8| < \epsilon$ (Simplification)

$2|x - 4| < \epsilon$ (Factoring)

$|x - 4| < \frac{\epsilon}{2}$ (Divide by 2)

The desired δ is $\delta = \frac{\epsilon}{2}$.

5) Find δ for an arbitrary ϵ in $\lim\limits_{x \to 3} (x^2 - 4) = 5$.

$|f(x) - L| = |x^2 - 4 - 5| = |x^2 - 9|$
$= |(x + 3)(x - 3)| = |x + 3|\,|x - 3|$
If $p = 1$ and $|x - 3| < 1$, then

$$-1 < x - 3 < 1$$
$$2 < x < 4$$
$$5 < x + 3 < 7$$

So $|x + 3| < 7$.

Thus $|x + 3|\,|x - 3| < 7|x - 3| < \epsilon$
if $|x - 3| < \frac{\epsilon}{7}$.

Let $\delta = \min\left\{1, \frac{\epsilon}{7}\right\}$.

Page 114

B. A proof that $\lim\limits_{x \to a} f(x) = L$ consists of a verification of the definition. Start by assuming you have an $\epsilon > 0$ and then determine a number δ so that the statement $|f(x) - L| < \epsilon$ can be deduced from the statement $0 < |x - a| < \delta$. Your proof then consists of

(1) assuming $\epsilon > 0$

(2) stating your choice for δ

(3) showing that $0 < |x - a| < \delta$ implies $|f(x) - L| < \epsilon$.

Example: Prove $\lim\limits_{x \to 3} (6x - 5) = 13$.

For $\epsilon > 0$, the scratchwork to determine a suitable $\delta > 0$ is in section A.

Proof:

Let $\epsilon > 0$. Choose $\delta = \frac{\epsilon}{6}$.

Then for $0 < |x - 3| < \delta$

$$|x - 3| < \frac{\epsilon}{6}$$
$$6|x - 3| < \epsilon$$
$$|6x - 18| < \epsilon$$
$$|6x - 5 - 13| < \epsilon$$

In summary, if $0 < |x - 3| < \delta$, then $|6x - 5 - 13| < \epsilon$.

Thus, $\lim\limits_{x \to 3} (6x - 5) = 13$.

6) Prove that $\lim\limits_{x \to 4} (2x + 1) = 9$.
(See question 4.)

For $\epsilon > 0$, we know from question 4 that
$\delta = \frac{\epsilon}{2}$ is suitable.

Proof:

Let $\epsilon > 0$. Choose $\delta = \frac{\epsilon}{2}$.

Then for $0 < |x - 4| < \delta$

$$|x - 4| < \frac{\epsilon}{2}$$

$$2|x - 4| < \epsilon$$

$$|2x - 8| < \epsilon$$

$$|(2x + 1) - 9| < \epsilon$$

In summary, if $0 < |x - 4| < \delta$,

then $|(2x + 1) - 9| < \epsilon$.

Thus $\lim\limits_{x \to 4} (2x + 1) = 9$.

7) Prove that $\lim\limits_{x \to 3} (x^2 - 4) = 5$
(See question 5.)

From question 5, we see $\delta = \min\left\{1, \frac{\epsilon}{7}\right\}$.

Proof:

Let $\epsilon > 0$. Choose $\delta = \min\left\{1, \frac{\epsilon}{7}\right\}$.

Then for $0 < |x - 3| < \delta$, we first have

$$|x - 3| < 1,$$

$$-1 < x - 3 < 1$$

$$5 < x + 3 < 7$$

Thus $|x + 3| < 7$.

We also have $|x - 3| < \frac{\epsilon}{7}$,

$$7|x - 3| < \epsilon$$

Therefore $|x + 3| \, |x - 3| < \epsilon$

$$|x^2 - 9| < \epsilon$$

$$|(x^2 - 4) - 5)| < \epsilon.$$

In summary, if $0 < |x - 3| < \delta$, then
$|(x^2 - 4) - 5| < \epsilon$.

Thus, $\lim\limits_{x \to 3} (x^2 - 4) = 5$.

Page 119

C. Recall that $\lim\limits_{x \to a} f(x) = \infty$ means that as x is assigned values that approach a, the corresponding $f(x)$ values grow larger without bound. The definition to make this precise is

$\lim\limits_{x \to a} f(x) = \infty$, means for all positive M there is a $\delta > 0$, such that $f(x) > M$ whenever $0 < |x - a| < \delta$.

Just like with ordinary limits, you may need to work "backwards" from $f(x) > M$ to determine a proper δ for a given M.

8) Prove that $\lim\limits_{x \to 2} \dfrac{1}{|4 - 2x|} = \infty$.

For $M > 0$, $f(x) > M$ is $\dfrac{1}{|4 - 2x|} > M$.

$\dfrac{1}{M} > |4 - 2x|$

$|4 - 2x| < \dfrac{1}{M}$

$|(-2)(x - 2)| < \dfrac{1}{M}$

$2|x - 2| < \dfrac{1}{M}$

$|x - 2| < \dfrac{1}{2M}$

Choose $\delta = \dfrac{1}{2M}$.

Proof:

Let $M > 0$; choose $\delta = \dfrac{1}{2M}$.

Then if $0 , |x - 2| < \delta$,

$|x - 2| < \dfrac{1}{2M}$

$2|x - 2| < \dfrac{1}{M}$

$|2x - 4| < \dfrac{1}{M}$

$|4 - 2x| < \dfrac{1}{M}$

$M < \dfrac{1}{|4 - 2x|}$

$\dfrac{1}{|4 - 2x|} > M$.

Thus, if $0 < |x - 2| < \delta$, then $\dfrac{1}{|4 - 2x|} > M$.

Therefore $\lim\limits_{x \to 2} \dfrac{1}{|4 - 2x|} = \infty$.

Section 2.5   Continuity

Suppose you could draw a graph of a function without lifting your pencil as you move from one end of an interval domain to the other; we would say that the function is "continuous"—meaning that the graph is unbroken over its domain. This section precisely gives the concept of an unbroken graph.

Concepts to Master

A. Continuity of a function at a point and on an interval

B. Intermediate Value Theorem

Summary and Focus Questions

Page 122

A. A function f is *continuous at a* means both numbers $f(a)$ and $\lim\limits_{x \to a} f(x)$ exist and are equal, that is $\lim\limits_{x \to a} f(x) = f(a)$.

If f is continuous at a, then the calculation of $\lim\limits_{x \to a} f(x)$ is easy; just use $x = a$ in $f(x)$. Continuity at a means the graph of f is unbroken as it passes through $(a, f(a))$.

f is *continuous on an interval* $[a, b]$ means f is continuous at each point in the interval. For the point a we have $\lim\limits_{x \to a^+} f(x) = f(a)$ (*right continuity*) and for the point b we have $\lim\limits_{x \to b^-} f(x) = f(a)$ (*left continuity*). The graph of $f(x)$ must be unbroken at each point in the interval.

All these types of functions are continuous at every number in their domains:

Type	Example	Continuous on Domain
Polynomial	$f(x) = 6x^2 - 5x$	all reals
Rational	$f(x) = \dfrac{x}{(x + 1)(x - 2)}$	all reals except $x = -1$, $x = 2$
Root	$f(x) = \sqrt{x + 1}$	all real x, $x \geq -1$
Trigonometric	$f(x) = \cot x$	all reals except multiples of π
Inverse Trigonometric	$f(x) = \cos^{-1} x$	$-1 \leq x \leq 1$
Exponential	$f(x) = 3^x$	all reals
Logarithmic	$f(x) = \ln x$	$x > 0$

1)

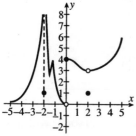

Determine the points where the function f above is not continuous (is discontinuous).

Discontinuous at

$x = -2$, because $\lim\limits_{x \to -2} f(x)$ does not exist.

$x = 0$, because $\lim\limits_{x \to 0} f(x)$ does not exist.

$x = 2$, because $f(2) = 1$, but $\lim\limits_{x \to 2} f(x) = 3$.

2) For the graph in problem 1 determine whether f is continuous on:

   a) $[-4, -3]$

   Yes.

   b) $[.5, 1.5]$

   Yes.

   c) $[-1, 1]$

   No, not continuous at 0.

   d) $[-2, -1]$

   No, not continuous at -2.

   e) $[0, 1]$

   Yes.

   f) $(-2, -1]$

   Yes.

3) Where is $f(x) = \dfrac{x}{(x - 1)(x + 2)}$ continuous?

Since f is rational, f is continuous everywhere it is defined: f is continuous for all x except $x = 1, x = -2$.

4) $\lim\limits_{x \to 0} \ln(\cos x)$.

As $x \to 0$, $\cos x \to 1$.
Thus $\ln(\cos x) \to \ln(1) = 0$.

5) Is $f(x) = 4x^2 + 10x + 1$ continuous at $x = \sqrt{2}$?

Yes. Since f is a polynomial, it is continuous everywhere.

6) Where is $f(x) = \sqrt{\dfrac{1-x}{|x|}}$ continuous?

$(-\infty, 0) \cup (0, 1]$, f is algebraic and this is the domain of f.

7) Where is $f(x) = \tan\!\left(\frac{\pi}{2}x\right)$ discontinuous?

$f(x)$ is not continuous where it is not defined: at $x = \ldots, -5, -3, -1, 1, 3, 5, \ldots.$

B. *The Intermediate Value Theorem:*
If f is continuous on $[a, b]$ and N is between $f(a)$ and $f(b)$, then there is at least one c between a and b so that $f(c) = N$.

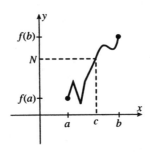

The theorem does not say where c is located other than somewhere between a and b. Nor does it say c is unique—more than one value for c is possible. Another way to state the Intermediate Value Theorem is that as x varies from a to b, $f(x)$ must attain all values between $f(a)$ and $f(b)$.

Example:

The function $f(x) = x + \dfrac{1}{x}$ is continuous on $\left[\frac{1}{2}, 3\right]$ and takes on all values between $2\frac{1}{2}$ and $3\frac{1}{3}$ as x ranges from $\frac{1}{2}$ to 3. (It also takes on other values but that is not important here.) The theorem does not apply for the interval $[-2, 2]$ because f is not continuous at $x = 0$. There is no x between -2 and 2 for which $f(x) = 0$, even though 0 is between $-2\frac{1}{2}$ and $2\frac{1}{2}$.

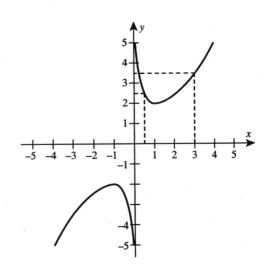

Page 129

8) Does $f(x) = x^2 - x - 3$ have a root (a value w such that $f(w) = 0$) in $[1, 3]$?

Yes. Since f is a polynomial f is continuous on $[1, 3]$. Since 0 is between $-3 = f(1)$ and $3 = f(3)$, by the Intermediate Value Theorem, there exists w between 1 and 3 with $f(w) = 0$.

9) Is the hypothesis of the Intermediate Value Theorem satisfied for

$$f(x) = \begin{cases} \sin \frac{\pi}{x} & x \neq 0 \\ 0 & x = 0 \end{cases}$$

with domain $[-2, 2]$?

No. $f(x)$ is not continuous at 0, hence not continuous on $[-2, 2]$.

10) For $f(x) = x^2 - 3x$, is there a number $t \in [-1, 5]$ such that $f(t) = -2$?

The answer is "yes", but we cannot use the Intermediate Value Theorem to support the answer. $f(-1) = 4$ and $f(5) = 10$, but -2 is not between 4 and 10. Solving $x^2 - 3x = -2$ gives $x = 1$, $x = 2$, both in $[-1, 5]$.

11) Find an interval $[a, b]$ for which $f(x) = x^5 - 10x^2 + 6x - 1$ has a root.

Since f is a polynomial f is continuous everywhere. We simply need to find an a with $f(a) < 0$ and b with $f(b) > 0$. $a = 0$ and $b = 3$ will do. By the Intermediate Value Theorem f has a root in $[0, 3]$.

12) Mark all the possible values for c in the Intermediate Value Theorem.

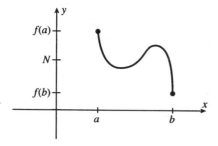

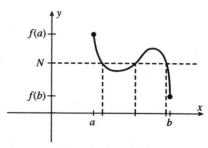

There are 3 possible values for c.

Section 2.6   Limits at Infinity; Horizontal Asymptotes

Limits at infinity describe the behavior of a graph as $y = f(x)$ as x gets larger and larger without bound. Just as $\lim_{x \to a} f(x) = \infty$ defined a vertical asymptote, $\lim_{x \to \infty} f(x) = L$ will define a horizontal asymptote.

Concepts to Master

A. Limits at infinity

B. Horizontal asymptote

C. Precise definition of $\lim_{x \to \infty} f(x) = L$

Summary and Focus Questions

Page
133

A. $\lim_{x \to \infty} f(x) = L$ means that as x is assigned larger values without bound, the corresponding values of $f(x)$ approach L.

$\lim_{x \to \infty} f(x) = L$ describes what happens to the graph of f out on the "far right side:" the graph approaches the line $y = L$.

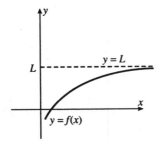

$\lim_{x \to -\infty} f(x) = L$ means that $f(x)$ approaches L as x takes more negative values without bound.

Some of the techniques for evaluating ordinary limits may be used for limits at infinity. In addition, for a limit involving a rational function, dividing the numerator and denominator by the highest power of x that appears in the denominator may help, as in:

$$\lim_{x \to \infty} \frac{4x^2 + 7x + 5}{2x^2 + 3} = \lim_{x \to \infty} \frac{\dfrac{(4x^2 + 7x + 5)}{x^2}}{\dfrac{(2x^2 + 3)}{x^2}} = \lim_{x \to \infty} \frac{4 + \dfrac{7}{x} + \dfrac{5}{x^2}}{2 + \dfrac{3}{x^2}} = \frac{4 + 0 + 0}{2 + 0} = 2$$

All of the usual limit theorems hold for limits at infinity.

$\lim_{x \to \infty} f(x) = \infty$ means that as x increases without bound, then so do the corresponding $f(x)$ values.

1) Evaluate $\lim\limits_{x\to\infty} f(x)$ and $\lim\limits_{x\to-\infty} f(x)$, where $y = f(x)$ has the given graph.

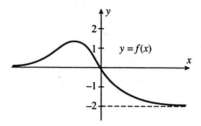

$$\lim_{x\to\infty} f(x) = -2.$$
$$\lim_{x\to-\infty} f(x) = 0.$$

2) Evaluate

a) $\lim\limits_{x\to\infty} \dfrac{5x^3 + 7x + 1}{3x^3 + 2x^2 + 3}$

Divide both numerator and denominator by x^3.

$$\lim_{x\to\infty} \frac{5 + \dfrac{7}{x^2} + \dfrac{1}{x^3}}{3 + \dfrac{2}{x} + \dfrac{3}{x^3}} = \frac{5 + 0 + 0}{3 + 0 + 0} = \frac{5}{3}.$$

Limits of this type can also be "eye-balled" by observing that for very large positive x, $5x^3 + 7x + 1 = 1 \approx 5x^3$ and $3x^3 + 2x^2 + 3 \approx 3x^3$. Thus the limit is $\lim\limits_{x\to\infty} \dfrac{5x^3}{3x^3} = \dfrac{5}{3}.$

b) $\lim\limits_{x\to-\infty} \dfrac{2x + 5}{\sqrt{x^2 + 4}}$

$$\lim_{x\to-\infty} \frac{\dfrac{2x + 5}{(-x)}}{\dfrac{\sqrt{x^2 + 4}}{(-x)}} = \lim_{x\to-\infty} \frac{\dfrac{2x + 5}{(-x)}}{\dfrac{\sqrt{x^2 + 4}}{\sqrt{x^2}}}$$

(Remember, since $x\to-\infty$, $x < 0$, so $\sqrt{x^2} = -x$.)

$$= \lim_{x\to-\infty} \frac{-2 - \dfrac{5}{x}}{\sqrt{1 + \dfrac{4}{x^2}}} = \frac{-2 - 0}{\sqrt{1 + 0}} = -2.$$

c) $\lim\limits_{x\to\infty} \dfrac{\cos x}{x}$

Since $-1 \le \cos x \le 1$, $-\dfrac{1}{x} \le \dfrac{\cos x}{x} \le \dfrac{1}{x}.$
Both $\lim\limits_{x\to\infty} \dfrac{-1}{x} = 0$ and $\lim\limits_{x\to\infty} \dfrac{1}{x} = 0.$
Therefore, $\lim\limits_{x\to\infty} \dfrac{\cos x}{x} = 0$ by the limits at infinity version of the Squeeze Theorem.

d) $\lim\limits_{x \to \infty} \dfrac{x^3 + 6x + 1}{2x^2 - 5x}$

Dividing the numerator and denominator by x^2, we have $\lim\limits_{x \to \infty} \dfrac{x + \frac{6}{x} + \frac{1}{x^2}}{2 - \frac{5}{x}}$.

As x grows larger, the denominator approaches 2 but the numerator grows larger, so

$$\lim\limits_{x \to \infty} \frac{x^3 + 6x + 1}{2x^2 - 5x} = \infty.$$

3) True or False: If all three limits exist, then
$$\lim\limits_{x \to \infty} (f(x)g(x)) = \left(\lim\limits_{x \to \infty} f(x)\right)\left(\lim\limits_{x \to \infty} g(x)\right)$$

True.

4) $\lim\limits_{x \to -\infty} \left(\dfrac{1}{2}\right)^x =$ _____.

∞

5) Find $\lim\limits_{x \to \infty} e^{-x}$.

As $x \to \infty$, $-x \to -\infty$. Thus $\lim\limits_{x \to \infty} e^{-x} = 0$.

6) Sometimes, Always, or Never:
$$\lim\limits_{x \to \infty} a^x = \infty.$$

Sometimes. True for $a < 1$. False for $0 < a < 1$.

Page 135

B. A *horizontal asymptote* for a function f is a line $y = L$ such that $\lim\limits_{x \to \infty} f(x) = L$, $\lim\limits_{x \to -\infty} f(x) = L$, or both. Here are the graphs of three examples:

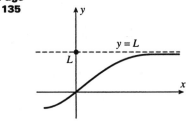 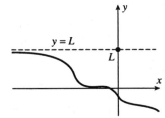 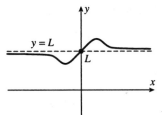

A function may have at most two horizontal asymptotes, one in each direction.

7) Find the horizontal asymptotes for the graph:

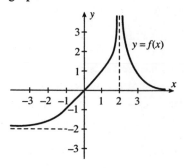

f has two horizontal asymptotes, $y = -2$ and $y = 0$. (Note: $x = 2$ is a vertical asymptote.)

8) Find the horizontal asymptotes for $f(x) = \dfrac{x}{|x| - 1}$.

$y = 1$, because $\lim\limits_{x \to \infty} \dfrac{x}{|x| - 1} = 1$ and $y = -1$, because $\lim\limits_{x \to -\infty} \dfrac{x}{|x| - 1} = -1$.

9) Find the asymptotes and sketch the graph of $f(x) = \dfrac{x}{x^2 - 1}$.

$\lim\limits_{x \to \infty} \dfrac{x}{x^2 - 1} = 0$ and $\lim\limits_{x \to -\infty} \dfrac{x}{x^2 - 1} = 0$, so $y = 0$ is the only horizontal asymptote.

$\lim\limits_{x \to 1^+} \dfrac{x}{x^2 - 1} = \infty$ and $\lim\limits_{x \to 1^-} \dfrac{x}{x^2 - 1} = -\infty$ so $x = 1$ is a vertical asymptote. Likewise $x = -1$ is a vertical asymptote because

$\lim\limits_{x \to -1^+} \dfrac{x}{x^2 - 1} = \infty$ and $\lim\limits_{x \to -1^-} \dfrac{x}{x^2 - 1} = -\infty$.

$\dfrac{x}{x^2 - 1} = 0$ only at $x = 0$, so $x = 0$ is the only intercept.

$$f'(x) = \frac{(x^2 - 1) - x(2x)}{(x^2 - 1)^2} = -\frac{x^2 + 1}{(x^2 - 1)^2}$$

which is always negative when it exists. Thus $f(x)$ is always decreasing.

$$f''(x) = -\frac{(x^2 - 1)^2 2x - (x^2 + 1)(2(x^2 - 1)2x)}{(x^2 - 2)^4}$$

$$= \frac{2x(x^2 + 3)}{(x^2 - 1)^3}$$

$f''(x) = 0$ at $x = 0$; 0 is a point of inflection.

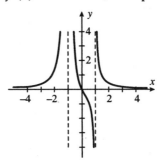

10) Find the vertical and horizontal asymptotes of $f(x) = \dfrac{1 - \cos x}{x}$.

$f(x)$ is not defined for $x = 0$, so $x = 0$ could be a vertical asymptote. However, $\lim\limits_{x \to 0} \dfrac{1 - \cos x}{x} = 0$, so f has no vertical asymptotes. To find limits at infinity we use the Squeeze Theorem. $-1 \le \cos x \le 1$. Thus, $-1 \le -\cos x \le 1$ and $0 \le 1 - \cos x \le 2$. Thus $\dfrac{0}{x} \le \dfrac{1 - \cos x}{x} \le \dfrac{2}{x}$, so $0 \le \dfrac{1 - \cos x}{x} \le \dfrac{2}{x}$ for $x > 0$. Since $\lim\limits_{x \to \infty} \dfrac{2}{x} = 0$, $\lim\limits_{x \to \infty} \dfrac{1 - \cos x}{x} = 0$. In a similar manner we can see that $\lim\limits_{x \to -\infty} \dfrac{1 - \cos x}{x} = 0$. Thus, $y = 0$ is the only horizontal asymptote.

Page 141

C. The formal definition of $\lim\limits_{x\to\infty} f(x) = L$ is:

for each $\epsilon > 0$ there exists a number N such that $|f(x) - L| < \epsilon$ whenever $x > N$.

11) $\lim\limits_{x\to\infty} f(x) = L$ means for all _____ there exists _____ such that _____ whenever _____.

$\epsilon > 0$
N
$|f(x) - L| < \epsilon$
$x > N$

12) Illustrate the definition of $\lim\limits_{x\to\infty} \dfrac{x}{x+1} = 1$ for $\epsilon = .01$.

We work "backwards:"

$|f(x) - L| < .01$

$\left|\dfrac{x}{x+1} - 1\right| < .01$

$\left|\dfrac{-1}{x+1}\right| < .01$

$\dfrac{1}{x+1} < .01$

$x + 1 > 100$

$x > 99$

For $N = 99$, if $x > N$ then $\left|\dfrac{x}{x+1} - 1\right| < .01$

Section 2.7   Tangents, Velocities, and Other Rates of Change

This section returns us to the beginning of the chapter for another look at a special kind of limit—instantaneous rate of change as a limit of average rates of change.

Concepts to Master

A. Slope of the tangent line to the graph of a function

B. Rates of change; instantaneous velocity

Summary and Focus Questions

Page 147

A. Let $m = \lim\limits_{x \to a} \dfrac{f(x) - f(a)}{x - a}$. If the number m exists, we say that m is the *slope of the tangent line to $y = f(x)$ at the point $(a, f(a))$.*

If we use $a + h = x$, then as $x \to a$, we have $h \to 0$.
Thus an alternate way to write the above limit is

$$m = \lim\limits_{h \to 0} \frac{f(a + h) - f(a)}{h}.$$

Either form of this limit may be used to calculate the slope of the tangent line. Notice that the direct substitution $x = a$ or $h = 0$ in these limits always yields $\frac{0}{0}$, so you can never simply use limit laws to evaluate a limit that is the slope of the tangent line.

1) Find the slope of the tangent line to $f(x) = x^2 + 2x$ at $x = 3$.

At $x = 3$, $f(3) = 3^2 + 2(3) = 15$.

$$\frac{f(x) - f(a)}{x - a} = \frac{x^2 + 2x - 15}{x - 3} = \frac{(x - 3)(x + 5)}{x - 3}$$

$$= x + 5 \text{ for } x \neq 3.$$

Thus, $\lim\limits_{x \to 3} \dfrac{f(x) - f(3)}{x - 3} = \lim\limits_{x \to 3} (x + 5) = 8.$

2) Find $\lim\limits_{h\to 0}\dfrac{f(a+h)-f(a)}{h}$ where $f(x)=\dfrac{1}{x-1}$ and $a=2$.

$$f(2+h)=\frac{1}{2+h-1}=\frac{1}{h+1}.$$

$$f(2+h)-f(2)=\frac{1}{h+1}-\frac{1}{2-1}$$

$$=\frac{1}{h+1}-\frac{1}{1}=\frac{1}{h+1}-\frac{h+1}{h+1}$$

$$=\frac{1-(h+1)}{h+1}=\frac{-h}{h+1}.$$

Thus, $\lim\limits_{h\to 0}\dfrac{f(a+h)-f(a)}{h}$

$$=\lim_{h\to 0}\frac{-h}{h+1}\cdot\frac{1}{h}=\lim_{h\to 0}\frac{-1}{h+1}=-1.$$

B. The expression $\dfrac{f(x)-f(a)}{x-a}$ is the *average rate of change* of $y=f(x)$ on the interval $[a,x]$. Thus $\lim\limits_{x\to a}\dfrac{f(x)-f(a)}{x-a}$ is the *instantaneous rate of change* of $y=f(x)$ at $x=a$.

In particular, if $y=f(x)$ is the position of an object at time x, $\dfrac{f(x)-f(a)}{x-a}$ is the *average velocity* and $\lim\limits_{x\to a}\dfrac{f(x)-f(a)}{x-a}$ is the *instantaneous velocity* at $x=a$.

3) Find the instantaneous velocity at time 3 seconds if a particle's position at time t is $f(t)=t^2+2t$ feet.

8 ft/s. This question is asking for the same information as question 1. Here the rate of change is interpreted as velocity instead of slope.

4) a) For a particle whose position at time t is $f(t)=6t^2-4t+1$ feet, find the average velocity over each of these intervals of time:

[1, 4]

$$\frac{f(4)-f(1)}{4-1}=\frac{81-3}{3}=26\ \text{ft/s}$$

[1, 2]

$$\frac{f(2)-f(1)}{2-1}=\frac{17-3}{1}=14\ \text{ft/s}$$

[1, 1.2]

$$\frac{f(1.2)-f(1)}{1.2-1}=\frac{4.84-3}{0.2}=9.2\ \text{ft/s}$$

[1, 1.01]

$$\frac{f(1.01)-f(1)}{1.01-1}=\frac{3.0806-3}{0.01}=8.06\ \text{ft/s}$$

Page 150

b) Find the instantaneous velocity of the particle at $t = 1$ second.

From part (a) it appears as though the answer is 8 ft/s. Let's find out:

$$\frac{f(t) - f(1)}{t - 1} = \frac{6t^2 - 4t + 1 - 3}{t - 1}$$

$$= \frac{6t^2 - 4t - 2}{t - 1}$$

$$= \frac{(6t + 2)(t - 1)}{t - 1}$$

$$= 6t + 2 \text{ for } t \neq 1.$$

Thus,

$$\lim_{t \to 1} \frac{f(t) - f(1)}{t - 1} = \lim_{t \to 1} (6t + 2) = 8 \text{ ft/s}.$$

Section 2.8   Derivatives

A good understanding of this section is critical to your success in calculus. Make sure you know the definition of a derivative and can interpret it as a rate of change.

Concepts to Master

A. Definition of the derivative; calculate and estimate the derivative of a function

B. Interpretations of the derivative

Summary and Focus Questions

Page 156

A. The derivative of $y = f(x)$ at a point a is

$$f'(a) = \lim_{h \to 0} \frac{f(a + h) - f(a)}{h}.$$

An equivalent way to write this without using the variable h is

$$f'(a) = \lim_{x \to a} \frac{f(x) - f(a)}{x - a}$$

If the function is given algebraically, such as $f(x) = x^2 + 2x + 1$, then $f'(a)$ is calculated in the same manner as the limits in section 2.6.

1) Find $f(a + h)$.

2) Find $f(a + h) - f(a)$.

3) Determine $\dfrac{f(a + h) - f(a)}{h}$ and simplify (if possible) when $h \neq 0$.

4) Find the limit of the result of part 3).

Example: Find $f'(a)$, where $f(x) = 3 + x^2$.

1) $f(a + h) = 3 + (a + h)^2 = 3 + a^2 + 2ah + h^2$

2) $f(a + h) - f(a) = (3 + a^2 + 2ah + h^2) - (3 + a^2) = 2ah + h^2$

3) $\dfrac{f(a + h) - f(a)}{h} = \dfrac{2ah + h^2}{h} = \dfrac{h(2a + h)}{h} = 2a + h$, for $h \neq 0$.

4) $f'(a) = \lim_{h \to 0} (2a + h) = 2a.$

If the function is given in table form, then the derivative is estimated by first calculating the difference quotients $\dfrac{f(a + h) - f(a)}{h}$, as in this example:

Estimate the derivative at $a = 4$ for the function given at the right.

x	f(x)
0	1
2	5
4	8
6	11
8	13
10	16

First compute a table of difference quotients:

h	a + h	f(a + h)	f(a + h) − f(a)	$\dfrac{f(a+h) - f(a)}{h}$
−4	0	1	−7	1.75
−2	2	5	−3	1.50
2	6	11	3	1.50
4	8	13	5	1.25
6	10	16	8	1.33

The table suggests that $f'(4)$ is somewhere between 1.25 and 1.75, and probably near 1.50 (since smaller values of h give difference quotients of 1.50). We estimate that $f'(4) \approx 1.50$.

1) Find $f'(3)$ for $f(x) = x^2 + 10x$.

$f'(3)$ is found in steps:

(1) $f(3 + h) = (3 + h)^2 + 10(3 + h)$
$$= 9 + 6h + h^2 + 30 + 10h$$
$$= 39 + 16h + h^2.$$

(2) $f(3 + h) - f(3) = (39 + 16h + h^2) - 39$
$$= 16h + h^2$$
$$= h(16 + h).$$

(3) $\dfrac{f(3 + h) - f(3)}{h} = \dfrac{h(16 + h)}{h} = 16 + h$

for $h \neq 0$.

(4) Finally, $f'(3) = \lim_{h \to 0}(16 + h) = 16.$

2) Find $f'(a)$ for $f(x) = 2x - x^2$.

(1) $f(a + h) = 2(a + h) - (a + h)^2$
$$= 2a + 2h - a^2 - 2ah - h^2.$$

(2) $f(a + h) - f(a)$
$$= 2a + 2h - a^2 - 2ah - h^2 - (2a - a^2)$$
$$= 2h - 2ah - h^2$$
$$= h(2 - 2a - h).$$

(3) $\dfrac{f(a + h) - f(a)}{h} = \dfrac{h(2 - 2a - h)}{h}$
$$= 2 - 2a - h \text{ for } h \neq 0.$$

(4) $f'(x) = \lim\limits_{h \to 0}(2 - 2a - h)$
$$= 2 - 2a - 0 = 2 - 2a.$$

3) The population of the village of Rosebush is given by the table below. Estimate the derivative for 1992.

t	P(t)
1989	841
1990	832
1991	821
1992	810
1993	801
1994	793

h	a + h	f(a + h)	f(a + h) − f(a)	$\frac{f(a + h) - f(a)}{h}$
−3	1989	841	31	−10.33
−2	1990	832	22	−11.00
−1	1991	821	11	−11.00
1	1993	801	−9	−9.00
2	1994	793	−17	−8.50

The difference quotients range from −8.50 to −11.00; we estimate $P'(1992) \approx -10.00$.

Page
157

B. If $f'(a)$ exists, the tangent line to the graph of f at a is the line through $(a, f(a))$ with slope $f'(a)$. If $f'(a)$ does not exist, then the tangent line might not exist, might be a vertical line, or might not be unique.

The slope of the tangent line (measured by $f'(a)$) is the same as the *instantaneous rate of change* of $y = f(x)$ with respect to x at $x = a$. The notion of instantaneous rate of change is crucial; the derivative measures how fast $f(x)$ is changing per unit change of x. Thus the derivative has many applications, such as velocity, metabolic rates, etc.

4) The slope of the tangent line to the graph of f at the point a is _____.

$f'(a)$, provided it exists.

5) True or False:
$f'(x)$ measures the average rate of change of $y = f(x)$ with respect to x.

False. $f'(x)$ measures instantaneous rate of change.

6) Given that $f'(x) = 3x^2 + 2$ for $f(x) = x^3 + 2x + 1$, find:

a) the equation of the tangent line to f at the point corresponding to $x = 2$.

The point of tangency has x coordinate 2 and y coordinate

$$f(2) = 2^3 + 2(2) + 1 = 13.$$

The slope is $f'(2) = 3(2)^2 + 2 = 14.$

Hence the tangent line has equation

$$y - 13 = 14(x - 2) \text{ or } y = 14x - 15.$$

b) the instantaneous rate of change of $f(x)$ with respect to x at $x = 3$.

$f'(3) = 3(3)^2 + 2 = 29.$

c) If $f(x)$ represents the distance in feet of a particle from the origin at time x, find the velocity at time $x = 4$ seconds.

$f'(4) = 3(4)^2 + 3 = 51$ ft/s.

7) An arrow is shot straight up into the air and falls back to earth after 8 seconds. Its height above the archer after t seconds is $H(t)$.

a) What is the meaning of $H'(2)$?

$H'(2)$ is the instantaneous rate of change of the height of the arrow above the archer after 2 seconds. The rate of change of height (distance) is velocity, so $H'(2)$ is the instantaneous velocity after 2 seconds.

b) What is $H'(9)$?

Since the arrow falls to earth after 8 seconds, it is not moving at time $t = 9$. Thus $H'(9) = 0$.

c) Is there a value a between 0 and 8 seconds for which $H'(a) = 0$?

Yes. $H'(a) = 0$ means that the arrow height does not change at the time a; so there would be an instant when the velocity is 0. At its highest point, before the arrow starts returning to earth, there is an instant when $H'(a) = 0$. Since the entire flight takes 8 seconds, you might guess that at $a = 4$, $H'(4) = 0$.

8)

For the function f given above, which is largest: $f'(a), f'(b)$, or $f'(c)$?

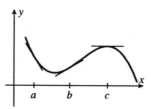

$f'(b)$ is the largest. $f'(b)$ is positive while $f'(a) < 0$ and $f'(c) = 0$.

Section 2.9   The Derivative as a Function

By letting the number a vary, $f'(a)$ itself varies and so f' is a function. The function f' provides useful information about the function f. You will also learn other notations for the derivative and how to determine values of a for which $f'(a)$ does not exist.

Concepts to Master

A. Differentiability; Other differentiation notation
B. Relationship of continuity to differentiability
C. Relationship between the graphs of f and f'.
D. Determining where a function is not differentiable.

Summary and Focus Questions

Page 163

A. By varying $x, f'(x)$ defines a function called *the derivative of f with respect to x:*

$$f'(x) = \lim_{h \to 0} \frac{f(x + h) - f(x)}{h}.$$

The function f is *differentiable at x* means that $f'(x)$ exists.
f is *differentiable on an interval* if it is differentiable at every number in the interval.
All of these notations are used to refer to the derivative of f at x:

$$f'(x), y', \frac{df}{dx}, \frac{dy}{dx}, \frac{d}{dx}f(x), Df(x), D_x f(x)$$

1) True or False:

$$f'(x) = \lim_{\Delta x \to 0} \frac{f(x + \Delta x) - f(x)}{\Delta x}$$

True. This is the definition of $f'(x)$ using the symbol Δx instead of h.

2) Is $f(x) = \sqrt{x}$ differentiable at $x = 0$?

No. $f'(0) = \lim_{h \to 0} \dfrac{f(0 + h) - f(0)}{h}$

$= \lim_{h \to 0} \dfrac{\sqrt{h} - \sqrt{0}}{h}$

$= \lim_{h \to 0} \dfrac{\sqrt{h}}{h} = \lim_{h \to 0} \dfrac{1}{\sqrt{h}}$,

which does not exist.

3) True or False:

For $y = f(x)$ the notation $f'(x)$ and $\frac{dx}{dy}$ have the same meaning.

False. $f'(x) = \frac{dy}{dx}$.

Page 169

B. If f is differentiable at a point c, then f is also continuous at c. (Roughly speaking, this means that if you can draw a tangent line to a graph, then the graph must be unbroken.) The converse is *false*: continuity *does not* imply differentiability.

4) Can a function be continuous at $x = 3$ but not differentiable at $x = 3$?

Yes. $f(x) = |x - 3|$ is an example. $f'(3)$ does not exist.

5) Can a function be differentiable at $x = 3$ and not continuous at $x = 3$?

No. This can never happen.

6) True or False:

For the function f graphed below:

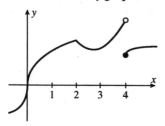

a) $f(x)$ is continuous at 0.

True.

b) $f(x)$ is differentiable at 0.

False.

c) $f(x)$ is continuous at 2.

True.

d) $f(x)$ is differentiable at 2.

False.

e) $f(x)$ is continuous at 3.

True.

f) $f(x)$ is differentiable at 3.

True.

g) $f(x)$ is continuous at 4.

False.

h) $f(x)$ is differentiable at 4.

False.

Page 164

C. The graph of f' may be determined from the graph of f by remembering that $f'(x)$ is the slope of the tangent line at $(x, f(x))$:

> If c is the slope of the tangent line at $(x, f(x))$, then (x, c) is on the graph of $y = f'(x)$.

> Conversely, if (x, c) is on the graph of f', then $f'(x) = c$ is the slope of the tangent line to f at $(x, f(x))$.

In the first graph below of $y = f(x)$, we have labeled some tangent line slope values. Beneath the graph is the corresponding graph of $y = f'(x)$.

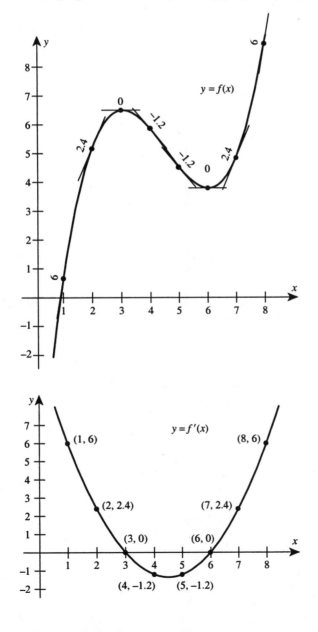

7) From the graph of *f* below, estimate $f'(1)$, $f'(2)$, $f'(3)$, and $f'(4)$, and then sketch a graph of $y = f'(x)$.

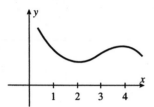

We estimate that $f'(1) = -2$, $f'(2) = 0$, $f'(3) = 1$, and $f'(4) = 0$. For these estimates, the graph of f' is:

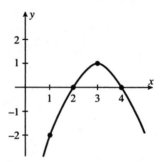

8) From the given graph of $y = f'(x)$, sketch a graph of $y = f(x)$.

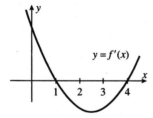

One such graph of $y = f(x)$ is:

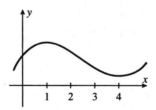

9) Dan is on a ladder, 2.5 meters tall, above a trampoline which is 1 meter off the ground. He jumps onto the trampoline which propels him 3 meters into the air before he lands on the ground. Let $f(t)$ be his height after t seconds. Sketch a graph of $f(t)$ and $f'(t)$.

At $t = 0$, $f(0) = 2.5$. Some time later, say $t = a$, he hits the trampoline and $f(a) = 1$. For a very short time he is below the trampoline rim but then ($t = b$) the springs of the trampoline begin to propel him upward. Later ($t = c$), he reaches his apex, $f(c) = 3$. Finally, he lands on the ground at $t = d$, so $f(d) = 0$.

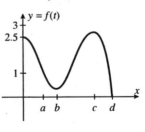

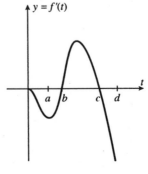

Page
170

D. There are three common ways for a function to fail to be differentiable at a point.

1) The graph of the function has a corner or "kink" in it.

Example:

$$f(x) = \begin{cases} x^2 & x \le 2 \\ (x - 4)^2 & x > 2 \end{cases}$$

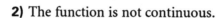

2) The function is not continuous.

Example:

$$f(x) = \begin{cases} x^2 & x \le 2 \\ 5 & x = 2 \\ 10 - x^2 & x > 2 \end{cases}$$

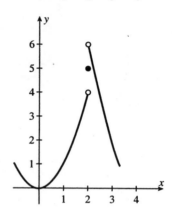

3) The graph of the function has a tangent line, but it is a vertical line.
Example: $f(x) = \sqrt[3]{x - 2}$

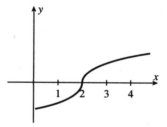

10) Where is $y = f(x)$ not differentiable?

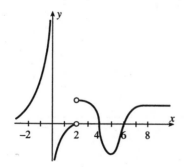

It appears that f is not differentiable at $x = 0, 2, 4$.

Technology Plus for Chapter 2

1) a) Use a graphing device to graph

$$f(x) = \frac{x^4 - 1}{x - 1}.$$

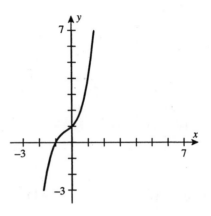

b) As you zoom in on the graph near $x = 1$, what does the limit appear to be? Describe the nature of the graph.

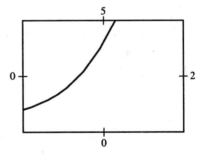

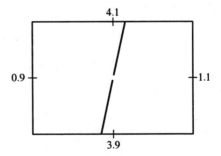

It appears that $\lim\limits_{x \to 1} f(x) = 4$.

The graph appears more linear.

c) Is $f(x)$ continuous at $x = 1$? What does the graphing device appear to tell you versus what you know about

$f(x) = \dfrac{x^4 - 1}{x - 1}$ at $x = 1$?

$f(x)$ is not continuous at $x = 1$. Some devices will produce a continuous looking line; others will show a gap, as indicated above.

2) A manufacturer of computer memory chips knows that selling x (hundred) chips will result in a revenue of $700x - .2x^2$.

a) Find $\lim\limits_{x \to 100} (700x - .2x^2)$.

$$\lim\limits_{x \to 100} (700x - .2x^2) = 700(100) - .2(100)^2$$
$$= 68{,}000.$$

b) If x varies by 1 unit, what is the range of the revenue?

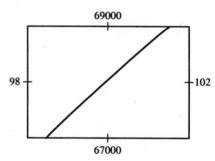

At $x = 99$, the revenue is \$67,339.80.
At $x = 101$, the revenue is \$68,659.80.
The range is from \$67,339.80 to \$68,659.80.

c) Question b) may be written in terms of the definition of a limit. What are each of these?

$f(x) =$ _____

$a =$ _____

$L =$ _____

$\epsilon =$ _____

$\delta =$ _____

$f(x) = 700x - .2x^2$

$a = 100$

$L = 68,000$

$\epsilon = \max\{|67,399.8 - 68,000|,$
$\qquad |68,659.8 - 68,000|\}$
$\quad = \max\{660.2, 659.8\} = 660.2$

$\delta = 1$

3) Use a graphing calculator to sketch the graph of $y = \left| \dfrac{1}{1 - \cos \pi x} \right|$. Where is the function discontinuous?

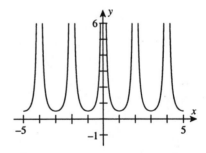

The function is not continuous at any even integer value of x.

4) Let $f(x) = 2^x$ and $x = 1$.

   a) Create the function

$$d(h) = \frac{f(1 + h) - f(1)}{h}.$$

   Use a calculator to compute several values for $d(h)$ and estimate $f'(1)$.

h	$1 + h$	$f(1 + h)$	$f(1 + h) - f(1)$	$\frac{f(1 + h) - f(x)}{h}$
1	2	4.00000	2.00000	2.00000
0.5	1.5	2.82843	0.82843	1.65685
0.1	1.1	2.14355	0.14355	1.43547
0.01	1.01	2.01391	0.01391	1.39111
0.001	1.001	2.00139	0.00139	1.38677
0.0001	1.0001	2.00014	0.00014	1.38634

We estimate $f'(1) \approx 1.386$.

   b) Do the same as part a) for $f'(2)$.

h	$2 + h$	$f(2 + h)$	$f(2 + h) - f(2)$	$\frac{f(2 + h) - f(2)}{h}$
1	3	8.00000	4.00000	4.00000
0.5	2.5	5.65685	1.65685	3.31371
0.1	2.1	4.28709	0.28709	2.87094
0.01	2.01	4.02782	0.02782	2.78222
0.001	2.001	4.00277	0.00277	2.77355
0.0001	2.0001	4.00028	0.00028	2.77268

We estimate $f'(2) \approx 2.772$.

Chapter 3 — Derivatives

© 1999 by Sidney Harris.

Section 3.1   Derivatives of Polynomials and Exponential Functions

This section is fundamental to all of calculus. It shows you how to calculate the derivative for many types of functions using formulas rather than using the limit definition.

Concepts to Master

A. The derivatives of $f(x) = c$ and $f(x) = x^n$

B. The derivatives of sums and differences

C. Definition of e; derivative of e^x

Summary and Focus Questions

Page 181

A. The derivative of a constant function $f(x) = c$ is 0; that is, $\frac{dc}{dx} = 0$.

(General Power Rule) The derivative of the function $f(x) = x^n$ is nx^{n-1}; that is, $\frac{d}{dx}(x^n) = nx^{n-1}$.

1) True, False:

   a) For $f(x) = -5, f'(x) = 0$.

       True.

   b) For $f(x) = x^6, f'(x) = 6x$.

       False, $f'(x) = 6x^5$.

   c) For $f(x) = 2^x, f'(x) = x2^{x-1}$.

       False. The variable x must be in the base of the expression and the exponent must be a constant; we will have to wait a while before we find out that the derivative is $(\ln 2)2^x$.

2) Find $f'(x)$ for each:

   a) $f(x) = 10$

       $f'(x) = 0$.

   b) $f(x) = x^{-3}$

       $f'(x) = -3x^{-4} = -\dfrac{3}{x^4}$.

Page
184

B. The derivative of an arithmetic combination of functions can be found by putting together the derivatives of the components using these rules:

$$[cf(x)]' = cf'(x) \text{ for any constant } c.$$

$$[f(x) \pm g(x)]' = f'(x) \pm g'(x).$$

3) Find $f'(x)$ for each:

a) $f(x) = 5x^4$

$$f'(x) = 20x^3.$$

b) $f(x) = x^4 + 4x^3$

$$f'(x) = 4x^3 + 12x^2.$$

c) $f(x) = 7x^3 + 6x^2 + 10x + 12$

$$f'(x) = 21x^2 + 12x + 10.$$

d) $f(x) = x^{-7} - x^{-6}$

$$f'(x) = -7x^{-8} + 6x^{-7}.$$

e) $f(x) = x^\pi$

$$f'(x) = \pi x^{\pi-1}.$$

4) Suppose the position of a particle along an axis at time t is $f(t) = t^2 + \frac{1}{t}$ ft. Find the velocity of the particle at time $t = 2$ seconds.

We can now use differentiation formulas to find velocities.
$$f(t) = t^2 + t^{-1}$$
$$f'(t) = 2t + (-1)t^{-2} = 2t - \frac{1}{t^2}$$
$$f'(2) = 2(2) - \frac{1}{2^2} = 3.75 \text{ ft/s}.$$

5) For $f(x) = mx + b$, find $f'(x)$. What does this say about the tangent line to $f(x)$ at any point?

$f'(x) = m$, a constant.
$f(x) = mx + b$ is a line with slope m. Thus, the tangent line is the line itself and the tangent line slope is always m.

Page 187

C. The number e is the constant (approximately 2.718) such that

$$\lim_{h \to 0} \frac{e^h - 1}{h} = 1.$$

In general, for $f(x) = a^x, f'(x) = f'(0) \cdot a^x$; that is, $(a^x)' = a^x \cdot \left(\lim_{h \to 0} \frac{a^h - 1}{h} \right)$.

Thus the number e was chosen in order to make this differentiation formula simple:

$$\text{If } f(x) = e^x, \text{ then } f'(x) = e^x.$$

6) For what number(s) k is $\lim_{h \to 0} \frac{k^h - 1}{h} = 1$?

$k = e$ is the only such number.

7) Find $f'(x)$ for:

   a) $f(x) = 3e^x$

$f'(x) = 3e^x.$

   b) $f(x) = e^3$

$f'(x) = 0.$ (Remember, e is a constant and therefore e^3 is a constant.)

   c) $f(x) = x^2 - 6x - e^x$

$f'(x) = 2x - 6 - e^x$

8) For what a is $(a^x)' = a^x$?

Only for $a = e$.

Section 3.2   The Product and Quotient Rules

This section contains rules for finding derivatives for functions formed from other functions by multiplication or division.

Concepts to Master

The derivatives of products and quotients

Summary and Focus Questions

Page 190

The derivative of a product or quotient may be found by putting together the derivatives of the components using these rules:

$$(f(x) \cdot g(x))' = f(x)g'(x) + g(x)f'(x).$$

$$\left[\frac{f(x)}{g(x)}\right]' = \frac{g(x)f'(x) - f(x)g'(x)}{(g(x))^2}$$

1) Find $f'(x)$ for each:

a) $f(x) = 5x^4(x + 1)$

$$f'(x) = 20x^3(x + 1) + 5x^4(1)$$
$$= 25x^4 + 20x^3.$$

b) $f(x) = (x^2 + x)(3x + 1)$

$$f'(x) = (x^2 + x)(3) + (3x + 1)(2x + 1)$$
$$= 9x^2 + 8x + 1.$$

c) $f(x) = \dfrac{x^3}{x^2 + 10}$

$$f'(x) = \frac{(x^2 + 10)(3x^2) - (x^3)(2x)}{(x^2 + 10)^2} = \frac{x^4 + 30x^2}{(x^2 + 10)^2}.$$

d) $f(x) = \dfrac{1 + x^{-1}}{2 - x^{-2}}$

$$f'(x) = \frac{(2 - x^{-2})(-x^{-2}) - (1 + x^{-1})(2x^{-3})}{(2 - x^{-2})^2}.$$

e) $f(x) = \dfrac{\sqrt{x} - 1}{\sqrt{x} + 1}$

$f(x) = \dfrac{x^{1/2} - 1}{x^{1/2} - 1}$, so $f'(x) =$

$\dfrac{(x^{1/2} + 1)\left(\frac{1}{2}x^{-1/2}\right) - (x^{1/2} - 1)\left(\frac{1}{2}x^{-1/2}\right)}{(x^{1/2} + 1)^2}$

$= \dfrac{2\left(\frac{1}{2}x^{-1/2}\right)}{(x^{1/2} + 1)^2} = \dfrac{1}{\sqrt{x}(\sqrt{x} + 1)^2}$.

f) $f(x) = x^2 e^x$

$f'(x) = x^2(e^x) + e^x(2x) = (x^2 + 2x)e^x$

2) Find $f'(x)$ for $f(x) = \dfrac{1}{x^2}$ two ways:

   a) by the Power Rule

$f(x) = \dfrac{1}{x^2} = x^{-2}$.

Thus, $f'(x) = -2x^{-3} = \dfrac{-2}{x^3}$.

   b) by considering $\dfrac{1}{x^2}$ as a quotient.

$f'(x) = \dfrac{x^2(0) - 1(2x)}{(x^2)^2} = \dfrac{-2x}{x^4} = \dfrac{-2}{x^3}$.

Section 3.3 Rates of Change in the Natural and Social Sciences

Rates of change play an important role in many fields, including the natural sciences (biology, chemistry, ...) and social sciences (economics, political science, ...).

Concepts to Master

Interpretation of a derivative as an instantaneous rate of change of a given quantity

Page 196

Summary and Focus Questions

For $y = f(x)$, $f'(x)$ measures the instantaneous rate of change of y with respect to x. If $f(x)$ represents some quantity at time x, $f'(x)$ will measure its instantaneous rate of change of the quantity at the given time x.

1) Water is flowing out of a water tower in such a fashion that after t minutes there are $10000 - 10t - t^3$ gallons remaining. How fast is the water flowing after 2 minutes?

Let $V(t) = 10000 - 10t - t^3$ be the volume at time t. The question asks for the rate of change of V.

$V'(t) = -10 - 3t^2$. At $t = 2$,

$V'(2) = -10 - 3(2)^2 = -22$ gal/min.

Note: The answer is negative, indicating that the volume of water is decreasing.

2) A space shuttle is $16t + t^3$ meters from its launch pad t seconds after liftoff. What is its velocity after 3 seconds?

$d(t) = 16t + t^3$, $d'(t) = 16 + 3t^2$

$d'(t) = 16 + 3(3)^2 = 43$ m/s.

3) A particle is moving along an axis so that at time t its position is

$f(t) = t^3 - 6t^2 + 6$ feet.

   a) What is its velocity at time t?

$$f'(t) = 3t^2 - 12t.$$

   b) What is the velocity at 3 seconds?

$$f'(3) = 3(3)^2 - 12(3) = -9 \text{ ft/s.}$$

   c) Is the particle moving left or right at 3 seconds?

Left, since $f'(3) = -9$ is negative.

   d) At what time(s) is the particle (instantaneously) motionless?

Motionless means its velocity $f'(t) = 0$
$3t^2 - 12t = 0$
$3t(t - 4) = 0$
At $t = 0$ and $t = 4$ the particle has zero velocity.

4) A stone is thrown upward from a 70 m cliff so that its height above ground is $f(t) = 70 + 3t - t^2$. What is the velocity of the stone as it hits the ground?

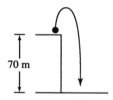

The time when the stone hits the ground is when $f(t) = 0$.
$70 + 3t - t^2 = 0$
$(10 - t)(7 + t) = 0$
$t = 10$ or $t = -7$ (Disregard $t = -7$.)
$f'(t) = 3 - 2t.$
$f'(10) = 3 - 20 = -17 \text{ m/s.}$

5) The numbers of yellow perch in a heavily fished portion of Lake Michigan have been declining rapidly. Using the data below estimate the rate of decline in 1996 by averaging the slopes of two secant lines.

t	P(t) (in millions)
1993	4.2
1994	4.0
1995	3.7
1996	3.6
1997	3.3
1998	3.1

The secant line for 1995–96 has slope

$$\frac{3.7 - 3.6}{1995 - 1996} = \frac{.1}{-1} = -0.1$$

The secant line for 1996–97 has slope

$$\frac{3.3 - 3.6}{1997-1996} = \frac{-.3}{1} = -0.3$$

We estimate $P'(1996) = \frac{-0.1 + (-0.3)}{2}$

$$= -0.2 \text{ per year.}$$

In 1996, the population was declining at 200,000 fish per year.

Section 3.4   Derivatives of Trigonometric Functions

This section contains formulas for derivatives of another type of function—trigonometric functions. Thus we are continuing to build up a base of basic derivative rules that is useful throughout calculus. Here again, after the basic rules are mastered, you need to use them to find derivatives of more complex functions.

Concepts to Master

A. Limits involving trigonometric functions
B. Derivatives of the six trigonometric functions

Summary and Focus Questions

Page 209

A. Two very important limits involving sine and cosine are:

$$\lim_{x \to 0} \frac{\sin x}{x} = 1 \text{ and } \lim_{x \to 0} \frac{\cos x - 1}{x} = 0$$

where x is the radian measure of an angle.

Many other limits involving trigonometric functions can be solved when transformed by algebra and limit laws into combinations of these two limits.

Example: Evaluate $\lim_{t \to 0} \frac{\sin 3t}{4t}$.

Since this resembles $\frac{\sin x}{x}$, we choose $x = 3t$. Then as $t \to 0$, $x \to 0$ and

$$\lim_{t \to 0} \frac{\sin 3t}{4t} = \lim_{t \to 0} \left(\frac{3}{4}\right)\frac{\sin 3t}{3t} = \frac{3}{4}\lim_{x \to 0} \frac{\sin x}{x} = \frac{3}{4}(1) = \frac{3}{4}.$$

1) $\lim_{x \to 0} \frac{1 - \cos x}{x} = $ _____.

$$\lim_{x \to 0} \frac{1 - \cos x}{x} = \lim_{x \to 0} -\frac{\cos x - 1}{x} = -0 = 0.$$

2) Find $\lim_{x \to 0} \frac{1}{x \cot x} = $ _____.

$$\frac{1}{x \cot x} = \frac{1}{x \frac{\cos x}{\sin x}} = \frac{\sin x}{x \cos x} = \frac{\sin x}{x} \cdot \frac{1}{\cos x}.$$

Then $\lim_{x \to 0} \frac{\sin x}{x} \cdot \frac{1}{\cos x} = 1 \cdot \frac{1}{1} = 1.$

3) Find $\lim\limits_{x \to 0} \dfrac{x^2}{2 \sin x}$.

$$\lim_{x \to 0} \frac{x^2}{2 \sin x} = \lim_{x \to 0} \frac{x}{2} \cdot \frac{x}{\sin x}$$

$$= \lim_{x \to 0} \frac{x}{2} \cdot \lim_{x \to 0} \frac{x}{\sin x}$$

$$= \lim_{x \to 0} \frac{x}{2} \cdot \lim_{x \to 0} \frac{1}{\frac{\sin x}{x}}$$

$$= (0)(1) = 0.$$

Page 211

B. These six differentiation formulas *must* be memorized.

$$\frac{d}{dx} \sin x = \cos x \qquad\qquad \frac{d}{dx} \cos x = -\sin x$$

$$\frac{d}{dx} \tan x = \sec^2 x \qquad\qquad \frac{d}{dx} \cot x = -\csc^2 x$$

$$\frac{d}{dx} \sec x = (\sec x)(\tan x) \qquad\qquad \frac{d}{dx} \csc x = -(\csc x)(\cot x)$$

4) Find y' for each:

a) $y = \sin x - \cos x$

$$y' = \cos x + \sin x.$$

b) $y = \dfrac{\tan x}{x + 1}$

$$y' = \frac{(x + 1)(\sec^2 x) - (\tan x)(1)}{(x + 1)^2}$$

$$= \frac{x \sec^2 x + \sec^2 x - \tan x}{(x + 1)^2}.$$

c) $y = \sin \dfrac{\pi}{4}$

$$y' = 0, \text{ since } y \text{ is a constant.}$$

d) $y = x^3 \sin x$

$$y' = x^3(\cos x) + (\sin x)3x^2$$
$$= x^3 \cos x + 3x^2 \sin x.$$

e) $y = x^2 + 2x \cos x$

$$y' = 2x + 2x(-\sin x) + 2 \cos x$$
$$= 2(x - x \sin x + \cos x).$$

f) $y = \dfrac{x}{\sec x + 1}$

$$y' = \frac{(\sec x + 1)1 - x(\sec x \tan x)}{(\sec x + 1)^2}$$

$$= \frac{\sec x + 1 - x \sec x \tan x}{(\sec x + 1)^2}.$$

g) $y = \dfrac{x}{\cot x}$

First rewrite as $y = x \tan x$.
$y' = x \sec^2 x + \tan x$.

5) Find where the graph of
$f(x) = \sqrt{3} \sin x + 3 \cos x$ has a horizontal
tangent.

Set $f'(x) = \sqrt{3} \cos x - 3 \sin x = 0$.
$\sqrt{3} \cos x = 3 \sin x$
$\dfrac{\sin x}{\cos x} = \dfrac{\sqrt{3}}{3}$; $\tan x = \dfrac{\sqrt{3}}{3}$.
There are horizontal tangents at
$x = \dfrac{\pi}{6} + k\pi$, where k is an integer.

Section 3.5 The Chain Rule

This section gives another rule for finding derivatives of functions built up from simpler ones. The Chain Rule determines the derivative of a composition of functions. The key to correct use of the Chain Rule comes from first recognizing that the function to be differentiated is a composite and determining the components.

Concepts to Master

A. Chain Rule for computing the derivative of a composition of functions
B. The derivative of a^x

Summary and Focus Questions

Page
215

A. If $y = f(u)$ and $u = g(x)$ (and thus $y = f \circ g(x)$) then the *Chain Rule* for computing the derivative of such a composition is:

$$\frac{dy}{dx} = \frac{dy}{du} \cdot \frac{du}{dx}$$

or equivalently

$$[f(g(x))]' = f'(g(x)) \cdot g'(x).$$

For example, if $y = (x^2 + 6x + 1)^4$, we recognize $f(x) = x^4$ and $g(x) = x^2 + 6x + 1$ as components such that $y = f(g(x))$.
Thus, $y' = 4(x^2 + 6x + 1)^3 \cdot (2x + 6)$.

$$\underbrace{f'(g(x))} \qquad \underbrace{g'(x)}$$

Recognizing components that are easily differentiated is an important first step. Composite functions often occur as expressions raised to a power, or as some function (such as sine or cosine) of an expression. Here are several examples. In each of these, y can be written as $f \circ g(x)$.

Example: $y = (x + 3)^4$

$y = f(u) = u^4$ (the "outer" function)

$u = g(x) = x + 3$ (the "inner" function)

Thus, $y' = 4(x + 3)^3 \cdot 1 = 4(x + 3)^3$.

Example: $y = \cos x^2$

$$y = f(u) = \cos u,$$
$$u = g(x) = x^2$$

Thus, $y' = (-\sin x^2) \cdot 2x = -2x \sin x^2$.

Example: $y = \sqrt{1 + \dfrac{1}{x}}$

$$y = f(u) = \sqrt{u},$$
$$u = g(x) = 1 + \frac{1}{x} = 1 + x^{-1}$$

Thus, $y' = \frac{1}{2}\left(1 + \frac{1}{x}\right)^{-1/2} \cdot (0 + (-1)x^{-2}) = \dfrac{-1}{2x^2\sqrt{1 + \frac{1}{x}}}$.

Sometimes, more than two functions are necessary, as in:

Example: $y = (3 + (x^3 - 2x)^5)^8$

$$y = f(u) = u^8 \text{ (the ``outer'' function)}$$
$$u = g(v) = 3 + v^5 \text{ (the ``middle'' function)}$$
$$v = h(x) = x^3 - 2x \text{ (the ``inner'' function)}$$

Here $y = f \circ g \circ h(x)$ and $\dfrac{dy}{dx} = \underbrace{8(3 + (x^3 - 2x)^5)^7}_{\frac{dy}{du}}\underbrace{(5(x^3 - 2x)^4)}_{\frac{du}{dv}}\underbrace{(3x^2 - 2)}_{\frac{dv}{dx}}$.

Example: $y = \tan^3 \sqrt{x}$

$$y = f(u) = u^3$$
$$u = g(v) = \tan(v)$$
$$v = h(x) = \sqrt{x}$$

Thus, $y' = \underbrace{(3 \tan^2 \sqrt{x})}_{\frac{dy}{du}}\underbrace{(\sec^2\sqrt{x})}_{\frac{du}{dv}}\underbrace{(\frac{1}{2}x^{-1/2})}_{\frac{dv}{dx}} = \frac{3}{2}x^{-1/2} \tan^2 \sqrt{x} \, \sec^2\sqrt{x}.$

The derivative of the exponential function may be generalized:

$$\frac{d}{dx}(e^{u(x)}) = e^{u(x)}u'(x).$$

For example, if $f(x) = e^{3x^2}$, then $f'(x) = e^{3x^2}(6x)$.

1) Suppose $h(x) = f(g(x)), f'(7) = 3$, $g(4) = 7$, and $g'(4) = 5$. Find $h'(4)$.

$$h'(4) = f'(g(4)) \cdot g'(4) = f'(7) \cdot 5$$
$$= 3 \cdot 5 = 15.$$

2) Find $\dfrac{dy}{dx}$ for:

a) $y = \sqrt{x^3 + 6x}$

Since $y = (x^3 + 6x)^{1/2}$,
$$y' = \tfrac{1}{2}(x^3 + 6x)^{-1/2}(3x^2 + 6) = \frac{3x^2 + 6}{2\sqrt{x^3 + 6x}}.$$

b) $y = \sec x^2$

$$y' = (\sec x^2)(\tan x^2)(2x).$$

c) $y = \sec^2 x$

Since $y = [\sec x]^2$,
$$y' = 2[\sec x]^1(\sec x \tan x) = 2 \sec^2 x \tan x.$$

d) $y = \cos^3 x^2$

Since $y = [\cos x^2]^3$,
$$y' = 3[\cos x^2]^2(-\sin x^2(2x))$$
$$= -6x(\cos^2 x^2)\sin x^2.$$

e) $y = \sin 2x \cos 3x$

y is a product of $\sin 2x$ and $\cos 3x$, so
$$y' = (\sin 2x)[-\sin 3x \cdot 3]$$
$$+ (\cos 3x)[\cos 2x \cdot 2]$$
$$= -3 \sin 2x \sin 3x + 2 \cos 3x \cos 2x.$$

f) $y = \sin(\sec x)$

$$y' = \cos(\sec x)(\sec x \tan x).$$

g) $y = \tan(x^2 + 1)^4$

$$y' = \sec^2(x^2 + 1)^4(4(x^2 + 1)^3(2x))$$
$$= 8x(x^2 + 1)^3 \sec^2(x^2 + 1)^4.$$

h) $y = \dfrac{1}{(x^2 - 1)^4}$

Since $y = (x^2 - 1)^{-4}$,
$$y' = -4(x^2 - 1)^{-5}(2x) = \frac{-8x}{(x^2 - 1)^5}.$$

i) $y = x^3 \sqrt{x^2 + 1}$

Since $y = x^3(x^2 + 1)^{1/2}$,
$$y' = x^3\left(\tfrac{1}{2}(x^2 + 1)^{-1/2} \cdot 2x\right)$$
$$+ (x^2 + 1)^{1/2} \cdot 3x^2$$
$$= \frac{x^4}{\sqrt{x^2 + 1}} + 3x^2\sqrt{x^2 + 1} = \frac{4x^4 + 3x^2}{\sqrt{x^2 + 1}}.$$

j) $f(x) = e^{\sqrt{x}}$

$$f(x) = e^{x^{1/2}}$$
$$f'(x) = e^{x^{1/2}}\left(\tfrac{1}{2}x^{-1/2}\right) = \frac{e^{\sqrt{x}}}{2\sqrt{x}}.$$

k) $f(x) = x^2 \cos e^x$

$$f'(x) = x^2(-\sin e^x \cdot e^x) + \cos e^x(2x)$$
$$= -x^2 e^x \sin e^x + 2x \cos e^x.$$

3) Does the following make sense?

$$\frac{dh}{dt} = \frac{dh}{dx}\frac{dx}{du}\frac{du}{dv}\frac{dv}{dt}$$

Yes. $h(t) = (h \circ x \circ u \circ v)(t)$. There are four "links" in this use of the Chain Rule.

4) The mean temperature at noon each day x at the top of Mt. Arthur is approximately

$$50 + 30 \sin\left(\frac{\pi(x - 200)}{180}\right)$$

degrees Fahrenheit. (Assume a 360 day year.) Find the rate of change of the daily temperature for January 15.

$$T(x) = 50 + 30 \sin\frac{\pi(x - 200)}{180}$$
$$T'(x) = 30 \cos\frac{\pi(x - 200)}{180} \cdot \frac{\pi}{180}$$
At $x = 15$, $T'(15) = 30 \cos\frac{\pi(-185)}{180} \cdot \frac{\pi}{180}$
$$\approx -.52 \text{ F}°/\text{day}.$$

For January 15, the noon temperature is decreasing about a half a degree per day.

B. If $f(x) = a^x$, then $f'(x) = a^x \ln a$.

Page 219

5) Find $f'(x)$ for each:

a) $f(x) = 4^x$

$$f'(x) = 4^x \ln 4$$

b) $f(x) = 3^{-x}$

$$f'(x) = -3^{-x} \ln 3$$

c) $f(x) = 2^x(x^2 + 1)$

$$f'(x) = 2^x(2x) + (x^2 + 1)2^x \ln 2$$
$$= 2^x(2x + (\ln 2)(x^2 + 1)).$$

Section 3.6   Implicit Differentiation

This section shows you a way to find the derivative of a function even when you are not given the rule for the function explicitly—the rule will be determined by an equation. This section also shows you how to find the derivatives of the inverse trigonometric functions.

Concepts to Master .

A. Define a functional relationship implicitly

B. Find $\frac{dy}{dx}$ where the function y of x is given implicitly

C. Definitions of the six inverse trigonometric functions

D. Derivatives involving inverse trigonometric functions

Summary and Focus Questions

Page 224

A. If y is a function of x defined by an equation not of the form $y = f(x)$, we say y as a function of x is defined *implicitly* by the equation.

Example: $x + \sqrt{y} = 1$ defines y as a function of x. If we solve for y (and remember that y must be nonnegative) we get $y = (1 - x)^2$ with domain $(-\infty, 1]$.

The graph of a relation is not necessarily the graph of a function. A portion of the graph may be the graph of a function. That function is defined implicitly by the equation.

For example, $x = y^2$ is a parabola opening to the right. It is not the graph of a function but the top half is the graph of the function $y = \sqrt{x}$ and the bottom half is the graph of $y = -\sqrt{x}$. Each function is defined implicitly by $x = y^2$.

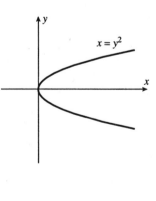

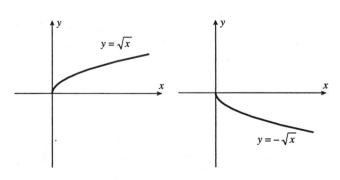

1) Sketch a graph of $\frac{x^2}{9} - \frac{y^2}{16} = 1$ and determine explicit forms for the functions that it defines.

We recognize the graph of the equation as a hyperbola opening left and right with asymptotes $y = \pm\frac{4}{3}x$.

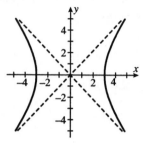

Solve for y:

$$-\frac{y^2}{16} = 1 - \frac{x^2}{9}$$
$$y^2 = \frac{16}{9}x^2 - 16$$

$$y = \sqrt{\frac{16}{9}x^2 - 16}, \quad y = -\sqrt{\frac{16}{9}x^2 - 16}.$$

The graphs of these functions, each with domain $(-\infty, -3] \cup [3, \infty)$ are the top half and bottom half, respectively, of the hyperbola.

2) The graph below is $x^{2/3} + y^{2/3} = 1$, which implicitly defines two functions y of x. Find the functions explicitly.

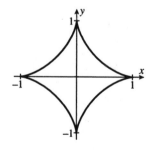

Solve for y:
$$x^{2/3} + y^{2/3} = 1$$
$$y^{2/3} = 1 - x^{2/3}$$
$$y^2 = (1 - x^{2/3})^3$$
$$y = (1 - x^{2/3})^{3/2} \text{ (the top half) and}$$
$$y = -(1 - x^{2/3})^{3/2} \text{ (the bottom half)}.$$

Page 225

B. Given an equation that defines y as an implicit function of x, $\frac{dy}{dx}$ may be found by implicit differentiation:

Step 1) Take the derivative with respect to x of both sides of the equation.

Step 2) Solve the result for $\frac{dy}{dx}$.

Be sure to remember that y is a function of x, so that when a term such as y^3 is differentiated, the result is $3y^2 y'$ (by the Chain Rule).

3) Find $\frac{dy}{dx}$ implicitly:

   a) $x^2 y^3 = 2x + 1$

Step 1) Differentiate with respect to x (using the Product Rule because both x^2 and y^3 are functions of x):
$$x^2(3y^2 y') + y^3(2x) = 2.$$

Step 2) Solve for y': $3x^2 y^2 y' = 2 - 2xy^3$
$$y' = \frac{2 - 2xy^3}{3x^2 y^2}.$$

   b) $3x^2 - 5xy + y^2 = 10$

Step 1) Differentiate:
$$6x - (5x(y') + y \cdot 5) + 2y \cdot y' = 0.$$

Step 2) Solve for y': $-5xy' + 2yy' = 5y - 6x$
$$(2y - 5x)y' = 5y - 6x$$
$$y' = \frac{5y - 6x}{2y - 5x}.$$

4) Find the slope of the tangent line to the curve defined by $x^2 + 2xy - y^2 = 1$ at the point $(5, 2)$.

The slope is $\frac{dy}{dx}$ at $(5, 2)$. First find $\frac{dy}{dx}$ implicitly:

Step 1) $2x + (2xy' + y \cdot 2) - 2yy' = 0$

Step 2) $2y'(x - y) = -2(x + y)$
$$y' = -\frac{x + y}{x - y} = \frac{x + y}{y - x}.$$
Now use $x = 5$, $y = 2$:
$$\frac{dy}{dx} = \frac{5 + 2}{2 - 5} = \frac{7}{-3} = -\frac{7}{3}.$$

5) For $x^2y = 1$ find y' both explicitly and implicitly.

Explicitly:
From $x^2y = 1$, $y = x^{-2}$ so $y' = -2x^{-3}$.
Implicitly:

Step 1) $x^2y' + 2xy = 0$

Step 2) $x^2y' = -2xy$

$$y' = -\frac{2y}{x}$$

These are the same because (remember $y = x^{-2}$)
$$\frac{-2y}{x} = \frac{-2x^{-2}}{x} = -2x^{-3}.$$

Page 229

C. Recall that for any function f for which f^{-1} is a function,

$$y = f^{-1}(x) \text{ if and only if } x = f(y)$$

None of the six trigonometric functions is one-to-one. However, when the domain of a trigonometric function is restricted to a certain interval, that restricted trigonometric function does have an inverse.

For example, the sine function is not one-to-one, but sine with domain $\left[-\frac{\pi}{2}, \frac{\pi}{2}\right]$ has an inverse function $\sin^{-1}$. Thus,

$$y = \sin^{-1}\left(\tfrac{1}{2}\right) \text{ means } \sin(y) = \tfrac{1}{2} \text{ for } -\tfrac{\pi}{2} \le y \le \tfrac{\pi}{2}; \text{ therefore } y = \tfrac{\pi}{6}.$$

The domains and ranges of the six inverse trigonometric functions are:

Function	Domain	Range
$\sin^{-1}$	$[-1, 1]$	$\left[-\frac{\pi}{2}, \frac{\pi}{2}\right]$
$\cos^{-1}$	$[-1, 1]$	$[0, \pi]$
$\tan^{-1}$	all reals	$\left(-\frac{\pi}{2}, \frac{\pi}{2}\right)$
$\cot^{-1}$	all reals	$(0, \pi)$
$\sec^{-1}$	$(-\infty, -1] \cup [1, \infty)$	$\left[0, \frac{\pi}{2}\right) \cup \left[\pi, \frac{3\pi}{2}\right)$
$\csc^{-1}$	$(-\infty, -1] \cup [1, \infty)$	$\left(0, \frac{\pi}{2}\right] \cup \left(\pi, \frac{3\pi}{2}\right]$

The first three functions are used frequently; their graphs are:

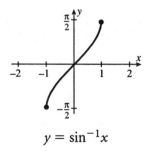

$y = \sin^{-1}x$

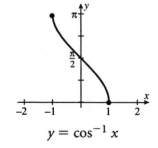

$y = \cos^{-1} x$

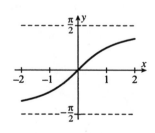

$y = \tan^{-1} x$

6) True or False:

a) $\sin^{-1}\left(\frac{3}{2}\right)$ is not defined.

True. $\frac{3}{2}$ is not in the domain of $\sin^{-1}$.

b) $\lim\limits_{x\to\infty} \tan^{-1} x = \frac{\pi}{2}$.

True.

7) Sometimes, Always, or Never:

a) $\cos(\cos^{-1} x) = x$

Sometimes. True for $-1 \le x \le 1$.

b) $\tan(\tan^{-1} x) = x$

Always.

c) $\tan^{-1}(\tan x) = x$

Sometimes. True for $-\frac{\pi}{2} \le x \le \frac{\pi}{2}$.

8) Determine each:

a) $\cos^{-1}\left(\frac{1}{2}\right)$

Let $x = \cos^{-1}\left(\frac{1}{2}\right)$. Then $0 \le y \le \pi$ and $\frac{1}{2} = \cos(y)$. From your knowledge of trigonometry, $y = \frac{\pi}{3}$.

b) $\tan^{-1}(-\sqrt{3})$

Let $y = \tan^{-1}(-\sqrt{3})$. Then $-\frac{\pi}{2} < y < \frac{\pi}{2}$ and $-\sqrt{3} = \tan y$. Thus $y = -\frac{\pi}{3}$.

c) $\sin^{-1}\left(\sin \frac{3\pi}{4}\right)$

Since $\sin \frac{3\pi}{4} = \frac{1}{\sqrt{2}}$, this problem asks you to find $y = \sin^{-1}\left(\frac{1}{\sqrt{2}}\right)$. Then $-\frac{\pi}{2} \le y \le \frac{\pi}{2}$ and $\sin y = \frac{1}{\sqrt{2}}$. Thus $y = \frac{\pi}{4}$.

d) $\sin^{-1}\left(\sin \frac{\pi}{3}\right)$

For $-\frac{\pi}{2} \le x \le \frac{\pi}{2}$, $\sin^{-1}(\sin x) = x$. Thus $\sin^{-1}\left(\sin \frac{\pi}{3}\right) = \frac{\pi}{3}$.

e) $\sec^{-1}\left(\sin\frac{\pi}{7}\right)$

This does not exist because $-1 \leq \sin\frac{\pi}{7} \leq 1$; $\sin\frac{\pi}{7}$ is not in the domain of $\sec^{-1}$.

9) For $0 < x < 1$, simplify $\tan(\cos^{-1} x)$.

Drawing and labeling a right triangle will help:

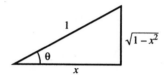

If $\cos\theta = x$ where $0 < \theta < \frac{\pi}{2}$ then $\theta = \cos^{-1} x$ and $\tan\theta = \frac{\sqrt{1-x^2}}{x}$.

Page 229

D. Here are the derivatives of the inverse trigonometric functions:

$$\frac{d}{dx}\sin^{-1} x = \frac{1}{\sqrt{1-x^2}} \qquad \frac{d}{dx}\cos^{-1} x = \frac{-1}{\sqrt{1-x^2}}$$

$$\frac{d}{dx}\tan^{-1} x = \frac{1}{1+x^2} \qquad \frac{d}{dx}\cot^{-1} x = \frac{-1}{1+x^2}$$

$$\frac{d}{dx}\sec^{-1} x = \frac{1}{x\sqrt{x^2-1}} \qquad \frac{d}{dx}\csc^{-1} x = \frac{-1}{x\sqrt{x^2-1}}$$

10) Find $f'(x)$ for each:

a) $f(x) = \sin^{-1}(x^3)$

$$f'(x) = \frac{1}{\sqrt{1-(x^3)^2}}(3x^2) = \frac{3x^2}{\sqrt{1-x^6}}$$

b) $f(x) = (\tan^{-1} x)^3$

$$f'(x) = 3(\tan^{-1} x)^2 \cdot \frac{1}{1+x^2} = \frac{3(\tan^{-1} x)^2}{1+x^2}$$

c) $f(x) = x\cos^{-1} x$

$$f'(x) = x\left(\frac{-1}{\sqrt{1-x^2}}\right) + (\cos^{-1} x)(1)$$
$$= \frac{-x}{\sqrt{1-x^2}} + \cos^{-1} x.$$

d) $f(x) = e^{\csc^{-1} x^2}$

$$f'(x) = e^{\csc^{-1} x^2}\left(\frac{-1}{x^2\sqrt{(x^2)^2-1}}\right)(2x)$$
$$= \frac{-2e^{\csc^{-1} x^2}}{x\sqrt{x^4-1}}.$$

11) Find y' for $y = \sin^{-1} x + \cos^{-1} x$. What can you conclude about $\sin^{-1} x + \cos^{-1} x$?

$y' = \dfrac{1}{\sqrt{1 - x^2}} + \dfrac{-1}{\sqrt{1 - x^2}} = 0$. Thus y is a constant. To determine the constant, choose

$x = 0$:

$y = \sin^{-1} 0 + \cos^{-1} 0 = 0 + \dfrac{\pi}{2} = \dfrac{\pi}{2}$.

Therefore $\sin^{-1} x + \cos^{-1} x = \dfrac{\pi}{2}$.

Section 3.7   Higher Derivatives

Since the derivative of a function is itself a function you may take the derivative of the derivative (called the second derivative). The same differentiation rules apply, use the processes twice.

Concepts to Master

A. Higher derivatives; Higher derivative notations
B. Acceleration as the derivative of velocity
C. Relationships between f, f', and f''

Summary and Focus Questions

Page 233

A. The second derivative of $y = f(x)$, denoted y'', is the derivative of y'. The third derivative, y''', is the derivative of y''. The nth derivative is denoted $y^{(n)}$.

For $y = x^6$,
$y' = 6x^5$
$y'' = 30x^4$
$y''' = 120x^3$
$y^{(4)} = 360x^2$.

Each of these are notations for the nth derivative of y with respect to x:

$$y^{(n)}, f^{(n)}(x), \frac{d^n y}{dx^n}, D^n f(x), D^n y$$

1) Find $y^{(2)}$ for:

   a) $y = 5x^3 + 4x^2 + 6x + 3$

$y' = 15x^2 + 8x + 6$, so $y'' = 30x + 8$.

   b) $y = \sqrt{x^2 + 1}$

$y = (x^2 + 1)^{1/2}$
$y' = \frac{1}{2}(x^2 + 1)^{-1/2}(2x) = x \cdot (x^2 + 1)^{-1/2}$
Thus, $y'' =$
$x\left[-\frac{1}{2}(x^2 + 1)^{-3/2}(2x)\right] + (x^2 + 1)^{-1/2}(1)$
$= -x^2(x^2 + 1)^{-3/2} + (x^2 + 1)^{-1/2}$.

2) True or False:
The notation for the sixth derivative of
$y = f(x)$ is $\dfrac{dy^6}{d^6x}$.

False. It is $\dfrac{d^6y}{dx^6}$. (Think of applying $\dfrac{d^6}{dx^6}$ to the function y.)

3) Find $f^{(4)}(x)$, where $f(x) = \dfrac{x^{10}}{90} + \dfrac{x^5}{60}$.

$$f'(x) = \frac{x^9}{9} + \frac{x^4}{12}$$
$$f^{(2)}(x) = x^8 + \frac{x^3}{3}$$
$$f^{(3)}(x) = 8x^7 + x^2$$
$$f^{(4)}(x) = 56x^6 + 2x.$$

4) Find a formula for $f^{(n)}(x)$, where $f(x) = \dfrac{1}{x^2}$.

$$f(x) = x^{-2}$$
$$f^{(1)}(x) = -2x^{-3}$$
$$f^{(2)}(x) = (-2)(-3)x^{-4}$$
$$f^{(3)}(x) = (-2)(-3)(-4)x^{-5}.$$
The pattern is
$$f^{(n)}(x) = (-2)(-3)...(-(n+1))x^{-(n+2)}$$
$$= \frac{(-1)^n(n+1)!}{x^{n+2}}.$$

5) If $y = f(x)$ is a polynomial of degree n, what is $y^{(n+1)}$?

$y^{(n+1)} = 0$. y' has degree $n - 1$, y'' has degree $n - 2$, and so on. $y^{(n)}$ has degree zero (is a constant). Thus, $y^{(n+1)} = 0$.

B. The rate of change of velocity is called *acceleration.* (The gas pedal in a car is sometimes called the "accelerator" because when you push your foot on it the car speeds up—the velocity changes.) Since y' measures the velocity of a particle whose position is given by $y = f(t)$, y'' measures the change in velocity; that is, y'' is the instantaneous acceleration of the particle.

Page
233

6) Find the velocity and acceleration after 2 seconds of a particle whose position (in meters) after t seconds is
$s = 3t^3 + 6t^2 + t + 1$.

$s' = 9t^2 + 12t + 1$. At $t = 2$, the velocity is
$s' = 9(2)^2 + 12(2) + 1 = 61$ m/s.
$s'' = 18t + 12$. At $t = 2$, the acceleration is
$s'' = 18(2) + 12 = 48$ m/s^2.

7) Find the acceleration at time $\frac{\pi}{6}$ for a
particle whose position in meters at time
t seconds is $s = \cos 2t$.

$s = \cos 2t,\ s' = -2 \sin 2t,$

$s'' = -4 \cos 2t.$ At $t = \frac{\pi}{6}$,

$s'' = -4 \cos \frac{\pi}{3} = -4\left(\frac{1}{2}\right) = -2$ m/s^2.

Page
233

C. The graph of $f''(x)$ has the same relationship to the graph of $f'(x)$ as the
graph of $f'(x)$ does to the graph of $f(x)$. Therefore, $f''(x)$ is the slope of the
tangent line to $y = f'(x)$.

8) Use your best judgement to determine
the graph of $y = f''(x)$ given this graph of
$y = f(x)$.

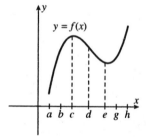

From the slopes of tangents to $y = f(x)$ the
graph of $y = f'(x)$ is:

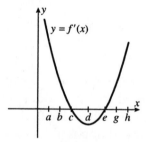

From the slopes of tangents to this graph of
$y = f'(x)$ the graph of $y = f''(x)$ is:

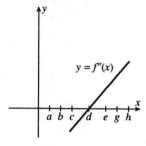

Section 3.8   Derivatives of Logarithmic Functions

This section completes the development of the derivative of logarithmic functions and provides a technique (logarithmic differentiation) for finding the derivative of certain complicated functions.

Concepts to Master

A. Derivatives of $\ln x$ and $\log_a x$

B. Logarithmic differentiation

C. The number e as a limiting value

Summary and Focus Questions

Page 240

A. If $f(x) = \ln x$, $f'(x) = \frac{1}{x}$. In general, $\frac{d}{dx} \ln u(x) = \frac{1}{u(x)} \cdot u'(x)$.

If $f(x) = \log_a x$, $f'(x) = \frac{1}{x \ln a}$.

1) Find $f'(x)$ for each:

a) $f(x) = \ln x^2$

$f(x) = 2 \ln x$, so $f'(x) = 2\left(\frac{1}{x}\right) = \frac{2}{x}$.
Alternatively, using the chain rule,
$f'(x) = \frac{1}{x^2}(2x) = \frac{2}{x}$.

b) $f(x) = \ln \cos x$

$f'(x) = \frac{1}{\cos x} \cdot (-\sin x) = -\tan x$.

c) $f(x) = x^2 \ln x$

$f'(x) - x^2 \left(\frac{1}{x}\right) + \ln x \,(2x) = x + 2x \ln x$.

d) $f(x) = \log_2(x^2 + 1)$

$f'(x) = \frac{1}{(\ln 2)(x^2 + 1)} \cdot (2x)$

$= \frac{2x}{(\ln 2)(x^2 + 1)}$.

e) $f(x) = \ln \sqrt{\dfrac{x^2 + 1}{2x^3}} \ (x > 0)$

$f(x) = \frac{1}{2} \ln \left(\frac{x^2 + 1}{2x^3}\right)$
$= \frac{1}{2}(\ln(x^2 + 1) - \ln 2 - 3 \ln x)$.
$f'(x) = \frac{1}{2}\left(\frac{2x}{x^2 + 1} - \frac{3}{x}\right)$.

Page 242

B. The derivative of a function $y = f(x)$ may be found using *logarithmic differentiation*—that is,

   1. Take ln of both sides of $y = f(x)$: $\ln y = \ln(f(x))$.

   2. Simplify $\ln(f(x))$ using properties of logarithms.

   3. Differentiate both sides implicitly and solve for y': $y' = y\frac{d}{dx}\ln(f(x))$.

Example: Find y', where $y = x^{2/x}$.

   1. $\ln y = \ln x^{2/x}$

   2. $\ln y = \frac{2}{x}\ln x$

   3. $\frac{1}{y}\cdot y' = \frac{2}{x}\left(\frac{1}{x}\right) + \ln\left(-\frac{2}{x^2}\right) = \frac{2}{x^2}(1 - \ln x)$

      $y' = x^{2/x}\left(\frac{2}{x^2}(1 - \ln x)\right) = 2x^{2/x\,-\,2}(1 - \ln x)$.

2) Find y' for each, using logarithmic differentiation.

   a) $y = \left(\frac{10x^3}{\sqrt{x+1}}\right)^4$

      1. $\ln y = \ln\left(\frac{10x^3}{\sqrt{x+1}}\right)^4$.

      2. $\ln y = 4(\ln 10x^3 - \ln\sqrt{x+1})$
            $= 4(\ln 10 + 3\ln x - \frac{1}{2}\ln(x+1))$

      3. $\frac{d}{dx}\ln y = 4\left(0 + \frac{3}{x} - \frac{1}{2}\frac{1}{x+1}(1)\right)$
              $= 4\left(\frac{3}{x} - \frac{1}{2(x+1)}\right)$.

      Thus, $y' = \left(\frac{10x^3}{\sqrt{x+1}}\right)^4\left[4\left(\frac{3}{x} - \frac{1}{2(x+1)}\right)\right]$.

   b) $y = 6^x$

      1. $\ln y = \ln 6^x$

      2. $\ln y = x\ln 6$.

      3. $\frac{d}{dx}\ln y = \ln 6$ (a constant).

      Thus, $y' = y\ln 6 = 6^x\ln 6$.

Page 244

C. Here is another way to determine the value of *e*:

$$\lim_{x \to 0^+} (1 + x)^{1/x} = e.$$

Equivalently, $\lim_{t \to \infty} \left(1 + \frac{1}{t}\right)^t = e.$

3) Find each limit:

a) $\lim_{h \to 0^+} (1 + h)^{1/h}$

e.

b) $\lim_{x \to \infty} \left(1 + \frac{1}{x}\right)^{3x}$

$$\lim_{x \to \infty} \left(1 + \frac{1}{x}\right)^{3x} = \lim_{x \to \infty} \left[\left(1 + \frac{1}{x}\right)^x\right]^3$$
$$= \left[\lim_{x \to \infty} \left(1 + \frac{1}{x}\right)^x\right]^3 = e^3.$$

c) $\lim_{x \to 1^+} x^{1/(x-1)}$

Let $t = x - 1$, hence $x = 1 + t$.

As $x \to 1^+$, $t \to 0^+$.

Thus $\lim_{x \to 1^+} x^{1/(x-1)} = \lim_{t \to 0^+} (1 + t)^{1/t} = e.$

4) Give three different definitions for the number *e*.

1. *e* is the unique number *a* for which the slope of the tangent line to $y = a^x$ at $(0, 1)$ is 1.

2. *e* is the unique number for which
$$\lim_{h \to 0^+} \frac{e^h - 1}{h} = 1.$$

3. $e = \lim_{x \to 0^+} (1 + x)^{1/x}.$

Section 3.9   **Hyperbolic Functions**

Certain combinations of e^x and e^{-x} give rise to functions called hyperbolic functions: sinh, cosh, etc. The reason they are called hyperbolic is because some of their properties are similar to ones for the trigonometric functions but are based on the hyperbola $x^2 - y^2 = 1$ rather than the circle $x^2 + y^2 = 1$. For example, the hyperbolic identity $\cosh^2 x - \sinh^2 x = 1$ may be remembered by noting that $(\cosh x, \sinh x)$ is on the hyperbola in the same manner that $\cos^2 x + \sin^2 x = 1$ may be remembered by noting that $(\cos x, \sin x)$ is on the circle.

Concepts to Master

A. Define the hyperbolic trigonometric functions

B. Derivatives involving hyperbolic functions

C. Definition and derivatives of inverse hyperbolic functions

Summary and Focus Questions

Page 246

A. The hyperbolic sine and cosine functions are:

$$\sinh x = \frac{e^x - e^{-x}}{2} \qquad\qquad \cosh x = \frac{e^x + e^{-x}}{2}$$

The other hyperbolic trigonometric functions are:

$$\tanh x = \frac{\sinh x}{\cosh x} \qquad\qquad \coth x = \frac{\cosh x}{\sinh x}$$

$$\operatorname{sech} x = \frac{1}{\cosh x} \qquad\qquad \operatorname{csch} x = \frac{1}{\sinh x}$$

1) Evaluate sinh 2.

$$\sinh 2 = \frac{e^2 - e^{-2}}{2} = \frac{e^4 - 1}{2e^2}.$$

2) Write coth x in terms of e^x.

$$\coth x = \frac{\cosh x}{\sinh x} = \frac{\frac{e^x + e^{-x}}{2}}{\frac{e^x - e^{-x}}{2}} = \frac{e^x + e^{-x}}{e^x - e^{-x}}.$$

3) Verify $\sinh 2x = 2 \sinh x \cosh x$.

$$2 \sinh x \cosh x = 2\frac{e^x - e^{-x}}{2} \cdot \frac{e^x + e^{-x}}{2}$$
$$= \frac{(e^x - e^{-x})(e^x + e^{-x})}{2}$$
$$= \frac{e^{2x} - e^{-2x}}{2} = \sinh 2x.$$

4) What is the domain and range of
$y = \cosh x$?

$$\cosh x = \frac{e^x + e^{-x}}{2} \text{ is defined for all real } x$$
and is always greater than or equal to one
(since e^x and e^{-x} are reciprocals). Thus the
domain is all reals and range is $[1, \infty)$.

5) Evaluate $\tanh(\ln 5)$

$$\tanh(\ln 5) = \frac{\sinh(\ln 5)}{\cosh(\ln 5)}$$
$$= \frac{\dfrac{e^{\ln 5} - e^{-\ln 5}}{2}}{\dfrac{e^{\ln 5} + e^{-\ln 5}}{2}} = \frac{5 - \frac{1}{5}}{5 + \frac{1}{5}} = \frac{12}{13}.$$

Page 248

B. Here are derivatives for the hyperbolics:

$$\frac{d}{dx}\sinh x = \cosh x \qquad\qquad \frac{d}{dx}\cosh x = \sinh x$$

$$\frac{d}{dx}\tanh x = \operatorname{sech}^2 x \qquad\qquad \frac{d}{dx}\coth x = -\operatorname{csch}^2 x$$

$$\frac{d}{dx}\operatorname{sech} x = -\operatorname{sech} x \tanh x \qquad\qquad \frac{d}{dx}\operatorname{csch} x = -\operatorname{csch} x \coth x$$

6) Find $f'(x)$ for each:

  a) $f(x) = \tanh\left(\frac{x}{2}\right)$

$$f'(x) = \operatorname{sech}^2\left(\tfrac{x}{2}\right)\left(\tfrac{1}{2}\right) = \tfrac{1}{2}\operatorname{sech}^2\left(\tfrac{x}{2}\right).$$

  b) $f(x) = \sqrt{\cosh x}$

$$f'(x) = \tfrac{1}{2}(\cosh)^{-1/2}(\sinh x) = \frac{\sinh x}{2\sqrt{\cosh x}}$$

  c) $f(x) = \sinh x + \cosh x$

$$f'(x) = \cosh x + \sinh x. \text{ Thus } f'(x) = f(x).$$
If this seems familiar it should since
$\sinh x + \cosh x = e^x$ and $(e^x)' = e^x$.

C. The inverses of the hyperbolic functions are:

Function	Domain	Range	Explicit Form
$\sinh^{-1} x$	all reals	all reals	$\ln(x + \sqrt{x^2 + 1})$
$\cosh^{-1} x$	$[1, \infty)$	$[0, \infty)$	$\ln(x + \sqrt{x^2 - 1})$
$\tanh^{-1} x$	$(-1, 1)$	all reals	$\frac{1}{2} \ln\left(\frac{1 + x}{1 - x}\right)$
$\coth^{-1} x$	$(-\infty, -1) \cup (1, \infty)$	all reals except 0	$\frac{1}{2} \ln\left(\frac{1 + x}{x - 1}\right)$
$\text{sech}^{-1} x$	$(0, 1]$	$[0, \infty)$	$\ln\left(\frac{1}{x} + \frac{\sqrt{1 - x^2}}{x}\right)$
$\text{csch}^{-1} x$	all reals except 0	all reals except 0	$\ln\left(\frac{1}{x} + \frac{\sqrt{1 + x^2}}{|x|}\right)$

The derivatives are:

$$\frac{d}{dx} \sinh^{-1} x = \frac{1}{\sqrt{1 + x^2}} \qquad \frac{d}{dx} \text{csch}^{-1} x = \frac{-1}{|x|\sqrt{x^2 + 1}}$$

$$\frac{d}{dx} \cosh^{-1} x = \frac{1}{\sqrt{x^2 - 1}} \qquad \frac{d}{dx} \text{sech}^{-1} x = \frac{-1}{x\sqrt{1 - x^2}}$$

$$\frac{d}{dx} \tanh^{-1} x = \frac{1}{1 - x^2} \qquad \frac{d}{dx} \coth^{-1} x = \frac{1}{1 - x^2}$$

7) Evaluate $\tanh^{-1} \frac{1}{2}$.

$$\tanh^{-1} \frac{1}{2} = y \text{ means } \tanh y = \frac{1}{2}.$$
$$\frac{e^y - e^{-y}}{e^y + e^{-y}} = \frac{1}{2}$$
$$2e^y - 2e^{-y} = e^y + e^{-y}$$
$$e^y = 3e^{-y}$$
$$e^{2y} = 3$$
$$2y = \ln 3$$
$$y = \frac{1}{2} \ln 3.$$

8) Show that $f(x) = \tanh^{-1} x$ is always positive.

$$f'(x) = \frac{1}{1 - x^2}$$
Since the domain is $(-1, 1)$ we have
$$-1 < x < 1, \text{ hence } \frac{1}{1 - x^2} > 0.$$

9) Verify the derivative $\frac{d}{dx}\coth^{-1}x = \frac{1}{1-x^2}$, using the explicit form for $\coth^{-1}x$.

$\coth^{-1}x = \frac{1}{2}\ln\left(\frac{1+x}{x-1}\right)$.

Therefore, $\frac{d}{dx}\coth^{-1}x$

$$= \frac{1}{2}\frac{1}{\frac{1+x}{x-1}}\frac{(x-1)(1)-(1-x)(1)}{(x-1)^2}$$

$$= \frac{1}{2}\frac{x-1}{1+x}\frac{-2}{(x-1)^2} = \frac{-1}{(1+x)(x-1)}$$

$$= \frac{1}{1-x^2}.$$

10) Find $f'(x)$ for:

 a) $f(x) = \sinh^{-1}x^2$

$$f'(x) = \frac{1}{\sqrt{1+(x^2)^2}}(2x) = \frac{2x}{\sqrt{1+x^4}}.$$

 b) $f(x) = \tanh^{-1}\sqrt{1-x^2}$

$$f'(x) = \frac{1}{1-(\sqrt{1-x^2})^2}\frac{1}{2}(1-x^2)^{-1/2}(-2x)$$

$$= \frac{1}{x^2}\frac{1}{2}\frac{1}{\sqrt{1-x^2}}(-2x)$$

$$= \frac{-1}{x\sqrt{1-x^2}}.$$

Section 3.10   Related Rates

If two quantities $a(t)$ and $b(t)$ are functions of time and are related by some equation, then by implicit differentiation their derivatives, $a'(t)$ and $b'(t)$, are also related. Thus, when one of the rates is given, it may be used to find the other rate. This is the idea behind "related rates" problems: applications that involve relating the derivatives of two or more quantities that change over time.

Concepts to Master

Solve related rates problems

Summary and Focus Questions

Page 253

A related rate problem is an application in which one or more rates of change are given and you are asked to find the rate of change of some other (related) quantity. The procedure to solve a related rate problem is:

1) Illustrate the problem with a picture, if possible.

2) Identify and label with constants all fixed quantities and with variables all quantities that vary with time. Any rates of change given in the problem are the rates of change of these variables. The unknown rate will also be the rate of change of one of these variables.

3) Find an equation that relates the constants, the variables whose rates are given, and the variable whose rate is unknown.

4) Differentiate the equation in 3) with respect to time.

5) Substitute all known quantities in the result from 4) and solve for the unknown rate.

Step 3) is often the hardest and will call upon your skills in remembering relationships from geometry, trigonometry and other branches of mathematics. Remember not to substitute known quantities until after differentiating in Step 4).

1) An observer, 300 meters from the launch pad of a rocket, watches it ascend vertically at 60 m/s. Find the rate of change of the distance between the rocket and the observer when the rocket is 400 meters high.

(1)

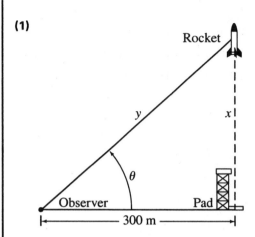

(2) Let y = distance between rocket and observer. Let x = distance between rocket and pad. $\frac{dx}{dt}$ is known (60); $\frac{dy}{dt}$ is unknown.

(3) Relate y, x, and 300: $y^2 = 300^2 + x^2$.

(4) Use implicit differentiation, remembering y and x are functions of time t:

$$2y\frac{dy}{dt} = 0 + 2x\frac{dx}{dt}$$
$$y\frac{dy}{dt} = x\frac{dx}{dt}.$$

(5) Substitute known values:

$\frac{dx}{dt} = 60$, $x = 400$. The value of y is known when $x = 400$; from

$$y^2 = 300^2 + (400)^2,$$

$$y = 500.$$

$$y\frac{dy}{dt} = x\frac{dx}{dt}$$
$$500\frac{dy}{dt} = 400(60)$$
$$\frac{dy}{dt} = 48 \text{ m/s}.$$

2) A spherical ball has its diameter increasing at 2 inches/s. How fast is the volume changing when the radius is 10 inches?

(1)

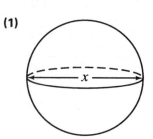

(2) Let x be the diameter V the volume. We are given $\frac{dx}{dt} = 2$.

(3) Relate V and x using the formula for the volume of a sphere:
$$V = \tfrac{4}{3}\pi\left(\tfrac{x}{2}\right)^3 = \tfrac{\pi}{6}x^3$$

(4) Differentiate:
$$\frac{dV}{dt} = \frac{3\pi}{6}x^2\frac{dx}{dt} = \frac{\pi x^2}{2}\frac{dx}{dt}.$$

(5) When the radius is 10, $x = 20$.
$$\frac{dV}{dt} = \frac{\pi(20)^2}{2} \cdot 2 = 400\pi \text{ in}^3/\text{s}.$$

3) A light is on top of a 12 ft vertical pole. A 6 ft woman walks away from the pole base at 4 ft/s. How fast is the angle made by the woman's head, the light, and the pole base changing 2 seconds after she starts walking?

(1)

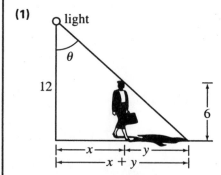

(2) Let θ be the desired angle. Let x be the distance between the pole base and the woman. Let y be the distance from the woman to the tip of her shadow.

(3) We are given $\frac{dx}{dt}$ ($= 4$) and we must find $\frac{d\theta}{dt}$.

Thus, we must relate x and θ. We can relate $x + y$ and θ so we first determine y in terms of x. By similar triangles, $\frac{x + y}{12} = \frac{y}{6}$. Thus,

$6x + 6y = 12y$

$6x = 6y$, or $x = y$.

Thus, $x + y = x + x = 2x$.

By trigonometry, we obtain the desired relation between x and θ:

$\tan \theta = \frac{2x}{12}$, or $6 \tan \theta = x$.

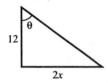

(4) Differentiate the relation:

$6 \sec^2 \theta \cdot \frac{d\theta}{dt} = \frac{dx}{dt}$.

(5) After 2 seconds, $x = 2(4) = 8$ so $2x = 16$ ft.

$\sec \theta = \frac{d}{12}$, so $\sec^2 \theta = \frac{d^2}{144}$.

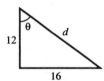

From the figure, $12^2 + 16^2 = d^2$, so $d^2 = 400$ and $\sec^2 \theta = \frac{400}{144}$. Finally, substituting into the result from (4):

$6\left(\frac{400}{144}\right)\frac{d\theta}{dt} = 4$

$\frac{d\theta}{dt} = 0.24$ radians/s.

4)

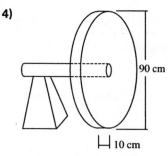

90 cm

10 cm

A 10 cm thick mill wheel is initially 90 cm in diameter and is wearing away (uniformly) at 55 cm³/hr. How fast is the diameter changing after 20 hours?

(1)

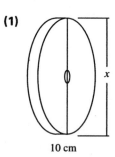

x

10 cm

(2) Let V be the volume and x the diameter; both are functions of time.

$\frac{dV}{dt}$ is known (-55), $\frac{dx}{dt}$ is unknown.

(3) From the formula for the volume of a cylinder: $V = 10\pi\left(\frac{x}{2}\right)^2$, so $V = 2.5\pi x^2$.

(4) Differentiate with respect to time:
$\frac{dV}{dt} = 5\pi x \cdot \frac{dx}{dt}$.

(5) 20 hours later the amount of material worn away is $(55)(20) = 1100$ cm³.
The volume after 20 hours is
$2.5\pi(90)^2 - 1100 \approx 62517$ cm³.
Thus the diameter x after 20 hours is
$62517 = 2.5\pi x^2$.

$x = \sqrt{\frac{62517.25}{2.5\pi}} = 89.2$ cm.

Finally, from $\frac{dV}{dt} = 5\pi x\frac{dx}{dt}$

$-55 = 5\pi(89.2)\frac{dx}{dt}$

$\frac{dx}{dt} = -0.039$ cm/hr.

Section 3.11   Linear Approximations and Differentials

The tangent line to a function $y = f(x)$ at a point a can be used to estimate a value for $f(x)$ when x is near a. This section shows how that is done.

Concepts to Master

A. Linear approximation of a function value

B. Differentials; approximations with differentials

Summary and Focus Questions

Page 259

A. Let $L(x)$ be the equation of the tangent line to $y = f(x)$ at $x = a$:

$$L(x) = f(a) + f'(a)(x - a).$$

Both $L(x)$ and $f(x)$ have the same first derivative so they both "head in the same direction" ($f'(a)$) at $x = a$.) For x near a, $L(x)$ may be used as a linear approximation to $f(x)$:

$$L(x) \approx f(x).$$

$L(x)$ is called a *linearization* of f at a.

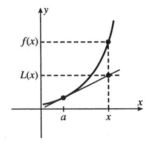

1) Find the linear approximation to $f(x) = 5x^3 + 6x$ at $x = 2$.

What we are asked to do is find the equation of the tangent line:
$f(2) = 5(2)^3 + 6(2) = 52$.
$f'(x) = 15x^2 + 6$, so
$f'(2) = 15(2)^2 + 6 = 66$.
Thus, $L(x) = 52 + 66(x - 2)$.

2) Approximate $f(1.98)$ for the function in question 1.

We use $f(1.98) \approx L(1.98)$.
$L(1.98) = 52 + 66(1.98 - 2) = 50.68$.
(*Note:* $f(1.98) = 50.69196$, so $L(1.98)$ is rather close and "easier" to calculate.)

3) Approximate $\sqrt{66}$ using a linear approximation.

Choose $f(x) = \sqrt{x}$ and $a = 64$ (64 is near 66). Then $f(64) = 8$, $f'(x) = \dfrac{1}{2\sqrt{x}}$, and $f'(64) = \dfrac{1}{2\sqrt{64}} = \dfrac{1}{16}$.

The linear approximation is:

$L(x) = 8 + \dfrac{1}{16}(x - 64)$.

At $x = 66$,

$L(66) = 8 + \dfrac{1}{16}(66 - 64) = 8.125$.

(Note: $\sqrt{66} \approx 8.1240384$).

4) Find the linear approximation of $f(x) = \sin x$ at $a = 0$. Use it to estimate $\sin\left(\dfrac{\pi}{15}\right)$.

$f(x) = \sin x$
$f(0) = \sin 0 = 0$.
$f'(x) = \cos x$
$f'(0) = \cos 0 = 1$.
So $L(x) = f(0) + f'(0)(x - 0)$
$\qquad = 0 + 1(x - 0) = x$.
Thus, $\sin x \approx x$, for x near 0.

For $x = \dfrac{\pi}{15}$, $\sin\left(\dfrac{\pi}{15}\right) \approx \dfrac{\pi}{15}$. (Using a calculator, $\sin \dfrac{\pi}{15} \approx .2079$ and $\dfrac{\pi}{15} \approx .2094$.)

Page 262

B. Let $y = f(x)$ be a differentiable function. The differential dx is an independent variable. The *differential dy* is defined as $dy = f'(x)\, dx$.

Note that dy is a function of both x (because of $f'(x)$) and dx.

If we let $dx = \Delta x$, then for small values of dx, the change in the function (Δy) is approximately the same as the change in the tangent line dy:

$\qquad dy \approx \Delta y$, when dx is small.

This is handy since dy may be easier to calculate than Δy.

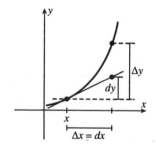

dy may be thought of as the *error* in calculating a value for y if an error of dx is made in estimating x. $\dfrac{dy}{y}$ is the *relative error*.

Example: The sides of a square field are measured and found to be 50 m with a possible error of .02 m. Estimate the maximum error and relative error of the area of the field.

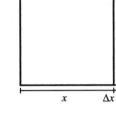

Let x = the side of the field and A be the area. We are given $\Delta x = dx = .02$ when $x = 50$.

$A = x^2$, so $dA = 2x \, dx$.

We estimate the maximum error ΔA with dA:

$$dA = 2(50)(.02) = 2 \text{ m}^2$$

and the relative error with $\frac{dA}{A}$:

$$\frac{dA}{A} = \frac{2 \text{ m}^2}{(50 \text{ m})^2} = .0008 = .08\%$$

5) True or False:

  a) $\Delta x = dx$

  True.

  b) $\Delta y = dy$

  False.

6) Compute dy and Δy for $f(x) = x^2 + 3x$ at $x = 2$ with $\Delta x = dx = 0.1$.

$f(x) = x^2 + 3x$.
$f' = 2x + 3$.
At $x = 2, f'(2) = 2(2) + 3 = 7$.
Thus, $dy = f'(x) \, dx = f'(2)(0.1)$
    $= 7(0.1) = 0.7$.
At $x = 2, y = f(2) = 2^2 + 3(2) = 10$.
At $x = 2 + \Delta x = 2 + 0.1 = 2.1$,
$y = f(2.1) = (2.1)^2 + 3(2.1) = 10.71$.
Thus, $\Delta y = f(2.1) - f(2)$
    $= 10.71 - 10 = 0.71$.
Note that $dy = 0.7$ and $\Delta y = 0.71$ are quite close but dy is easier to calculate.

7) A circle has a radius of 20 cm with a possible error of 0.02 cm. Use differentials to estimate the maximum error and relative error for the area of the circle.

Let x = radius of circle and A = area of circle. We are given $\Delta x = dx = 0.02$ for $x = 20$. The error is ΔA which we estimate with dA: $A = \pi x^2$, so
$dA = 2\pi x \, dx = 2\pi(20) \cdot (0.02) = 0.8\pi$.
The relative error is
$\frac{dA}{A} = \frac{0.8\pi}{\pi(20)^2} \approx .002 = .2\%$.

Technology Plus for Chapter 3

1) a) Use a computer or calculator to graph
$f(x) = x - \sin x$, for $-20 \leq x \leq 20$.

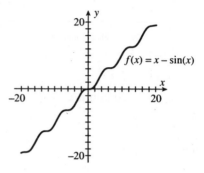

b) Compute and graph $f'(x)$.

$f'(x) = 1 - \cos x.$

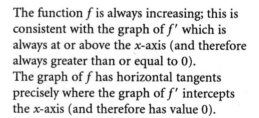

c) What can you conclude about the graph
of f in relation to the graph of f'?

The function f is always increasing; this is
consistent with the graph of f' which is
always at or above the x-axis (and therefore
always greater than or equal to 0).
The graph of f has horizontal tangents
precisely where the graph of f' intercepts
the x-axis (and therefore has value 0).

2) Use a calculator to graph $f(x) = |9 - x^2|$
for $-5 \leq x \leq 5$. Where is f not differentiable?

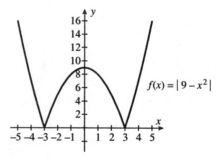

The function is not differentiable at $x = -3$
and $x = 3$.

3) Graph the function $f(x) = \dfrac{1}{5\sin x + \cos x}$ for $-5 \le x \le 5$. For how many points in $[-5, 5]$ is f not continuous?

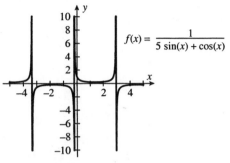

There are three values for x where $f(x)$ is not continuous. By zooming in, we see that they are approximately -3.334, -0.189, and 2.963.

4) Let $f(x) = \sqrt{1 + x}$

    a) Find the equation of the tangent line at $x = 3$.

$f(x) = (1 + x)^{1/2};\ f(3) = (1 + 3)^{1/2} = 2$

$f'(x) = \frac{1}{2}(1 + x)^{-1/2} = \dfrac{1}{2\sqrt{1 + x}}$

$f'(3) = \dfrac{1}{2\sqrt{1 + 3}} = \dfrac{1}{4}$

The tangent line is $y = 2 + \frac{1}{4}(x - 3)$.

    b) Graph $f(x)$ and use the tangent line on the same screen. Use a window of $[-1, 6]$ by $[-1, 6]$.

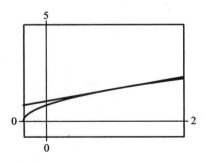

c) Use a calculator to complete this table of values for *x* near 3:

x	f(x)	tan line	difference
2.5			
2.9			
2.95			
2.99			
2.999			
2.9999			

x	f(x)	tan line	difference
2.5	1.870829	1.87500	−0.12500
2.9	1.974842	1.97500	−0.02500
2.95	1.987461	1.98750	−0.01250
2.99	1.997498	1.99750	−0.00250
2.999	1.99975	1.99975	−0.00025
2.9999	1.999975	1.99998	−0.00002

Chapter 4   Applications of Differentiation

"GRANTZ IS CHARTING HIS LIFE BASED ON GENETIC VS. ENVIRONMENTAL FACTORS."

© 1999 by Sidney Harris.

Section 4.1 Maximum and Minimum Values

This section gives you some tools for finding the largest and smallest values of $f(x)$ for a given function f, where x is from a set D. The largest and smallest values of $f(x)$ for all x in D are called absolute extrema; those places where $f(x)$ is largest or smallest in a region around the point x are called relative extrema. Extreme values are important in applications (viz., maximum profit, minimum force, etc.)

Concepts to Master

A. Absolute maxima, minima, and extrema; the Extreme Value Theorem

B. Relative (or local) maxima, minima, and extrema

C. Critical Numbers; Fermat's Theorem about local extrema

D. The Closed Interval Method for finding extrema for a continuous function on a closed interval

Summary and Focus Questions

A. Let f be a function with domain D. f has an *absolute maximum* at c (and $f(c)$ is the *maximum value*) means $f(x) \leq f(c)$ for all $x \in D$. Thus, a highest point on the graph of f occurs at $(c, f(c))$.

f has an *absolute minimum* at c means $f(x) \geq f(c)$ for all $x \in D$, so $(c, f(c))$ is a lowest point on the graph.

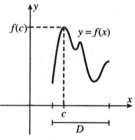

The *extreme values* of f are the maximum of f (if there is one) and the minimum of f (if there is one). The following three graphs from left to right have two, one, and no extreme values on D respectively.

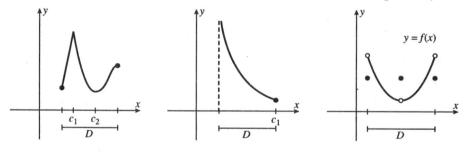

The Extreme Value Theorem: The extreme values must always exist for a continuous function whose domain is a closed interval.

Page 277

1) If $f(x) \geq f(c)$ for all x in the domain of the function f, then f has an _____ at c.

absolute minimum.

2) True or False:
A function may have more than one absolute minimum value on a set D.

False. There is only one absolute minimum value if it exists. However, this minimum could occur at more than one point.

3) Where does the absolute minimum value of $f(x) = 9 - x^2$ with domain $[-2, 2]$ occur?

$f(x) = 9 - x^2$ with domain $[-2, 2]$ has a minimum value of 5 which occurs at both $x = 2$ and $x = -2$.

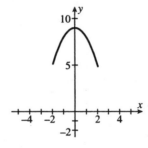

4) Does the Extreme Value Theorem guarantee that the extreme values exist for the function graphed?

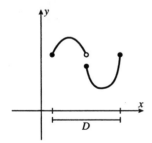

No. The function is not continuous on D. For this function, the extreme values exist, but their existence is not guaranteed by the Extreme Value Theorem.

Page 278

B. If there is some open interval I containing c such that $f(c) \geq f(x)$ for all
$x \in I$, then f has a *local (or relative) maximum* at c. In case $f(c) \leq f(x)$ for all
$x \in I$, we say f has a *local (or relative) minimum* at c.

The following graph is a function with domain $[a, b]$ with four local
maxima at $x = c, p, r, t$ and three local minima at $x = d, q, s$.

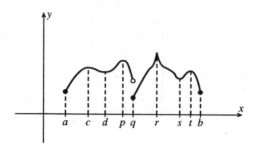

5) True or False:
If f has domain $[a, b]$, $c \in (a, b)$, and f has
an absolute minimum at $x = c$ then f has a
local minimum at $x = c$.

True.

6) How many local maxima and minima does
the following function with domain $[a, b)$
have?

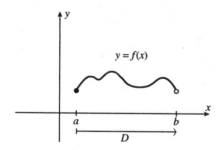

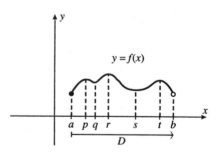

f has 3 local maxima (at $x = p, r$, and t) and
2 local minima (at $x = q$ and s).

C. A *critical number c* is a number in the domain of f for which either $f'(c) = 0$ or $f'(c)$ does not exist.

Page 281

Determining where $f'(x) = 0$ is a matter of solving the equation $f'(x) = 0$. The method of solution depends greatly on the nature of the function. You can use a computer algebra system or a calculator to approximate a solution. If solving by hand, factoring may work, as in this example.

Example:

Solve $f'(x) = 0$ where $f(x) = 3x^4 + 20x^3 - 36x^2 + 7$.
$f'(x) = 12x^3 + 60x^2 - 72x = 12x(x - 1)(x + 6)$.
So $f'(x) = 0$ at $x = 0$, $x = 1$, $x = -6$.

Determining where $f'(c)$ does not exist is often (but not always!) a matter of finding points where a denominator is 0 or where an even root of a negative number occurs.

Examples:

Find all points in the domain of f at which $f'(x)$ does not exist.

a) $f(x) = 9\sqrt{x} + x^{3/2}$. Here the domain of f is $[0, \infty]$,
$f'(x) = \dfrac{9}{2\sqrt{x}} + \dfrac{3\sqrt{x}}{2}$ and $f'(x)$ does not exist for $x = 0$.

b) $f(x) = \sqrt{4 - x}$. The domain of f is $(-\infty, 4]$ and $f'(x) = \dfrac{-1}{2\sqrt{4 - x}}$ does not exist at $x = 4$.

Fermat's Theorem: If f has a local maximum or minimum at c and $f'(c)$ exists, then $f'(c) = 0$.

In other words, if $f(c)$ is a local extremum for f, then c is a critical number of f. The converse is false—a critical number does not have to be where a local maximum or local minimum occurs.

7) Suppose $f'(c) = 0$. Is c a local maximum or minimum?

Not necessarily; for example, $f(x) = x^3$ with $c = 0$ has $f'(c) = 0$ but $c = 0$ is not a local extremum of $f(x) = x^3$.

8) Find the critical numbers for:

a) $f(x) = \sqrt[5]{x}$

$f(x) = x^{1/5}$

$f'(x) = \frac{1}{5}x^{-4/5} = \dfrac{1}{5\sqrt[5]{x^4}}$. This does not exist for $x = 0$. $f'(x)$ is never 0, so the only critical number is $x = 0$.

b) $f(x) = x + \frac{1}{x}$

$f(x) = x + x^{-1}$

$f'(x) = 1 + (-1)x^{-2} = 1 - \frac{1}{x^2}$

$f'(x) = 0$ at $x = 1$, $x = -1$.

$f'(x)$ does not exist at $x = 0$ (but 0 is not in the domain of f). The critical numbers are $x = 1$ and $x = -1$.

Page 282

D. The extreme values of a continuous function f on a closed interval $[a, b]$ always exist; they occur either at a, at b, or at a critical number of f in (a, b). Thus, to find the absolute maximum and minimum values of f on a closed interval $[a, b]$, use the *Closed Interval Method:*

1) Find $f(x)$ at the critical numbers x in (a, b).

2) Find $f(a)$ and $f(b)$.

3) Choose the largest and smallest from the results of Steps 1 and 2.

9) Find the extreme values of
$f(x) = \sqrt{10x - x^2}$ on $[2, 10]$.

Since f is continuous on $[2, 10]$, the extreme values occur at 2, 10, or some critical number between 2 and 10. $f(x) = (10x - x^2)^{1/2}$

$f'(x) = \frac{1}{2}(10x - x^2)^{-1/2}(10 - 2x)$

$= \frac{5 - x}{\sqrt{10x - x^2}}.$

On $[2, 10]$, $f'(x) = 0$ at $x = 5$. $f'(x)$ does not exist at $x = 10$. Thus 5 is the only critical number between 2 and 10. Computing function values:

x	5	2	10
y	5	4	0

The absolute minimum is 0 at $x = 10$ and the absolute maximum is 5 at $x = 5$.

10) Do the extreme values of
$$f(x) = 6 + 12x - x^2 \text{ on } (4, 10] \text{ exist?}$$

f is a polynomial and, therefore, continuous on (4, 10]. However, (4, 10] is not a closed interval so we cannot immediately conclude the extreme values of *f* exist. From the graph of *f* below (it is a portion of a parabola) we see that the extreme values do exist: the absolute maximum is 42 at $x = 6$ and the absolute minimum is 26 at $x = 10$.

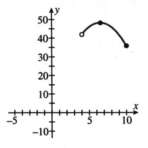

11) Find the extreme values of
$$f(x) = 2x + \sin x \text{ on } [0, 2\pi].$$

$f'(x) = 2 + \cos x = 0$
$\cos x = -2$ has no solution.
There are no critical numbers.

$f(0) = 2(0) + \sin 0 = 0$
$f(2\pi) = 2(2\pi) + \sin 2\pi = 4\pi.$

The absolute minimum is 0 and the absolute maximum is 4π.

Section 4.2   The Mean Value Theorem

The Mean Value Theorem is very important because it contains an equation that relates function values $f(x)$ to derivative values $f'(x)$. Rolle's Theorem is a special case of the Mean Value Theorem.

Concepts to Master

A. Rolle's Theorem; Mean Value Theorem

B. Relationship between functions f and g when $f' = g'$

Summary and Focus Questions

Page 288

A. *Rolle's Theorem:*
  If the function f
  1) is continuous on $[a, b]$
  2) is differentiable on (a, b)
  3) has $f(a) = f(b)$ then there is at least
      one number $c \in (a, b)$ such that $f'(c) = 0$.

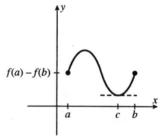

In terms of the graph of f, Rolle's Theorem says that if the three hypotheses are true, then there is at least one horizontal tangent.

The Mean Value Theorem is similar to Rolle's Theorem except we do not assume that $f(a)$ equals $f(b)$. (If $f(a) = f(b)$, then the conclusion of the Mean Value Theorem becomes the conclusion of Rolle's Theorem.)
Mean Value Theorem:
If the function f
  1) is continuous on $[a, b]$
  2) is differentiable on (a, b) then there is at least
      one $c \in (a, b)$ such that $f'(c) = \dfrac{f(b) - f(a)}{b - a}$.

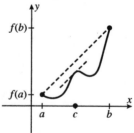

Since $\dfrac{f(b) - f(a)}{b - a}$ is the slope of the line joining the endpoints of the graph,

the Mean Value Theorem simply says there is at least one point on the graph somewhere between the endpoints where the tangent line has the same slope as the line joining the endpoints. The Mean Value Theorem is important because

$$f'(c) = \frac{f(b) - f(a)}{b - a}$$

is an equation involving both f and f'.

1) For $f(x) = 1 - x^2$ on $[-2, 1]$, do the hypotheses and conclusion of Rolle's Theorem hold?

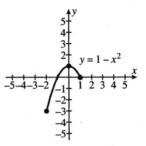

f is continuous on $[-2, 1]$ and differentiable on $(-2, 1)$ but since $f(-2) = -3 \neq 0 = f(1)$ one hypothesis fails. The conclusion holds because $0 \in (-2, 1)$ and $f'(0) = 0$.

2) Do the hypotheses and conclusion of the Mean Value Theorem hold for
$f(x) = 3x - \dfrac{x^3}{3}$ on $[-1, 4]$?

The hypotheses hold because f is a polynomial and therefore continuous and differentiable everywhere. Since the hypotheses are true, the conclusion must be true as well.

$$\frac{f(4) - f(-1)}{4 - (-1)} = \frac{\dfrac{-28}{3} - \dfrac{-8}{3}}{5} = -\frac{4}{3}.$$

$$f'(x) = 3 - x^2 = -\frac{4}{3};\ x^2 = \frac{13}{3},\ \text{so}$$

$$x = \sqrt{\frac{13}{3}} \cdot \left(-\sqrt{\frac{13}{3}}\ \text{is not in } [-1, 4]\right).$$

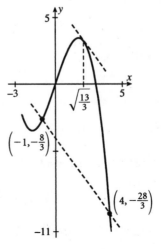

3) Mark on the *x*-axis the point(s) *c* in the conclusion of the Mean Value Theorem for the function below.

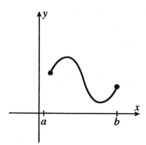

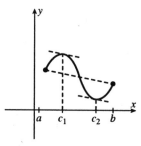

There are two points (c_1 and c_2) that satisfy the conclusion of the Mean Value Theorem.

4) David is driving on an interstate highway which has a speed limit of 55 mi/hr. At 2 p.m. he is at milepost 110 and at 5 p.m. he is at milepost 290. Is this enough evidence to prove that David is guilty of speeding?

Yes. Let $f(t)$ be his position at time t. $f(2) = 110$ and $f(5) = 290$. The mean value on $[2, 5]$ is $\frac{f(5) - f(2)}{5 - 2} = \frac{290 - 110}{5 - 2} = 60$. By the Mean Value Theorem there is a time $c \in (2, 5)$ such that $f'(c) = 60$. So at least once (at time c) David's velocity was 60 mi/hr.

5) Find all numbers *c* that satisfy the conclusion of the Mean Value Theorem for $f(x) = x^3 - 3x^2 + x$ on $[0, 3]$.

The mean value on $[0, 3]$ is
$$\frac{f(3) - f(0)}{3 - 0} = \frac{3 - 0}{3 - 0} = 1.$$
$f'(x) = 3x^2 + 6x + 1.$
$3x^2 - 6x + 1 = 1$
$3x^2 - 6x = 0$
$3x(x - 2) = 0.$
Thus, $f'(x) = 1$ at $x = 0$, $x = 2$. However, $0 \notin (0, 3)$ so $c = 2$ is the only point satisfying the Mean Value Theorem.

6) Suppose $f'(x) \leq 2$ for all x. If $f(1) = 8$ what is the largest possible value that $f(5)$ could be?

Since $f'(x)$ exists everywhere we can use the Mean Value Theorem for f on $[1, 5]$. For some $c \in (1, 5)$,

$$f'(c) = \frac{f(5) - f(1)}{5 - 1} = \frac{f(5) - 8}{4}$$

Since $f'(c) \leq 2$

$$\frac{f(5) - 8}{4} \leq 2$$

$$f(5) - 8 \leq 8$$

$$f(5) \leq 16$$

So $f(5)$ can be no more than 16.

Page 292

B. If $f'(x) = g'(x)$ for all $x \in (a, b)$, then $f(x) = g(x) + c$ where c is a constant.

7) True or False: If $f'(x) = g'(x)$, then $f(x) = g(x)$.

False. For example, $f(x) = x^2$ and $g(x) = x^2 + 6$ have the same derivative, $2x$.

8) If $f'(x) = 3x^2$ then what is $f(x)$? (Hint: If $g(x) = x^3$ then $g'(x) = 3x^2$.)

$f'(x) = x^3 + c$, where c is some constant. Each of these are valid expressions for $f(x)$:

$x^3 + 1$

$x^3 - 10$

$x^3 + \pi$

Section 4.3   How Derivatives Affect the Shape of a Graph

This section shows how values for f' and f'' determine the way the graph of f increases and decreases, as well as curves upward and downward.

Concepts to Master

A. The Increasing/Decreasing Test

B. The First Derivative Test

C. Concave upward and downward; Concavity Test

D. Point of inflection; Second Derivative Test

Summary and Focus Questions

Page 294

A. Recall that a function is increasing on an interval I if the graph rises as you move left to right. Therefore the tangent lines to an increasing function must also increase and thus the derivative (slope of the tangent) is positive.

Increasing/Decreasing Test:
If f is differentiable on an interval I

a) $f'(x) > 0$ for all $x \in I$ implies f is increasing on I.

b) $f'(x) < 0$ for all $x \in I$ implies f is decreasing on I.

1) Is $f(x) = x^2 + 6x$ increasing on $[-1, 2]$?

Yes. $f'(x) = 2x + 6$. $2x + 6 > 0$ when $2x > -6$ or $x > -3$. Thus for all $x \in [-1, 2]$, $f'(x) > 0$. Therefore f is increasing on $[-1, 2]$.

2) Where is $f(x) = x^3 - 3x^2$ increasing and where is it decreasing?

$f'(x) = 3x^2 - 6x = 3x(x - 2) = 0$ at $x = 0$, $x = 2$. At other values of x, $f'(x)$ will be either positive or negative.

	3x	x − 2	3x(x − 2)
x < 0	−	−	+
0 < x < 2	+	−	−
2 < x	+	+	+

f is increasing on $(-\infty, 0)$ and $(2, \infty)$; f is decreasing on $(0, 2)$.

3) Sketch a graph of a differentiable function having all these properties:
$f(0) = 1, f(2) = 3, f(5) = 0, f'(x) > 0$ for $x \in (0, 2)$ and $x \in (5, \infty), f'(x) < 0$ for $x \in (-\infty, 0)$ and $x \in (2, 5)$.

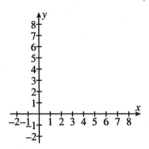

One such graph is:

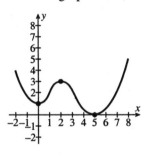

Page 295

B. *The First Derivative Test:*

Let f be continuous on $[a, b]$. Let c be a critical number for f in (a, b).

a) If $f'(x)$ is positive on (a, c) and negative in (c, b), then f has a local maximum at c.

b) If $f'(x)$ is negative in (a, c) and positive in (c, b), then f has a local minimum at c.

c) If $f'(x)$ has the same sign (either positive or negative) in both (a, c) and (c, b), then $f(c)$ is not a local extrema.

4) Suppose 8 is a critical number for a function f with $f'(x) > 0$ for $x \in (8, 9)$ and $f'(x) < 0$ for $x \in (7, 8)$. Then 8 is:

a) a local maximum

b) a local minimum

c) not a local extremum

b), by the First Derivative Test.

5) What can you conclude about the local extrema of a differential function f from the following? Critical numbers for f are $x = 1, 3, 4, 6, 8$.

Interval	Sign of $f'(x)$
$(-\infty, 1)$	positive
$(1, 3)$	negative
$(3, 4)$	negative
$(4, 6)$	positive
$(6, 8)$	positive
$(8, \infty)$	negative

f has a local maximum at $x = 1$ and $x = 8$; f has a local minimum at $x = 4$; there are no local extrema at $x = 3$ or $x = 6$.

6) Is $x = 5$ a local minimum for $f(x) = 10x - x^2 + 1$?

No. $f'(x) = 10 - 2x > 0$ for all $x < 5$ while $f'(x) < 0$ for all $x > 5$. f has a local *maximum* at $x = 5$.

7) Find where the local maximum and minimum values of $f(x) = 4x^3 - x^4$ occur.

$f'(x) = 12x^2 - 4x^3 = 4x^2(3 - x)$

$f'(x) = 0$ at $x = 0$, $x = 3$, so there are two critical numbers. Since $f'(x) \geq 0$ for $x > 3$ and $f'(x) \leq 0$ for $x > 3$, 3 is a local maximum. No extreme value occurs at $x = 0$.

8) Find where the local maximum and minimum values of $f(x) = x + \sin x$ occur.

$f'(x) = 1 + \cos x$. Since $-1 \leq \cos x \leq 1$, $f'(x) \geq 0$ for all x. Therefore, f has no local extreme values.

C. A function f is *concave upward* on an interval I if f lies above all tangent lines to f in I. f is *concave downward* on I if the graph of f is below all tangent lines.

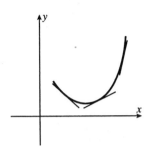

concave upward

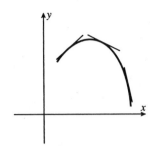

concave downward

The test for concavity involves the second derivative:

Concavity Test: If f is twice differentiable on an interval I (meaning $f''(x)$ exists for all $x \in I$) then:

a) If $f''(x) > 0$ for all $x \in I$, f is concave *upward* on I.

b) If $f''(x) < 0$ for all $x \in I$, f is concave *downward* on I.

9) Use the graph below to answer true or false to each.

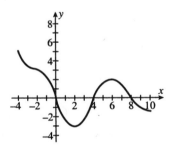

a) $f''(x) > 0$ for $x \in (2, 4)$ True.

b) $f''(x) < 0$ for $x \in (-4, -2)$ False.

c) $f''(6) = 0$ False. $f''(6)$ is negative.

d) $f''(2) > 0$ True.

e) f is concave upward on $(0, 2)$ True.

10) What can be said about $f'(x)$ and $f''(x)$ for each?

a)

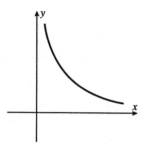

$f'(x) < 0$ and $f''(x) > 0$

b)

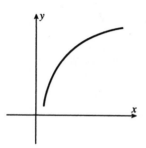

$f'(x) > 0$ and $f''(x) < 0$

c)

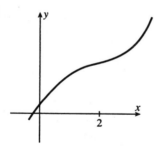

$f'(x) \geq 0$ for all x,
$f''(x) > 0$ for $x > 2$,
$f''(x) < 0$ for $x < 2$

Page
298

D. A *point of inflection* for *f* is a point on the graph of *f* where concavity changes from concave downward to concave upward or from concave upward to concave downward.

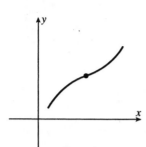 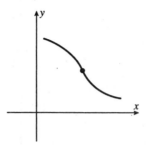

Concave downward to concave upward

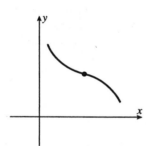 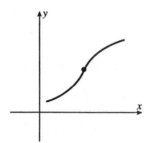

Concave upward to concave downward

The Second Derivative Test:

If f'' is continuous on (a, b) containing c and $f'(c) = 0$, then

a) if $f''(c) < 0$, f has a local maximum at c.

b) if $f''(c) > 0$, f has a local minimum at c.

If it should happen that both $f'(c) = 0$ and $f''(c) = 0$ then no conclusion can be made about an extremum at c.

11) What are the points of inflection for the function in question 9)?

$$x = -2, 0, 4, 8$$

12) Find the points of inflection for $f(x) = 8x^3 - x^4$.

$f'(x) = 24x^2 - 4x^3$,
$f''(x) = 48x - 12x^2 = 12x(4 - x) = 0$
at $x = 0$, $x = 4$. Since f is a polynomial the only points of inflection are at 0 and 4.

13) Find all local extrema of

   a) $f(x) = x^3 - 6x^2 - 36x$.

$f'(x) = 3x^2 + 12x - 36 = 3(x - 2)(x + 6)$.
The critical points are at $x = 2, -6$.
$f''(x) = 6x + 12$.
At $x = 2, f''(2) = 24 > 0$ so f has a local minimum at 2.
At $x = -6, f''(-6) = -24 < 0$ so f has a local maximum at -6.

   b) $f(x) = x^2 + 2 \cos x$.

$f'(x) = 2x - 2 \sin x = 0$ when $x = \sin x$.
The only solution to this equation is $x = 0$.
$f''(x) = 2 - 2 \cos x$ but $f''(x) = 0$ so the Second Derivative Test fails to give additional information. However $f''(x) \geq 0$, for all x, means f is concave upward and thus f has a local minimum at 0.

14) Sketch a graph of a function f having all these properties:
$f(-1) = 4, f(0) = 2, f(2) = 1, f(3) = 0$
$f'(x) \leq 0$ for $x < 3$ and
$f'(x) \geq 0$ for $x > 3$.
$f''(x) < 0$ for $0 < x < 2$ and
$f''(x) \geq 0$ elsewhere.

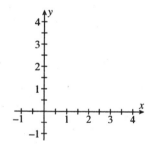

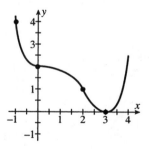

15) Orlando's car is stopped at a stop sign. He accelerates to 30 mi/hr and maintains this speed. He then sees a traffic light ahead and lets up on the gas pedal until his speed is 15 mi/hr. At that instant the traffic light turns green and he resumes his cruising speed of 30 mi/hr. Sketch a graph of the position function $p(t)$ of Orlando's car from the stop sign.

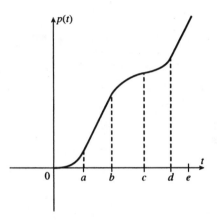

At $t = 0$, Orlando is stopped. He increases his speed ($p''(t) > 0$) on the interval $[0, a]$, so the graph is concave upward. In the interval $[a, b]$ he is cruising at 30 mi/hr ($p'(t) = 30$). At $t = b$ he begins to slow down and for $t \in [b, c]$, $p''(t) < 0$, so the graph is concave downward. At $t = c$ the light turns green, he accelerates in $[c, d]$ and resumes cruising at 30 mi/hr in time interval $[d, e]$.

Section 4.4 Indeterminate Forms and L'Hospital's Rule

Sometimes blithely applying the limit theorems of chapter 2 to a limit results in nonsensical expressions such as $\frac{0}{0}$ even though the limit exists. For example, applying the limit theorems to $\lim\limits_{x \to 2} \frac{x^2 - 4}{x - 2}$ yields $\frac{0}{0}$; we know the limit exists and is 4 (you need to resort to factoring or some other technique to find the limit.) L'Hospital's Rule is a method for evaluating limits that have indeterminate forms such as $\frac{0}{0}$. There are seven such forms discussed in this section.

Concepts to Master

Indeterminate forms; L'Hospital's Rule

Summary and Focus Questions

Page 305

Indeterminate forms $\frac{0}{0}$ and $\frac{\infty}{\infty}$.

L'Hospital's Rule says that if $\lim\limits_{x \to a} f(x) = 0$ and $\lim\limits_{x \to a} g(x) = 0$, then $\lim\limits_{x \to a} \frac{f(x)}{g(x)}$ may be determined by finding the limit of the slopes of the tangent lines near $x = a$.

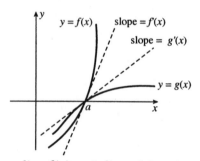

$\lim\limits_{x \to a} f(x) = 0, \lim\limits_{x \to a} g(x) = 0.$

For x near a, $\dfrac{f(x)}{g(x)} \approx \dfrac{f'(x)}{g'(x)}$

L'Hospital's Rule for $\frac{0}{0}$:

If $\lim\limits_{x \to a} f(x) = 0$ and $\lim\limits_{x \to a} g(x) = 0$, then $\lim\limits_{x \to a} \dfrac{f(x)}{g(x)} = \lim\limits_{x \to a} \dfrac{f'(x)}{g'(x)}$, provided $g'(x)$ is not zero near a.

L'Hospital's Rule for $\frac{\infty}{\infty}$:

If $\lim\limits_{x \to a} f(x) = \infty$ and $\lim\limits_{x \to a} g(x) = \infty$, then $\lim\limits_{x \to a} \dfrac{f(x)}{g(x)} = \lim\limits_{x \to a} \dfrac{f'(x)}{g'(x)}$.

Example: Although $\lim\limits_{x \to a} \dfrac{x^2 - 4}{x - 2}$ can be evaluated by factoring, canceling, and applying the limit theorems, L'Hospital's Rule is easier:

$$\lim\limits_{x \to a} \frac{x^2 - 4}{x - 2} = \lim\limits_{x \to 2} \frac{2x}{1} = 4.$$

Indeterminate form $0 \cdot \infty$:

To evaluate $\lim\limits_{x \to a} f(x) \cdot g(x)$ which has indeterminate form $0 \cdot \infty$, first rewrite $f(x) \cdot g(x)$ as either $\dfrac{f(x)}{1/g(x)}$ or $\dfrac{g(x)}{1/f(x)}$ to get either $\dfrac{0}{0}$ or $\dfrac{\infty}{\infty}$. Now apply L'Hospital's Rule.

Example: $\lim\limits_{x \to \infty} xe^{-x} \,(\text{form } \infty \cdot 0) = \lim\limits_{x \to \infty} \dfrac{x}{e^x} \left(\text{form } \dfrac{\infty}{\infty}\right) = \lim\limits_{x \to \infty} \dfrac{1}{e^x} = 0.$

Indeterminate form $\infty - \infty$:

For $\lim\limits_{x \to a} f(x) - g(x) = \infty - \infty, f(x) - g(x)$ must be rewritten as a single term. When the trigonometric functions are involved, switching to all sines and cosines may help.

Example: $\lim\limits_{x \to \pi/2-} (\sec x - \tan x) = \lim\limits_{x \to \pi/2-} \left(\dfrac{1}{\cos x} - \dfrac{\sin x}{\cos x}\right)$
$$= \lim\limits_{x \to \pi/2-} \dfrac{1 - \sin x}{\cos x} \left(\text{form } \dfrac{0}{0}\right) = \lim\limits_{x \to \pi/2-} \dfrac{-\cos x}{-\sin x} = 0.$$

Indeterminate forms 0^0, 1^∞, and ∞^0:

Finally, for limits of the type $\lim\limits_{x \to a} (f(x))^{g(x)}$, indeterminate forms of type 0^0, 1^∞, and ∞^0 are possible. In these cases,

    1. Let $y = f(x)^{g(x)}$.

    2. Then $\ln y = g(x) \ln f(x)$.

    3. If $\lim\limits_{x \to a} g(x) \ln f(x)$ (form $0 \cdot \infty$) exists and equals L, then
        $\lim\limits_{x \to a} (f(x))^{g(x)} = e^L$.

1) True or False:
All limits having one of the seven indeterminate forms are solved by reducing (if necessary) to limits of the form $\dfrac{0}{0}$ or $\dfrac{\infty}{\infty}$.

True. L'Hospital's Rule only applies in $\dfrac{0}{0}$ and $\dfrac{\infty}{\infty}$ cases. The other five forms reduce to one of these two forms.

2) Evaluate each:

a) $\lim\limits_{x \to 3} \dfrac{x^2 - 7x + 12}{x^2 - 9}$

$$\lim\limits_{x \to 3} \dfrac{x^2 - 7x + 12}{x^2 - 9} \left(\text{form } \tfrac{0}{0}\right) = \lim\limits_{x \to 3} \dfrac{2x - 7}{2x}$$

$$= -\dfrac{1}{6}.$$

b) $\lim\limits_{x \to 0} \dfrac{\sin x}{x}$

$$\lim\limits_{x \to 0} \dfrac{\sin x}{x} \text{ (a familiar limit)} = \lim\limits_{x \to 0} \dfrac{\cos x}{1}$$

$$= \dfrac{1}{1} = 1.$$

c) $\lim\limits_{x \to \infty} \dfrac{e^x}{3x}$

$$\lim\limits_{x \to \infty} \dfrac{e^x}{3x} \left(\text{form } \tfrac{\infty}{\infty}\right) = \lim\limits_{x \to \infty} \dfrac{e^x}{3} = \infty.$$

d) $\lim\limits_{x \to 0^+} x \cot x$

This has form $0 \cdot \infty$. Rewrite $x \cot x$ as

$$x \cdot \dfrac{\cos x}{\sin x} = \dfrac{x}{\sin x} \cdot \cos x.$$

$$\lim\limits_{x \to 0^+} x \cot x = \lim\limits_{x \to 0^+} \dfrac{x}{\sin x} \cdot \cos x$$

$$= \left(\lim\limits_{x \to 0^+} \dfrac{x}{\sin x}\right) \cdot \left(\lim\limits_{x \to 0^+} \cos x\right) \left(\text{form } \tfrac{0}{0}\right)$$

$$= \lim\limits_{x \to 0^+} \dfrac{1}{\cos x} \cdot \lim\limits_{x \to 0^+} \cos x = \lim\limits_{x \to 0^+} 1 = 1.$$

e) $\lim\limits_{x \to \infty} \left(\dfrac{1}{x^2 - 1} - \dfrac{1}{x - 1}\right)$

This limit is $0 - 0 = 0$. (Not an indeterminate form.)

f) $\lim\limits_{x \to 0^+} \left(\dfrac{1}{x} - \dfrac{1}{e^x - 1}\right)$

$$\lim\limits_{x \to 0^+} \left(\dfrac{1}{x} - \dfrac{1}{e^x - 1}\right) (\text{form } \infty - \infty)$$

$$= \lim\limits_{x \to 0^+} \dfrac{e^x - 1 - x}{x(e^x - 1)} \left(\text{form } \tfrac{0}{0}\right)$$

$$= \lim\limits_{x \to 0^+} \dfrac{e^x - 1}{xe^x + e^x - 1} \left(\text{form } \tfrac{0}{0} \text{ again}\right)$$

$$= \lim\limits_{x \to 0^+} \dfrac{e^x}{xe^x + 2e^x} = \dfrac{1}{0 + 2} = \dfrac{1}{2}.$$

g) $\lim\limits_{x \to 0^+} \left(\dfrac{1}{\sin^2 x} - \dfrac{\cot x}{x} \right)$

$\lim\limits_{x \to 0^+} \left(\dfrac{1}{\sin^2 x} - \dfrac{\cot x}{x} \right)$

$= \lim\limits_{x \to 0^+} \dfrac{x - \cos x \sin x}{x \sin^2 x} \left(\text{form } \dfrac{0}{0} \right)$

$= \lim\limits_{x \to 0^+} \dfrac{1 - \cos^2 x + \sin^2 x}{\sin^2 x + 2x \sin x \cos x}$

$= \lim\limits_{x \to 0^+} \dfrac{2 \sin^2 x}{\sin x(\sin x + 2x \cos x)}$

$= \lim\limits_{x \to 0^+} \dfrac{2 \sin x}{\sin x + 2x \cos x} \left(\text{form } \dfrac{0}{0} \right)$

$= \lim\limits_{x \to 0^+} \dfrac{2 \cos x}{3 \cos x - 2x \sin x} = \dfrac{2}{3}.$

h) $\lim\limits_{x \to 0^+} (\csc x)^x$

This has form ∞^0. Let $y = (\csc x)^x$.

$\ln y = x \ln(\csc x).$

$\lim\limits_{x \to 0^+} x(\ln \csc x) \, (\text{form } 0 \cdot \infty)$

$= \lim\limits_{x \to 0^+} \dfrac{\ln \csc x}{x^{-1}} \left(\text{form } \dfrac{\infty}{\infty} \right)$

$= \lim\limits_{x \to 0^+} \dfrac{\frac{1}{\csc x} \cdot (-\csc x \cot x)}{(-1)x^{-2}}$

$= \lim\limits_{x \to 0^+} x^2 \cot x \, (\text{form } 0 \cdot \infty)$

$= \lim\limits_{x \to 0^+} \dfrac{x^2}{\tan x} \left(\text{form } \dfrac{0}{0} \right)$

$= \lim\limits_{x \to 0^+} \dfrac{2x}{\sec^2 x} = \dfrac{2(0)}{1} = 0.$

Thus $\lim\limits_{x \to 0^+} (\csc x)^x = e^0 = 1.$

Section 4.5   Summary of Curve Sketching

This section reviews and applies the curve sketching calculus concepts of the previous sections. It also introduces the notion of a slant asymptote of $y = f(x)$—a non-vertical line that the graph of a function approaches for large positive or negative values of x.

Concepts to Master

A. Curve sketching using information obtained through calculus concepts

B. Slant asymptotes

Summary and Focus Questions

Page 314

A. You should develop a checklist of features to consider before drawing a graph. Questions to answer in sketching a graph of $y = f(x)$ include:

A. Domain of f	For what x is $f(x)$ defined?
B. x-intercept(s)	What are the solution(s) (if any) to $f(x) = 0$?
y-intercept	What value (if any) is $f(0)$?
C. Symmetry about y-axis	Is $f(-x) = f(x)$?
Symmetry about origin	Is $f(-x) = -f(x)$?
Periodic	Is there a number p such that $f(x + p) = f(x)$ for all x in the domain?
D. Horizontal asymptote(s)	Does $\lim\limits_{x \to \infty} f(x)$ or $\lim\limits_{x \to -\infty} f(x)$ exist?
Vertical asymptote(s)	For what a is $\lim\limits_{x \to a^+} f(x) = \infty$ or $-\infty$?
	For what a is $\lim\limits_{x \to a^-} f(x) = \infty$ or $-\infty$?
E. Increasing	On what intervals is $f'(x) \geq 0$?
Decreasing	On what intervals if $f'(x) \leq 0$?
F. Critical numbers	Where does $f'(x) = 0$ or not exist?
Local extrema	Where are the local maxima or minima (if any)? (Use the First and Second Derivative Tests.)
G. Concave upward	On what intervals is $f''(x) \geq 0$?
Concave downward	On what intervals is $f''(x) \leq 0$?
Inflection points	Where does f change concavity?
	Where does $f''(x) = 0$ or not exist?

1) Sketch a graph of $y = f(x)$ that has all these properties:

domain $= (-\infty, -1) \cup (1, \infty)$

$f(2) = 0$

$f(x) = -f(-x)$

$\lim\limits_{x \to \infty} f(x) = 4$ and $\lim\limits_{x \to 1+} f(x) = -\infty$

$f'(3) = 0, f'(5) = 0$

$f'(x) \geq 0$ on $(1, 3) \cup (5, \infty)$

$f'(x) \leq 0$ on $(3, 5)$

local maximum at 3

local minimum at 5

$f''(x) \geq 0$ on $(4, 6)$

$f''(x) \leq 0$ on $(1, 4) \cup (6, \infty)$

$f''(4) = 0, f''(6) = 0$

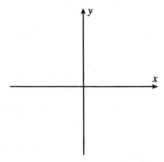

2) Sketch a graph of $f(x) = \dfrac{1}{x^2 - 4}$.

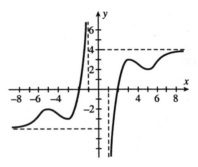

A. The domain is all x except where $x^2 - 4 = 0$; all x except $x = 2, x = -2$.

B. $\dfrac{1}{x^2 - 4}$ is never 0, so no x-intercepts.

At $x = 0$, $y = \dfrac{1}{0^2 - 4} = -\dfrac{1}{4}$, the y-intercept.

C. $f(-x) = \dfrac{1}{(-x)^2 - 4} = \dfrac{1}{x^2 - 4} = f(x)$, so there is symmetry about the y-axis. But $f(-x) \neq -f(x)$, so there is no symmetry about $(0, 0)$.

D. $\lim\limits_{x \to \infty} \dfrac{1}{x^2 - 4} = 0$ and $\lim\limits_{x \to -\infty} \dfrac{1}{x^2 - 4} = 0, y = 0$ is the only horizontal asymptote.

$\lim\limits_{x \to 2+} \dfrac{1}{x^2 - 4} = \infty, \lim\limits_{x \to 2-} \dfrac{1}{x^2 - 4} = -\infty,$

$\lim\limits_{x \to -2+} \dfrac{1}{x^2 - 4} = -\infty,$ and $\lim\limits_{x \to -2-} \dfrac{1}{x^2 - 4} = \infty.$

Therefore, $x = 2$ and $x = -2$ are vertical asymptotes.

E. $f'(x) = (-1)(x^2 - 4)^{-2}(2x)$

$\quad = -2x(x^2 - 4)^{-2} = \dfrac{-2x}{(x^2 - 4)^2}.$

For $x > 0$, $x \neq 2$, $f'(x) < 0$ so f is decreasing on $(0, 2)$ and $(2, \infty)$. For $x < 0$, $x \neq -2$, $f'(x) > 0$ so f is increasing on $(-\infty, -2)$ and $(-2, 0)$.

F. $f'(x) = \dfrac{-2x}{(x^2 - 4)^2} = 0$ at $x = 0$.

$f''(x) = (-2x)[-2(x^2 - 4)^{-3} \cdot 2x]$
$\quad\quad + (x^2 - 4)^{-2}(-2)$
$\quad = (x^2 - 4)^{-3}[8x^2 - 2(x^2 - 4)]$
$\quad = \dfrac{6x^2 + 8}{(x^2 - 4)^3}.$

$f''(0) = \dfrac{8}{(-4)^3} = -\dfrac{1}{8}.$ Thus f has a local maximum at $x = 0$.

G. Since $f''(x) = \dfrac{6x^2 + 8}{(x^2 - 4)^3}$, the sign of $f''(x)$ is determined by $(x^2 - 4)^3$ and thus by $x^2 - 4$.

$x^2 - 4 > 0$ for $x > 2$ and $x < -2$ and $x^2 - 4 < 0$ for $-2 < x < 2$. f is concave upward on $(-\infty, -2)$ and $(2, \infty)$ and concave downward on $(-2, 2)$.

H.

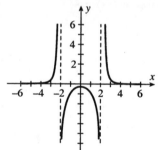

3) Sketch a graph of $f(x) = x - \ln x$.

A The domain is $(0, \infty)$.

B. $x - \ln x$ is never 0, so no x-intercept.

C. Since $x > 0$ there is no symmetry.

D. $\lim_{x \to 0^+} (x - \ln x) = \infty$

$\lim_{x \to \infty} (x - \ln x) = \infty$

E. $f'(x) = 1 - \frac{1}{x}$.

For $0 < x < 1, f'(x) < 0$ and f is decreasing. For $x > 1, f'(x) > 0$ and f is increasing.

F. $f'(x) = 1 - \frac{1}{x} = 0$ at $x = 1$.

$f''(x) = x^{-2}$, so $f''(x) = 1 > 0$.

Thus f has a local minimum at $x = 1$.

G. Since $f''(x) > 0$ the graph is always concave upward.

H. Here is a sketch:

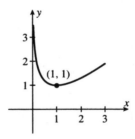

Page 320

B. The line $y = mx + b$ is a *slant asymptote* of $y = f(x)$ means that

$$\lim_{x \to \infty} [f(x) - (mx + b)] = 0 \text{ or } \lim_{x \to -\infty} [f(x) - (mx + b)] = 0.$$

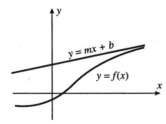
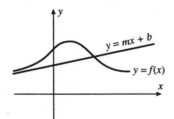

A slant asymptote is a nonvertical, nonhorizontal line that the curve approaches for large (positive or negative) values of x.

4) Find a slant asymptote for
$$f(x) = \frac{x^2 + x + 1}{x}.$$

$$f(x) = \frac{x^2 + x + 1}{x} = x + 1 + \frac{1}{x}.$$
The line $y = x + 1$ is a slant asymptote
because
$$\lim_{x \to \infty} (f(x) - (x + 1))$$
$$= \lim_{x \to \infty} \left((x + 1 + \frac{1}{x}) - (x + 1)\right)$$
$$= \lim_{x \to \infty} \frac{1}{x} = 0.$$

5) Can a polynomial of degree greater than
one have a slant asymptote?

No. If $f(x)$ is a polynomial whose degree is
two or more then $f(x) - (mx + b)$ will also
have degree two or more. Thus,
$\lim_{x \to \infty} [f(x) - (mx + b)]$ will not be zero (it
will be ∞ or $-\infty$).

6) Show that $y = \sqrt{x^2 - 4}$ has two slant
asymptotes.

For large x, $\sqrt{x^2 - 4}$ is approximately
$\sqrt{x^2} = x$, so we guess that $y = x$ is a slant
asymptote.
$$\lim_{x \to \infty} (\sqrt{x^2 - 4} - x)$$
$$= \lim_{x \to \infty} (\sqrt{x^2 - 4} - x)\frac{(\sqrt{x^2 - 4} + x)}{(\sqrt{x^2 - 4} + x)}$$
$$= \lim_{x \to \infty} \frac{(x^2 - 4) - x^2}{(\sqrt{x^2 - 4} + x)}$$
$$= \lim_{x \to \infty} \frac{-4}{(\sqrt{x^2 - 4} + x)} = 0$$
Similarly for large negative x,
$\sqrt{x^2 - 4} \approx \sqrt{x^2} = -x$. Thus, $y = -x$, is
the other slant asymptote.
We note that $y = \sqrt{x^2 - 4}$ is the top half of
the hyperbola $x^2 - y^2 = 4$ with asymptotes
$y = x$ and $y = -x$.
A graph of $y = \sqrt{x^2 - 4}$ is given here:

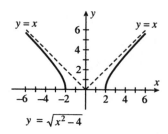

Section 4.6 Graphing with Calculus and Calculators

This section shows you how some knowledge of calculus will help you use your calculator to draw more accurate graphs.

Concepts to Master

A. Use calculus to improve information from graphing calculator displays
B. Determine the graphs of families of functions

Summary and Focus Questions

Page 322

A. A graphing calculator can display an initial graph of a function. The concepts of calculus can then be used to refine it so that details such as relative extrema and concavity stand out.

1) Use graphing calculator to graph $f(x) = 9x^4 - 76x^3 + 180x^2$. Then use calculus to refine it.

An initial graph from a calculator with window $[-1, 4]$ by $[-1, 330]$ will look something like:

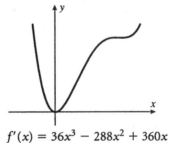

$$f'(x) = 36x^3 - 288x^2 + 360x$$
$$= 12x(3x^2 - 19x + 30)$$
$$= 12x(x - 3)(3x - 10) = 0$$
at $x = 0$, $x = 3$, $x = \frac{10}{3}$.
$$f''(x) = 108x^2 - 456x + 360$$

x	f	f'	f''	Type
0	0	0	360	relative minimum
3	297	0	−36	relative maximum
$\frac{10}{3}$	296.3	0	40	relative minimum

The graph of f with its critical numbers 0, 3, and $\frac{10}{3}$ is:

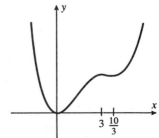

Zooming in, using a window of [2.8, 3.5] by [295, 300] we see

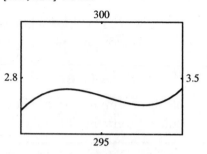

B. A *family* of functions is a collection of functions defined by a formula with one or more arbitrary constants. Varying the values of the constants results in changes in the graphs of the functions.

Page 327

For example, the family described by $f(x) = x^2 + bx + 1$ has graphs which are parabolas passing through (0, 1) with vertex (relative minimum) at $x = -\dfrac{b}{2}$.

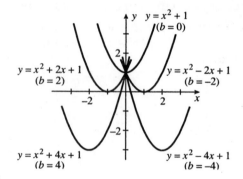

2) Describe the graphs of the family of functions $f(x) = x^3 - 3ax$.

Each function passes through (0, 0). $f'(x) = 3(x^2 - a)$. For $a > 0$, f has a relative maximum at $x = -\sqrt{a}$ and relative minimum at $x = \sqrt{a}$. For $a < 0$, f has no extrema.

$f''(x) = 6x$ so f is concave up for $x > 0$ and concave down for $x < 0$. Thus $x = 0$ is an inflection point. The family of graphs is:

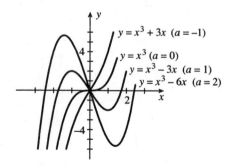

Section 4.7   Optimization Problems

An optimization problem seeks to find the largest (or smallest) value of a quantity (such as maximum revenue or minimum surface area) given certain limits to the problem. These problems arise in many areas of application. There is a particular method to follow that will help set up and solve the problems.

Concepts to Master

Solve applied extrema problems

Summary and Focus Questions

Page 329

An optimization problem can usually be expressed as "find the maximum (or minimum) value of some quantity Q under a certain set of given conditions." For example, suppose we wish to find the dimensions of the largest rectangular field enclosed by 100 feet of fencing. The quantity to be maximized is the area, subject to the condition that the perimeter is 100.

A procedure to help solve problems of this type is:

(1) Introduce notation identifying the quantity to be maximized (or minimized), Q, and the other variable(s). Here a diagram may help.

(2) Write Q as a function of the other variables. Also, express any relationships or conditions among the other variables with equations.

(3) Rewrite, if necessary, Q as a function of just one of the variables, using given relationships and determine the domain D of Q.

(4) Find the absolute extrema of Q on domain D.

In our example above, the steps in the solution are:

(1) Let A = area, x = width, y = height

(2) $A = xy$, $2x + 2y = 100$

(3) From $2x + 2y = 100$, $y = 50 - x$.
Thus, $A = x(50 - x) = 50x - x^2$.
The domain is [0, 50].

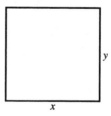

(4) $A'(x) = 50 - 2x = 0$ at $x = 25$. Thus, $y = 50 - 25 = 25$. So the largest area enclosed by 100 feet of fencing is a square 25 ft by 25 ft.

For steps 2 and 3 in the procedure you may need to recall facts and relationships from geometry, trigonometry, etc.

1) A gardener wishes to enclose a rectangular area with 300 feet of fencing and fence it down the middle to divide it into two equal subareas. What is the largest rectangular area that may be enclosed?

(1) Let A be the area of the enclosed rectangle.

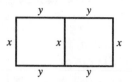

Label one side as x and the half side of the other as y (since that side is bisected).

(2) $A = x(2y) = 2xy$. The total amount of fence is $3x + 4y = 300$. Thus, $y = \frac{300 - 3x}{4}$.

(3) Therefore,
$$A = 2xy = 2x\frac{300 - 3x}{4} = 150x - \frac{3}{2}x^2.$$
Since $3x + 4y = 300$ and both $x \geq 0$ and $y \geq 0$, $x \in [0, 100]$.

(4) Maximize $A = 150x - \frac{3}{2}x^2$ on $[0, 100]$. Clearly 0 and 100 produce minimum values, so A is maximum when $A' = 0$.
$A' = 150 - 3x = 0$ at $x = 50$.
Then $y = \frac{300 - 3(50)}{4} = 37.5$.
Thus the maximum value of A is $2xy = 2(50)(37.5) = 3750$ ft^2.

2) A cylindrical aluminum cup is to be made from 12 square inches of pressed aluminum. What is the largest possible volume of such a cup?

(1) Maximize the volume V of a cup whose circular base has radius r and whose height is h.

(2) $V = \pi r^2 h$. The total surface area (bottom plus side) is $12 = \pi r^2 + 2\pi rh$.
Thus $h = \frac{12 - \pi r^2}{2\pi r}$.

(3) Thus $V = \pi r^2 \frac{12 - \pi r^2}{2\pi r}$.
$V = 6r - \frac{\pi}{2}r^3$, where $r \in \left[0, \sqrt{\frac{12}{\pi}}\right]$.

(4) $V' = 6 - \frac{3\pi}{2}r^2 = 0$
$r^2 = \frac{4}{\pi}$
$r = \frac{2}{\sqrt{\pi}}$
$V'' = -3\pi r$ so $r = \frac{2}{\sqrt{\pi}}$ is a maximum. At $r = \frac{2}{\sqrt{\pi}}$ the maximum volume is
$V = 6\left(\frac{2}{\sqrt{\pi}}\right) - \frac{\pi}{2}\left(\frac{2}{\sqrt{\pi}}\right)^3 = \frac{8}{\sqrt{\pi}}$ in^3.

3) Find the dimensions of the largest rectangular peg that can be put into a round hole with diameter 10 cm.

(1) "Largest peg" means one with the largest cross sectional area. The cross section of the peg in the hole is given in this diagram.

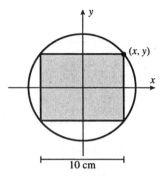

(2) The quantity to be maximized is area
$A = (2x)(2y) = 4xy$.

The equation for the hole is $x^2 + y^2 = 25$. Therefore, $y = \sqrt{25 - x^2}$.

(3) $A(x) = 4xy = 4x\sqrt{25 - x^2}$ where $x \in [0, 5]$.

(4) Maximize $A(x) = 4x\sqrt{25 - x^2}$ on $[0, 5]$. Both 0 and 5 produce a minimum area of zero, so the maximum occurs when $A'(x) = 0$.

$$A'(x) = (4x)\tfrac{1}{2}(25 - x^2)^{-1/2}(-2x)$$
$$+ (25 - x^2)^{1/2}(4)$$
$$= 4\left(-\frac{x^2}{\sqrt{25 - x^2}} + \sqrt{25 - x^2}\right)$$
$$= 4\left(\frac{-x^2 + (25 - x^2)}{\sqrt{25 - x^2}}\right)$$
$$= \frac{4(25 - 2x^2)}{\sqrt{25 - x^2}}$$

$A'(x) = 0$ when $25 - 2x^2 = 0$

$2x^2 = 25$, $x^2 = \dfrac{25}{2}$

$x = \dfrac{5}{\sqrt{2}} = \dfrac{5\sqrt{2}}{2}$ cm

$y = \sqrt{25 - x^2} = \sqrt{25 - \dfrac{25}{2}} = \sqrt{\dfrac{25}{2}}$
$= \dfrac{5\sqrt{2}}{2}$.

As you may well have guessed, a square peg.

Section 4.8   Applications to Economics

Quantities in economics such as revenue and costs are functions of the number of items sold or produced. Thus, calculus may be used to determine optimization problems in economics. The derivatives are referred to as marginal rates of change; thus, for example, the derivative of revenue is called marginal revenue.

Concepts to Master

A. Marginal cost; minimize average cost

B. Marginal revenue; maximize net revenue (profit).

Summary and Focus Questions

Page 340

A. If $C(x)$ represents the (total) cost function to produce x items, the *average cost* function is:

$$c(x) = \frac{C(x)}{x}.$$

Marginal cost is the derivative $C'(x)$ of total cost.

Average cost $c(x)$ is minimum when $c(x) = C'(x)$.

1) Find the minimum average cost for a company which produces x items at a total cost of $3x^2 + 10x + 1875$ dollars.

$C(x) = 3x^2 + 10x + 1875$
$C'(x) = 6x + 10$
$c(x) = \frac{3x^2 + 10x + 1875}{x} = 3x + 10 + \frac{1875}{x}$
Average cost is minimum when
$c(x) = C'(x)$:
$3x + 10 + \frac{1875}{x} = 6x + 10$
$\frac{1875}{x} = 3x$
$x^2 = 625$
$x = 25$
($x = -25$ is not a meaningful answer).
Thus the minimum average cost is
$3(25) + 10 + \frac{1875}{25} = \$160$ per unit.

B. The *price (or demand) function* $d(x)$ is the price per item a firm must charge to sell x items. The (total) *revenue function* $R(x) = x\,d(x)$ is the revenue obtained by selling x items at $d(x)$ dollars each.

Marginal revenue is the derivative $R'(x)$ of total revenue.
The *profit function* $P(x) = R(x) - C(x)$ is a maximum when marginal revenue equals marginal cost.

2) Suppose a laundry determines that to attract x customers per day its price of service must be $3.20 - 0.02x$ dollars. If their total cost to serve x customers is $0.05x^2 + 1.10x + 120$ dollars, how many customers should be served to maximize their profit?

(1) The demand is $p(x) = 3.20 + 0.02x$, so the revenue is $R(x) = p \cdot x = 3.20x + 0.02x^2$. Thus marginal revenue is
$R'(x) = 3.20 + 0.04x$.

(2) Total cost is $C(x) = 0.05x^2 + 1.10x + 120$, so marginal cost is $C'(x) = 0.10x + 1.10$.

(3) Equate $R'(x)$ and $C'(x)$:
$3.20 + 0.04x = 0.10x + 1.10$
$2.10 = 0.06x$
$x = 35$ customers.

Section 4.9   Newton's Method

Suppose you make a guess at the solution to $f(x) = 0$. Newton's method is a means to obtain a more accurate estimate from your initial guess. Usually the calculator is used to make sure your initial guess is somewhat near the actual root.

Concepts to Master

Approximations of solutions to $f(x) = 0$ using Newton's Method

Summary and Focus Questions

Page 345

Let f be a continuously differentiable function on an open interval with a real root. If x_1 is an estimate of the root, then

$$x_2 = x_1 - \frac{f(x_1)}{f'(x_1)}$$

is often (but not always) a closer approximation to the root. The point x_2 is where the tangent line to f at x_1 crosses the x-axis.

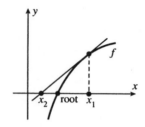

The process of obtaining x_2 may be repeated using x_2 in the role of x_1 to obtain another approximation x_3. In this way we can generate a sequence of approximations $x_1, x_2, x_3, \ldots$ which can approach the value of the root. In general, to obtain a desired degree of accuracy, repeat Newton's Method until the difference between successive x_i is within that accuracy.

1) Indicate the point x_2 determined by Newton's Method in each:

   a)

   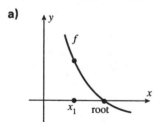

   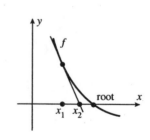

b)

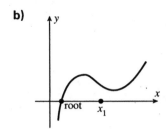

2) Use Newton's Method twice to approximate a root of $x^3 - 2x - 2 = 0$. Use an initial estimate of 1.0.

3) Approximate a solution to $\cos x = x$ to six decimals. Use Newton's Method with an initial estimate of 0.

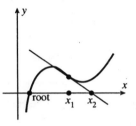

(Note: There is no guarantee that x_2 will be a better estimate than x_1.)

$f(x) = x^3 - 2x - 2, f'(x) = 3x^2 - 2$
$x_1 = 1.0$, so
$$x_2 = 1.0 - \frac{f(1.0)}{f'(1.0)} = 1.0 - \frac{-3.0}{1.0} = 4.0.$$

$$x_3 = 4.0 - \frac{f(4.0)}{f'(4.0)} = 4.0 - \frac{54}{46} \approx 2.826.$$

Let $f(x) = \cos x - x$ and $x_1 = 0$. This table shows successive values for x using Newton's Method.

	x	$x - \dfrac{f(x)}{f'(x)}$
$x_1 =$	0	1
$x_2 =$	1	0.750363868
$x_3 =$	0.750364	0.739112891
$x_4 =$	0.739113	0.739085133
$x_5 =$	0.739085	0.739085133
$x_6 =$	0.739085	0.739085133

To six decimals, the solution to $\cos x = x$ is 0.739085.

Section 4.10   Antiderivatives

In this section you begin to learn how to "undo" a derivative; that is, given a derivative, recover the function from where it came. We will see that there is not just one antiderivative function, but rather a whole collection of functions, all of which differ by a constant from one another.

Concepts to Master

General and particular antiderivatives of a function

Summary and Focus Questions

$F(x)$ is an *antiderivative* of the function $f(x)$ means $F'(x) = f(x)$. For example, $F(x) = 3x^4$ is an antiderivative of $f(x) = 12x^3$ because $(3x^4)' = 12x^3$.

Page 351

If F is an antiderivative of f then *all* other antiderivatives of f have the form $F(x) + C$, where C is a constant.

Here are some antidifferentiation formulas that you need to know well:

Function	Form of All Antiderivatives
x^n (except $n = -1$)	$\frac{x^{n+1}}{n+1} + C$
x^{-1}	$\ln |x| + C$
$\sin x$	$-\cos x + C$
$\cos x$	$\sin x + C$
$\sec^2 x$	$\tan x + C$
$\sec x \tan x$	$\sec x + C$
e^x	$e^x + C$
$\frac{1}{\sqrt{1 - x^2}}$	$\sin^{-1}x + C$
$\frac{1}{1 + x^2}$	$\tan^{-1}x + C$

The antiderivative of $f \pm g$ is the antiderivative of f plus or minus the antiderivative of g, and the antiderivative of $cf(x)$ is c times the antiderivative of $f(x)$ (c, any constant).

The general solution to the *differential equation* $\frac{dy}{dx} = f(x)$ is all antiderivatives of f. To find a particular solution, first find the general solution, then substitute the given values to determine the constant C.

Example:

The general solution to $\frac{dy}{dx} = 10x$ is $y = 5x^2 + C$. If we are also given that $y = 4$ when $x = 2$, then substituting we have $4 = 5(2)^2 + C$, so $C = -16$.

The particular solution is $y = 5x^2 - 16$.

1) If $g(x)$ is an antiderivative of $h(x)$, then
_____$'(x) =$ _____(x).

$g'(x) = h(x)$

2) Find all antiderivatives of:

a) $f(x) = x^7$

$\frac{x^8}{8} + C$

b) $f(x) = \cos x - \sec^2 x$

$\sin x - \tan x + C$

c) $f(x) = x + x^{-2}$

$\frac{x^2}{2} - \frac{1}{x} + C$

d) $f(x) = x - e^x$

$\frac{x^2}{2} - e^x$

e) $f(x) = \dfrac{4}{\sqrt{1 - x^2}}$

$4 \sin^{-1}x + C$

3) Find $f(x)$ where $f(2) = 3$ and
$f'(x) = 4x + 5$.

$f(x) = 2x^2 + 5x + C$
$f(2) = 2(2)^2 + 5(2) + C = 18 + C$
$18 + C = 3$ so $C = -15$
$f(x) = 2x^2 + 5x - 15$

4) A particle moves along a scale with velocity
$v = 3t + 7$. If the particle is at 4 on the
scale at time $t = 1$, find the position
function $s(t)$.

$s(t)$ is an antiderivative of $v(t) = 3t + 7$.
$s(t) = \frac{3}{2}t^2 + 7t + C$
$4 = \frac{3}{2}(1)^2 + 7(1) + C$
$C = -\frac{9}{2}$
$s(t) = \frac{3}{2}t^2 + 7t - \frac{9}{2}$

5) A company estimates that the marginal cost
to produce x items is $26 - 0.12x$ dollars per
item. If fixed costs are $4000, find the cost
to produce 50 items.

If $C(x)$ is the cost function, then
$C(0) = 4000$ and $C'(x) = 26 - 0.12x$.
Thus, $C(x) = 26x - 0.06x^2 + 4000$
$C(50) = \$5150$

6) Solve $\dfrac{dy}{dx} = 2 + x$, given that $y = 2$ when $x = 0$.

From $y' = 2 + x$, $y = 2x + \dfrac{x^2}{2} + C$.

At $x = 0$, $2 = 0 + 0 + C$. Thus, $2 = C$ and $y = \dfrac{x^2}{2} + 2x + 2$.

Technology Plus for Chapter 4

1) Let $f(x) = x^4 - 3x^3 + 5x$. Using a graph of the function estimate:

   a) the critical points of f.

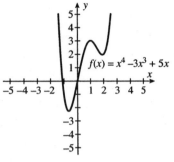

There are three critical points; they are (approximately) -0.656, 1.000, and 1.906.

   b) the absolute minimum and maximum of f on $[0, 2]$.

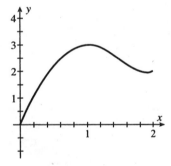

By using a window of $[0, 2]$ by $[-1, 4]$ we see that the absolute minimum occurs at $x = 0$ and is $f(0) = 0$ and the absolute maximum occurs at $x = 1$ and is $f(1) = 3$.

2) Let $f(x) = x^3 - x^2 + 2$.

   a) On the same screen, draw all three of these:

      $y = f(x)$, the secant line joining $(0, f(0))$ and $(3, f(3))$, the line tangent to f at $c = 1$.

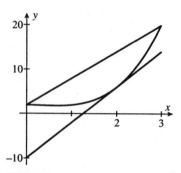

The line joining the endpoints has slope
$$\frac{f(3) - f(0)}{3 - 0} = \frac{20 - 2}{3 - 0} = 6 \text{ and equation}$$
$y = 6x + 2$.
$f'(x) = 3x^2 - 0\,2x;\, f'(2) = 8.\, f(2) = 6$.
The tangent line at $c = 1$ has equation
$y - 6 = 8(x - 2)$, or $y = 8x - 10$.

   b) Is $c = 2$ a value that satisfies the conclusion of the Mean Value Theorem?

No. Since $f'(2) = 8$, the tangent at $c = 2$ is too steep; it would need to have a slope of 6.

3) Let $f(x) = x^4 - 3x^3 - x^2 + 3x$. Draw the graph of f'' on a calculator with window $[-2, 4]$ by $[-12, 6]$. Where is f concave up and concave down?

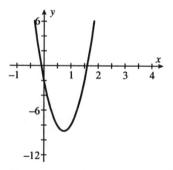

The function f'' crosses the x-axis at about -0.014 and 1.604. $f''(x) \geq 0$ on $(-\infty, -0.104] \cup [1.604, \infty)$, so f is concave upward there. On $[-0.104, 1.604]$ $f''(x) \leq 0$ and so f is concave downward there.

4) Use a calculator to make a table of values

of $f(x) = \dfrac{x}{\sqrt{4x^2 + 900}}$ for large values of x.

Use the table to estimate $\lim\limits_{x \to \infty} f(x)$.

x	f(x)
100	0.4944681764
1,000	0.4999437595
10,000	0.4999994375
100,000	0.4999999944
1,000,000	0.4999999999
10,000,000	0.5000000000

From the table, we see that $\lim\limits_{x \to \infty} f(x) = \frac{1}{2}$.

5) Let $f(x) = \dfrac{1}{x^2} - \dfrac{1}{(x-2)^2}$. Draw the graph of f on a calculator with window $[-2, 4]$ by $[-6, 6]$. Describe the graph.

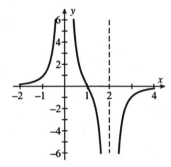

A. The domain is $(-\infty, 0) \cup (0, 2) \cup (2, \infty)$.

B. The x-intercept is 1; there is no y-intercept.

C. The graph is not symmetric about the y-axis or origin. (It is symmetric about the point $(1, 0)$.)

D. $y = 0$ is the horizontal asymptote. There are vertical asymptotes at $x = 0$ and $x = 2$.

E. Increasing on $(-\infty, 0) \cup (2, \infty)$; decreasing on $(0, 2)$.

F. The graph has no relative extrema.

G. $(1, 0)$ is an inflection point. The graph is concave up on $(-\infty, 0) \cup (0, 1)$ and concave down for $(1, 2] \cup [2, \infty)$.

6) Gordon is standing on one side of a canal that is one mile wide. A lighthouse is one mile away downstream from the opposite side of the canal (see the figure). Gordon can run 4 mi/hr and swim 1 mi/hr. Since Gordon needs to get to the lighthouse as fast as he can, he will run along the canal part way, then swim directly to the lighthouse.

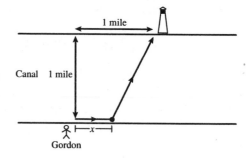

a) Write a function $T(x)$ for the time it takes Gordon to make the entire trip, where x is the distance that he runs. (Ignore any current in the water.)

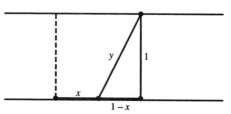

The time it takes to run x miles is $\frac{x}{4}$.
The distance swam (y) satisfies
$y^2 = (1 - x)^2 + 1^2$. Thus,
$y = \sqrt{(1 - x)^2 + 1}$
Therefore, the total trip time is

$$T(x) = \frac{x}{4} + \frac{\sqrt{(1 - x)^2 + 1}}{1}$$

$$= \frac{x}{4} + \sqrt{(1 - x)^2 + 1}.$$

b) Use a graphing calculator to estimate the distance that he should run to minimize his trip time.

With a window of [0, 1] by [1, 1.5], the graph of $T(x)$ is

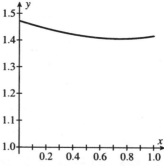

The minimum value of $T(x)$ occurs at approximately 0.7418. So Gordon should run about three quarters of the way and then swim the rest of the way.

Chapter 5 — Integrals

© 1999 by Sidney Harris.

Section 5.1   Areas and Distances

Chapters 5, 6, 8 and 9 cover the concepts and techniques of integration—that part of calculus that may be used to find areas, volumes, centers of mass, etc. (We will see later in this chapter how this relates to derivatives.) To this end, this section introduces two types of problems—finding areas under curves and finding distances given velocities. Just as limits were used in the beginning of Chapter 2 to find tangents and velocities, limits will be used here to determine areas and distances.

Concepts to Master

A. Approximation of area under the graph of a function

B. Area as a limit of sums

C. Determining distances from velocities

Summary and Focus Questions

Page 367

A. One way to approximate the area under a curve $y = f(x)$ that lies above the x-axis and between the vertical lines $x = a$ and $x = b$ is to cover it with several adjoining rectangles, estimate the area of these rectangles and add them up.

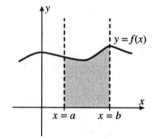

To approximate the area described above:

(1) Partition $[a, b]$ into n equal subintervals, each with width

$$\Delta x = \frac{b - a}{n}.$$

(2) Select a sample point x_i^* from each subinterval $[x_{i-1}, x_i]$. (Usually the choice is made in a systematic way, such as "choose the right endpoint of each subinterval.")

(3) Calculate

$$\sum_{i=1}^{n} f(x_i^*)\Delta x = f(x_1^*)\Delta x + f(x_2^*)\Delta x + f(x_3^*)\Delta x + \ldots + f(x_n^*)\Delta x.$$

Here is an example:
Use 3 subintervals to estimate the area under the curve $y = x^3 + x$ between the lines $x = 2$ and $x = 8$. Use the midpoint of each subinterval for sample points.

For $n = 3$, $\Delta x = \frac{8-2}{3} = 2$.

i	subint.	x_i^*	$f(x_i^*)$	$f(x_i^*)\Delta x$
1	[2, 4]	3	30	60
2	[4, 6]	5	130	260
3	[6, 8]	7	350	700

The total of the last column, 1020, is an approximation of the area.

1) Use midpoints as sample points to estimate the area under the function $f(x) = 20x - x^2$ that lies above the x-axis between $x = 4$ and $x = 16$. Use 3 subintervals of equal width. Sketch the area and approximating rectangles.

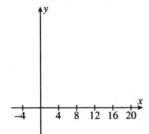

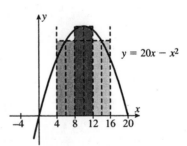

$y = 20x - x^2$

$\Delta x = \frac{16-4}{3} = 4$.

i	subint.	x_i^*	$f(x_i^*)$	$f(x_i^*)\Delta x$
1	[4, 8]	6	84	336
2	[8, 12]	10	100	400
3	[12, 16]	14	84	336

The area is approximately
$336 + 400 + 336 = 1072$.

2) Use the left endpoints as sample points to estimate the area under the function $f(x) = 1 + 4x^2$ that lies above the *x*-axis between $x = 2$ and $x = 8$. Use 4 subintervals of equal width.

$\Delta x = \dfrac{8 - 2}{4} = 1.5.$

i	subint.	x_i^*	$f(x_i^*)$	$f(x_i^*)\Delta x$
1	[2, 3.5]	2	17	25.5
2	[3.5, 5]	3.5	50	75
3	[5, 6.5]	5	101	151.5
4	[6.5, 8]	6.5	170	255

The area is approximately
$25.5 + 75 + 151.5 + 255 = 507.$

3) Estimate the area under the curve given below between $x = 1$ and $x = 6$. Use right endpoints for sample points and intervals of width 1.

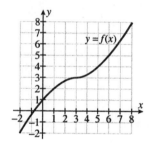

$\Delta x = 1$

i	subint.	x_i^*	$f(x_i^*)$	$f(x_i^*)\Delta x$
1	[1, 2]	2	2.7	2.7
2	[2, 3]	3	3	3
3	[3, 4]	4	3.3	3.3
4	[4, 5]	5	4	4
5	[5, 6]	6	5	5

The area is approximately
$2.7 + 3 + 3.3 + 4 + 5 = 18.$

B. For a continuous function f, the larger the value of n, the better the approximating sum

$$\sum_{i=1}^{n} f(x_i^*)\Delta x = f(x_1^*)\Delta x + f(x_2^*)\Delta x + f(x_3^*)\Delta x + \ldots + f(x_n^*)\Delta x$$

is to the exact area.

We define the area under a curve $y = f(x)$ that lies above the x-axis and between the vertical lines $x = a$ and $x = b$ to be:

$$\text{Area} = \lim_{n\to\infty} \sum_{i=1}^{n} f(x_i^*)\Delta x = \lim_{n\to\infty} [f(x_1^*)\Delta x + f(x_2^*)\Delta x + f(x_3^*)\Delta x + \ldots + f(x_n^*)\Delta x].$$

For some functions the area may be determined exactly by:

(1) Partition $[a, b]$ into n subintervals of equal length $\Delta x = \dfrac{b-a}{n}$.

(2) Select a point x_i^* in each subinterval and write each x_i^* in terms of n and i. Do so consistently. For example, choosing x_i^* to be the left endpoint of $[x_{i-1}, x_i]$ means $x_i^* = a + (i-1)\Delta x$.

(3) Write the approximating sum $\sum_{i=1}^{n} f(x_i^*)\Delta x$ as a function of n.

(4) Determine the limit of the expression found in step (3) as $n\to\infty$.

4) True or False:
The area of a region bounded by a non-negative continuous function $y = f(x)$, $y = 0$, $x = a$, $x = b$ is the limit of the sums of the areas of approximating rectangles.

True.

5) Find the area under $y = 10 - 2x$ between $x = 1$ and $x = 4$. Take x_i^* to be the right endpoint. Use subintervals of equal width.

With n subintervals we have
$$\Delta x = \frac{b - a}{n} = \frac{4 - 1}{n} = \frac{3}{n}.$$

$x_1 = 1 + \Delta x$, $x_2 = 1 + 2\Delta x$, and in general $x_i = 1 + i\Delta x$. Selecting x_i^* to be the right endpoint means $x_i^* = x_i$. Thus
$$x_i^* = 1 + i\Delta x = 1 + i\left(\frac{3}{n}\right) = 1 + \frac{3i}{n}.$$
Hence $f(x_i^*) = 10 - 2\left(1 + \frac{3i}{n}\right) = 8 - \frac{6i}{n}$.
The approximating sum is

$$\sum_{i=1}^{n} f(x_i^*)\Delta x = \sum_{i=1}^{n} \left(8 - \frac{6i}{n}\right)\left(\frac{3}{n}\right)$$
$$= \sum_{i=1}^{n} \left(\frac{24}{n} - \frac{18i}{n^2}\right)$$
$$= \sum_{i=1}^{n} \frac{24}{n} - \sum_{i=1}^{n} \frac{18i}{n^2}$$
$$= \frac{24}{n} \sum_{i=1}^{n} 1 - \frac{18}{n^2} \sum_{i=1}^{n} i$$
$$= \frac{24}{n}(n) - \frac{18}{n^2}\left(\frac{n(n+1)}{2}\right)$$
$$= 24 - 9\frac{n(n+1)}{n^2}.$$

Finally,
$$\lim_{n \to \infty} \left(24 - 9\frac{n(n+1)}{n^2}\right) = 24 - 9(1) = 15.$$

Page 374

C. If an object is moving in a positive direction and has a velocity of $v = f(t)$ at time t for $a \le t \le b$, then

$$\lim_{n \to \infty} \sum_{i=1}^{n} f(x_i^*)\Delta x = \lim_{n \to \infty} [f(x_1^*)\Delta x + f(x_2^*)\Delta x + f(x_3^*)\Delta x + \ldots + f(x_n^*)\Delta x]$$

is the distance the object travels during the time interval $[a, b]$.

6) Shu-Ping rides her bike from home to the park. She notes the velocity every 5 minutes in the table below. How far did she travel?

time	velocity (mi/hr)
10:00	0 (start)
10:05	12
10:10	13
10:15	16
10:20	8
10:25	0 (reaches park)

$\Delta x = .0833$ (5 minutes)
We use the left endpoints for sample points.

i	subint.	x_i^*	$f(x_i^*)$	$f(x_i^*)\Delta x$
1	[10:00, 10:05]	10:00	0	0.000
2	[10:05, 10:10]	10:05	12	1.000
3	[10:10, 10:15]	10:10	13	1.083
4	[10:15, 10:20]	10:15	16	1.333
5	[10:20, 10:25]	10:20	8	0.667

The distance traveled is approximately
$0 + 1 + 1.083 + 1.333 + 0.667$
$= 4.083$ miles.

Section 5.2　The Definite Integral

This section defines the definite integral as a limit of sums like those discussed in the previous section. Some properties of definite integrals (such as the integral of the sum of two functions is the sum of the integrals) are discussed; if you think of definite integrals as areas, then many of the properties can be visualized in graphs.

Concepts to Master

A. Definition of $\int_a^b f(x)\ dx$; Integrability; Riemann sum

B. Evaluating definite integrals

C. Properties of definite integrals

Summary and Focus Questions

Page 378

A. The *definite integral* of a function f on an interval $[a, b]$ is

$$\int_a^b f(x)\ dx = \lim_{n \to \infty} \sum_{i=1}^n f(x_i^*)\Delta x$$

$$= \lim_{n \to \infty} [f(x_1^*)\Delta x + f(x_2^*)\Delta x + f(x_3^*)\Delta x + \dots + f(x_n^*)\Delta x].$$

When this limit exists we say that f is *integrable on the interval* $[a, b]$. The definite integral always exists for continuous functions and for monotonic (increasing or decreasing) functions.

The sums may be generalized by not requiring that all subintervals $[x_{i-1}, x_i]$ be the same width. If $\Delta x_i = x_i - x_{i-1}$, then a *Riemann sum* for f has the form:

$$\sum_{i=1}^n f(x_i^*)\Delta x_i = [f(x_1^*)\Delta x_1 + f(x_2^*)\Delta x_2 + f(x_3^*)\Delta x_3 + \dots + f(x_n^*)\Delta x_n]$$

where each x_i^* is chosen from $[x_{i-1}, x_i]$, $i = 1, 2, 3, \dots, n$.

For $f(x) \geq 0$ on $[a, b]$, $\int_a^b f(x)\ dx$ is the area under the graph of f between $x = a$ and $x = b$.

1) If $\Delta x = \dfrac{3-1}{n}$ on the interval $[1, 3]$ and

$x_i^* \in [x_{i-1}, x_i]$ for each i then

$$\lim_{n\to\infty} \sum_{i=1}^{n} (x_i^*)^4 \Delta x = \underline{\hspace{1cm}}$$

$$\int_1^3 x^4 \, dx$$

2) Does $\displaystyle\int_1^6 [\![x]\!] \, dx$ exist?

($[\![x]\!]$ is the greatest integer function.)

Yes, because $f(x) = [\![x]\!]$ is increasing (monotone) on $[1, 6]$.

3) Construct a Riemann sum for $f(x) = x^2$ on the interval $[3, 10]$ using the partition $[3, 4], [4, 6], [6, 9], [9, 10]$.

We must select x_i^* points in $[x_{i-1}, x_i]$ and calculate Δx_i for each subinterval:

Select $x_1^* = 3 \in [3, 4]; \Delta x_1 = 1$
$x_2^* = 5 \in [4, 6]; \Delta x_2 = 2$
$x_3^* = 8 \in [6, 9]; \Delta x_3 = 3$
$x_4^* = 9 \in [9, 10]; \Delta x_4 = 1$

(Your x_i^* choices may be different.)
The resulting Riemann sum for these choices of x_i^* is:

i	subint.	Δx_i	x_i^*	$f(x_i^*)$	$f(x_i^*)\Delta x_i$
1	[3, 4]	1	3	9	9
2	[4, 6]	2	5	25	50
3	[6, 9]	3	8	64	192
4	[9, 10]	1	9	81	81

$$\sum_{i=1}^{4} f(x_i^*)\Delta x_i = 9 + 50 + 192 + 81 = 332.$$

4) Write a definite integral for the shaded area.

$$\int_{-1}^{3} \left(\frac{x^3}{3} - x^2 + 2\right) dx.$$

B. One of the simplest Riemann sums to calculate is the case where the partition is regular $\left(\text{all } \Delta x_i \text{ are equal to } \dfrac{b-a}{n}\right)$ and $x_i^* = x_i$, the right endpoint of the ith subinterval. In this case the Riemann sum is:

$$\frac{b-a}{n} \sum_{i=1}^{n} f\left(a + i\frac{b-a}{n}\right).$$

Midpoint Rule:

Using midpoints for x_i^* and regular partitions, the Riemann sum is

$$\frac{b-a}{n} \sum_{i=1}^{n} f(\overline{x_i}), \text{ where } \overline{x_i} = \frac{x_i + x_{i-1}}{2}.$$

Determining the exact value of $\displaystyle\int_a^b f(x)\, dx$ from the definition requires simplifying sigma notation sums. Some summations may be rewritten as a function of their upper limit which permits the summation notation to be replaced with a simple expression. Here are five such expressions:

$$\sum_{i=1}^{n} 1 = n \qquad\qquad \sum_{i=1}^{n} i = \frac{n(n+1)}{2}$$

$$\sum_{i=1}^{n} i^2 = \frac{n(n+1)(2n+1)}{6} \qquad \sum_{i=1}^{n} i^3 = \left(\frac{n(n+1)}{2}\right)^2$$

$$\sum_{i=1}^{n} i^4 = \frac{n(n+1)(2n+1)(3n^2 + 3n - 1)}{30}$$

These may be combined to simplify sums as in this example:

$$\sum_{i=1}^{n}(i^2 + 4i) = \sum_{i=1}^{n} i^2 + \sum_{i=1}^{n} 4i = \sum_{i=1}^{n} i^2 + 4\sum_{i=1}^{n} i$$

$$= \frac{n(n+1)(2n+1)}{6} + 4\frac{n(n+1)}{2} = \frac{n(n+1)(2n+13)}{6}$$

5) Use a partition of 4 equal sized subintervals and right endpoints for sample points to approximate

$$\int_{-2}^{4} 3x^2\, dx.$$

$$\Delta x = \frac{4 - (-2)}{4} = 1.5$$

i	subint.	x_i^*	$f(x_i^*)$	$f(x_i^*)\Delta x$
1	$[-2, -0.5]$	-0.5	0.750	1.125
2	$[-0.5, 1]$	1	3.000	4.500
3	$[1, 2.5]$	2.5	18.750	28.125
4	$[2.5, 4]$	4	48.000	72.000

$$\sum_{i=1}^{4}(3x_i^*)\Delta x =$$

$$1.125 + 4.500 + 28.125 + 72.000 = 105.75$$

6) Approximate $\int_1^4 x^3 \, dx$ using the Midpoint Rule with $n = 3$.

$$\Delta x = \frac{b - a}{n} = \frac{4 - 1}{3} = 1.$$

The Riemann sum is calculated as follows:

i	subint.	$\overline{x}_i$ mid pt	$f(\overline{x}_i)$
1	[1, 2]	1.5	3.375
2	[2, 3]	2.5	15.625
3	[3, 4]	3.5	42.875

$$\frac{b - a}{n} \sum_{i=1}^{3} f(\overline{x}_i)$$

$$= 1(3.375 + 15.625 + 42.875) = 61.875.$$

7) Write the following as an expression in n:

$$\sum_{i=1}^{n} (12i^2 - 4ni)$$

$$\sum_{i=1}^{n} (12i^2 - 4ni) = \sum_{i=1}^{n} 12i^2 - \sum_{i=1}^{n} 4ni$$

$$= 12 \sum_{i=1}^{n} i^2 - 4n \sum_{i=1}^{n} i$$

$$= 12 \frac{n(n + 1)(2n + 1)}{6} - (4n) \frac{n(n + 1)}{2}$$

$$= n(n + 1)(2n + 2) = 2n(n + 1)^2.$$

8) Find $\int_0^5 (x^3 + 2x^2) \, dx$ using the definition and right endpoints for x_i^*.

With n subintervals $\Delta x = \frac{5 - 0}{n} = \frac{5}{n}$.

$x_1 = 0 + \Delta x$, $x_2 = 0 + 2\Delta x$, ... and so in general $x_i = i\Delta x$. Right endpoints for x_i^* means $x_i^* = x_i = i\Delta x = i\left(\frac{5}{n}\right) = \frac{5i}{n}$.

The Riemann sum is

$$\sum_{i=1}^{n} f(x_i^*) \Delta x = \sum_{i=1}^{n} \left(\left(\frac{5i}{n}\right)^3 + 2\left(\frac{5i}{n}\right)^2 \right) \left(\frac{5}{n}\right)$$

$$= \sum_{i=1}^{n} \frac{625i^3}{n^4} + \frac{250i^2}{n^3}$$

$$= \frac{625}{n^4} \sum_{i=1}^{n} i^3 + \frac{250}{n^3} \sum_{i=1}^{n} i^2$$

$$= \frac{625}{n^4} \left(\frac{n(n + 1)}{2}\right)^2 + \frac{250}{n^3} \left(\frac{n(n + 1)(2n + 1)}{6}\right)$$

$$= \frac{625}{4} \left(\frac{n + 1}{n}\right)^2 + \frac{250}{6} \left(\frac{(n + 1)(2n + 1)}{n^2}\right).$$

Finally,

$$\lim_{n \to \infty} \left[\frac{625}{4} \left(\frac{n + 1}{n}\right)^2 + \frac{250}{6} \left(\frac{(n + 1)(2n + 1)}{n^2}\right) \right]$$

$$= \frac{625}{4} + \frac{250}{6}(2) = \frac{2875}{12} \approx 239.5833.$$

9) Suppose $f(x)$ is continuous and increasing on $[a, b]$. Let $\Re$ be the value of a Riemann sum for $\displaystyle\int_a^b f(x)\, dx$ using the left endpoint of each subinterval. Which is true?

a) $\Re \le \displaystyle\int_a^b f(x)\, dx$

b) $\Re = \displaystyle\int_a^b f(x)\, dx$

c) $\Re \ge \displaystyle\int_a^b f(x)\, dx$

a), since the left endpoint will have the smallest value of $f(x)$ in each subinterval.

10) Evaluate $\displaystyle\int_3^6 (12 - x)\, dx$.

(Hint: what area is it?)

The integral represents the shaded trapezoid area below.

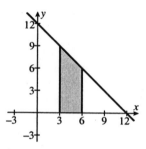

This area is
$\frac{1}{2}(9 + 6)3 = 22.5$

C. Several integral properties allow definite integrals to be evaluated and estimated. If f and g are integrable functions on an interval containing a, b, and c,

$$\int_a^b f(x)\, dx = -\int_b^a f(x)\, dx.$$

$$\int_a^b f(x)\, dx = \int_a^c f(x)\, dx + \int_c^b f(x)\, dx.$$

$$\int_a^b (f(x) \pm g(x))\, dx = \int_a^b f(x)\, dx \pm \int_a^b g(x)\, dx.$$

$$\int_a^b k\, f(x)\, dx = k\int_a^b f(x)\, dx \text{ (where k is a constant).}$$

If $f(x) \ge g(x)$ for all $x \in [a, b]$, then $\displaystyle\int_a^b f(x)\, dx \ge \int_a^b g(x)\, dx$.

If $m \le f(x) \le M$ for $x \in [a, b]$, then $m(b - a) \le \displaystyle\int_a^b f(x)\, dx \le M(b - a)$.

11) What are the missing limits of integration?

$$\int_4^? x^3 \, dx = \int_?^? x^3 \, dx + \int_5^7 x^3 \, dx$$

$$\int_4^7, \int_4^5, \int_5^7.$$

12) True or False:

a) $\int_0^1 (\sin x + \cos x) \, dx$
$$= \int_0^1 \sin x \, dx + \int_0^1 \cos x \, dx$$

True.

b) $\int_0^1 x(x^2 + 1) \, dx$
$$= \left(\int_0^1 x \, dx\right)\left(\int_0^1 (x^2 + 1) \, dx\right)$$

False.

c) $\int_0^1 x(x^2 + 1) \, dx = x\int_0^1 (x^2 + 1) \, dx$

False.

d) $\int_0^1 x(t^2 + 1) \, dt = x\int_0^1 (t^2 + 1) \, dt$

True. (x is a constant here because t is the variable of integration.)

e) $\int_0^{\pi/4} \tan x \, dx \geq 0$

True, because for $0 \leq x \leq \frac{\pi}{4}$, $\tan x \geq 0$.

f) $\int_0^1 2^x \, dx \leq \int_0^1 x \, dx$

False, for $0 \leq x \leq 1$, $x < 2^x$, hence
$$\int_0^1 x \, dx \leq \int_0^1 2^x \, dx.$$

g) $\int_1^3 \frac{1}{x} \, dx = \int_1^5 \frac{1}{x} \, dx + \int_5^3 \frac{1}{x} \, dx$

True.

13) Using $1 \leq x^2 + 1 \leq 10$ for $1 \leq x \leq 3$,
find upper and lower estimates for
$$\int_1^3 (x^2 + 1) \, dx.$$

Since $1 \leq x^2 + 1 \leq 10$ on $[1, 3]$,
$$1(3 - 1) \leq \int_1^3 (x^2 + 1) \, dx \leq 10(3 - 1).$$
Thus $2 \leq \int_1^3 (x^2 + 1) \, dx \leq 20.$

Section 5.3 The Fundamental Theorem of Calculus

Until now derivatives and definite integrals seem to have nothing to do with one another. The Fundamental Theorem of Calculus relates the two. The name "*The Fundamental Theorem of Calculus*" suggests that its importance cannot be overemphasized. It is stated in two equivalent versions.

Concepts to Master

Both forms of the Fundamental Theorem of Calculus.

Summary and Focus Questions

Page 391

The Fundamental Theorem of Calculus may be expressed in two different forms:

Part 1: If f is continuous on $[a, b]$ and if g is a function defined by

$$g(x) = \int_a^x f(t)\, dt,$$ then g is continuous on $[a, b]$ and $g'(x) = f(x)$ for all $x \in (a, b)$.

For example, if $g(x) = \int_2^x (t^2 + 2t)\, dt$, then $g'(x) = x^2 + 2x$.

Part 2: If f is continuous on $[a, b]$ and F is any antiderivative of f, then

$$\int_a^b f(x)\, dx = F(b) - F(a).$$

(Recall that F is an antiderivative of f means $f' = F$.)

For example, since an antiderivative of $f(x) = 6x^2 + 2x$ is $F(x) = 2x^3 + x^2$, we have $\int_1^3 (6x^2 + 2x)\, dx = F(3) - F(1) = [2(3)^3 + 3^2] - [2(1)^3 + 1^2] = 60$.

Both forms are useful. Part 1 will be useful in future sections when we define functions in that way. Part 2 provides a handy way to evaluate definite integrals —find an antiderivative and plug in the endpoints.

1) If $h(x) = \int_4^x \sqrt{t}\, dt$, then $h'(x) =$ _____.

$h'(x) = \sqrt{x}$. (Remember that h is a function of x, not t.)

2) If $f(x)$ is continuous on $[a, b]$, then for all

$x \in (a, b)$, $\dfrac{d}{dx}\displaystyle\int_a^x f(t)\, dt = $ _____.

$f(x)$. This is Part 1 of the Fundamental
Theorem of Calculus.

3) If $f'(x)$ is continuous on $[a, b]$, then for all

$x \in [a, b]$, $\displaystyle\int_a^x f'(x)\, dx = $ _____.

$f(x) - f(a)$. This is the second part of the
Fundamental Theorem of Calculus.

4) What do questions 2) and 3) say about
differentiation and integration.

Question 2) says differentiation will "undo"
integration while 3) says integration will
"undo" taking a derivative. Thus integration
and differentiation are inverse processes.

5) Evaluate $\displaystyle\int_0^1 (1 - x)\, dx$

An antiderivative of $1 - x$ is $F(x) = x - \dfrac{x^2}{2}$.

Thus $\displaystyle\int_0^1 (1 - x)\, dx = F(1) - F(0)$
$$= \left(1 - \tfrac{1}{2}\right) - \left(0 - \tfrac{0}{2}\right)$$
$$= \tfrac{1}{2}.$$

6) For each $x \in [1, 3]$, let $H(x)$ be the value of
the area shaded below:

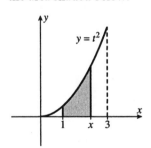

Find $H'(x)$.

$H(x) = \displaystyle\int_1^x t^2\, dt$. By Part 1 of the
Fundamental Theorem, $H'(x) = x^2$.

Section 5.4   Indefinite Integrals and the Total Change Theorem

This section shows you how to evaluate definite integrals by applying Part 2 of the Fundamental Theorem of Calculus. This is a basic technique that must be learned well. The section concludes by observing that the Fundamental Theorem of Calculus says that the total change of a function from a to b is the integral of the instantaneous rate of change of the function.

Concepts to Master

A. Indefinite integrals; Evaluation of definite integrals using the Fundamental Theorem

B. The Total Change Theorem

Summary and Focus Questions

Page 401

A. The *indefinite integral of f with respect to x* is $\int f(x)\ dx$ and is the antiderivative of f. That is $\int f(x)\ dx = F(x)$ means $F'(x) = f(x)$.
For example, $\int 2x\ dx = x^2 + C$, where C is an arbitrary constant, the *constant of integration.*

Here are some properties of indefinite integrals.

The derivative of $\int f(x)\ dx$ is $f(x)$, by definition.

$\int c f(x)\ dx = c \int f(x)\ dx$ (c, a constant)

$\int (f(x) \pm g(x))\ dx = \int f(x)\ dx \pm \int g(x)\ dx$

$\int x^n\ dx = \frac{x^{n+1}}{n+1} + C$ for any number n except -1.

$\int \frac{1}{x}\ dx = \ln |x| + C$

$\int \sin x\ dx = -\cos x + C \quad \int \sec x \tan x\ dx = \sec x + C$

$\int \cos x\ dx = \sin x + C \quad \int \csc^2 x\ dx = -\cot x + C$

$\int \sec^2 x\ dx = \tan x + C \quad \int \csc x \cot x\ dx = -\csc x + C$

$\int e^x\ dx = e^x + C$

$\int a^x\ dx = \frac{a^x}{\ln a} + C$

$\int \frac{1}{\sqrt{1 - x^2}}\ dx = \sin^{-1} x + C \qquad \int \frac{1}{1+x^2}\ dx = \tan^{-1} x + C$

Part 2 of the Fundamental Theorem of Calculus provides a handy procedure for evaluating some integrals without resorting to limits of Riemann sums:

1. Find $\int f(x)\, dx = F(x)$ (any antiderivative of f):

2. Compute $F(x)\Big]_a^b = F(b) - F(a)$.

Remember that the method may be applied when f is continuous on $[a, b]$.

1) Evaluate

a) $\int x^4\, dx$

$\dfrac{x^5}{5} + C$

b) $\int (6x + 12x^3)\, dx$

c) $\int (2 \cos x - \sec^2 x)\, dx$

$3x^2 + 3x^4 + C$

2) Determine

a) $\displaystyle\int_0^1 \sqrt{x}\, dx$

$2 \sin x - \tan x + C$

$\displaystyle\int_0^1 x^{1/2}\, dx = \dfrac{x^{3/2}}{3/2} = \dfrac{2}{3}x^{3/2}\Big]_0^1$

b) $\displaystyle\int_{-1}^3 x^{-3}\, dx$

$= \dfrac{2}{3}(1^{3/2} - 0^{3/2}) = \dfrac{2}{3}.$

The Fundamental Theorem does not apply because x^{-3} is not continuous at $x = 0$. We will have to wait until Chapter 8, section 8, to evaluate this integral.

c) $\displaystyle\int_{\pi/4}^{\pi/3} \sec x \tan x\, dx$

$\displaystyle\int_{\pi/4}^{\pi/3} \sec x \tan x = \sec x\Big]_{\pi/4}^{\pi/3}$

d) $\displaystyle\int_1^3 \dfrac{1}{x}\, dx$

$= \sec \dfrac{\pi}{3} - \sec \dfrac{\pi}{4} = 2 - \sqrt{2}.$

$\displaystyle\int_1^3 \dfrac{1}{x}\, dx = \ln |x|\Big]_1^3 = \ln 3 - \ln 1$

$= \ln 3 - 0 = \ln 3.$

3) True or False:

a) $\int 5 f(x) \, dx = 5 \int f(x) \, dx.$

True.

b) $\int [f(x)]^2 \, dx = \left(\int f(x) \, dx \right)^2.$

False.

c) $\int f(x) g(x) \, dx = \int f(x) \, dx \int g(x) \, dx.$

False. (Try $f(x) = x^3$ and $g(x) = x^4$.)

d) $\frac{d}{dx} \left(\int f(x) \, dx \right) = f(x).$

True.

Page 404

B. *The Total Change Theorem:*

By The Fundamental Theorem, Part 2, $\int_a^b f(x) \, dx = F(b) - F(a)$, when $\int f(x) \, dx = F(x)$. Thus, $F(b) - F(a)$, the total change in F, is $\int_a^b f(x) \, dx$.

For functions that are not nonnegative, the integral represents a kind of "net area."

For example, for the function graphed here, $\int_a^b f(x) \, dx = -A_1 + A_2 - A_3 + A_4$.

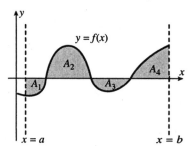

4) If $R(x)$ is the revenue obtained by selling x units, then the marginal revenue is $R'(x)$. What is $\int_{x_1}^{x_2} R'(x) \, dx$?

This is the total change in revenue when the number of units sold increases from x_1 to x_2.

5) Evaluate $\displaystyle\int_{-2}^{2} x^3\,dx$.

What does this number represent?

$$\int_{-2}^{2} x^3\,dx = \left.\frac{x^4}{4}\right]_{-2}^{2} = \frac{2^4}{4} - \frac{(-2)^4}{4} = 0.$$

The net area is 0 which means that the area above the x-axis and the area below the x-axis are the same.

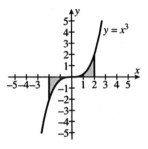

Section 5.5   The Substitution Rule

This section gives a technique for evaluating indefinite integrals that involve the composition of functions. It is the Chain Rule stated for indefinite integrals.

Concepts to Master

A. Evaluation of integrals using substitution

B. Integrals involving symmetric functions

Summary and Focus Questions

Page 410

A. The method of substitution is used to evaluate integrals of the form

$$\int f(g'(x))\, dx.$$

The integral is rewritten in the form $\int f(u)\, du$, where $u = g(x)$. The important step is to recognize a proper choice for u and the corresponding $du = g'(x)\, dx$.

For example, $\int (x^2 + 1)^3 x\, dx$ has the best choice $u = x^2 + 1$. For then $du = 2x\, dx$ and thus $x\, dx = \frac{du}{2}$. Now make the substitutions:

$$\int (x^2 + 1)^3 x\, dx = \int u^3 \frac{du}{2} = \frac{1}{2}\int u^3\, du = \frac{1}{2}\frac{u^4}{4} + C = \frac{(x^2 + 1)^4}{8} + C.$$

Frequently the integral will contain a multiple of $g'(x)\, dx$ with the final integral containing a constant $\left(\text{like the } \frac{1}{2} \text{ above}\right)$.

In the case of a definite integral, such as $\int_0^1 (x^2 + 1)^3 x\, dx$, you may either

1. evaluate using the limits 0 and 2 after substitution:

$$\int_0^2 (x^2 + 1)^3 x\, dx = \frac{(x^2 + 1)^4}{8}\Big|_0^2 = \frac{625}{8} - \frac{1}{8} = 78$$

   or

2. change the limits to fit u:

   Since $u(x) = x^2 + 1$, $u(2) = 5$ and $u(0) = 1$.

   Hence $\int_0^1 (x^2 + 1)^3 x\, dx = \frac{1}{2}\int_1^5 u^3\, du = \frac{u^4}{8}\Big|_1^5 = \frac{625}{8} - \frac{1}{8} = 78.$

1) Use the substitution $u = x^2 + 3$ to evaluate

$$\int x\sqrt{x^2 + 3} \; dx.$$

$u = x^2 + 3$, so $du = 2x \; dx$ and $x \; dx = \frac{du}{2}$.

$\int x(x^2 + 3)^{1/2} \; dx = \int u^{1/2} \frac{du}{2} = \frac{1}{2}\int u^{1/2} \; du$

$= \frac{1}{2} \cdot \frac{2}{3} u^{3/2} = \frac{1}{3}(x^2 + 3)^{3/2} + C$

2) State a substitution $u = g(x)$ that may be used for each of the following. Write the simplified integral and evaluate.

a) $\int (x^3 + 3x)^4 (x^2 + 1) \; dx$

$u = \underline{\hspace{2cm}}$

$du = \underline{\hspace{2cm}}$

$u = x^3 + 3x$

$du = (3x^2 + 3)dx = 3(x^2 + 1)dx.$

Thus $(x^2 + 1)dx = \frac{du}{3}$ and

$\int (x^3 + 3x)^4 (x^2 + 1) \; dx = \int u^4 \frac{du}{3}$

$= \frac{1}{3}\int u^4 \; du = \frac{1}{3}\frac{u^5}{5} = \frac{1}{15}u^5$

$= \frac{1}{15}(x^3 + 3x)^5 + C.$

b) $\int x^2 \sin x^3 \; dx$

$u = \underline{\hspace{2cm}}$

$du = \underline{\hspace{2cm}}$

$u = x^3$

$du = 3x^2 \; dx.$

Thus $x^2 \; dx = \frac{du}{3}$ and $\int x^2 \sin x^3 \; dx$

$= \int \sin u \frac{du}{3} = \frac{1}{3}\int \sin u \; du$

$= -\frac{1}{3}\cos u = -\frac{1}{3}\cos x^3 + C.$

c) $\int \tan^2 x \sec^2 x \; dx$

$u = \underline{\hspace{2cm}}$

$du = \underline{\hspace{2cm}}$

$u = \tan x$

$du = \sec^2 x \; dx.$

Thus $\int (\tan x)^2 \sec^2 x \; dx = \int u^2 \; du$

$= \frac{1}{3}u^3 = \frac{1}{3}\tan^3 x + C.$

3) Evaluate $\int_0^1 \sqrt{2x+1}\ dx$.

$u = 2x + 1,\ du = 2\ dx,\ dx = \frac{du}{2}$.

Thus $\int_0^1 (2x+1)^{1/2}\ dx = \int u^{1/2}\ \frac{du}{2}$.

$= \frac{1}{2}\int u^{1/2}\ du = \frac{1}{2} \cdot \frac{2}{3} u^{3/2}$

$= \frac{1}{3}(2x+1)^{3/2}\Big]_0^1 = \frac{1}{3}(3^{3/2} - 1) = \sqrt{3} - \frac{1}{3}$.

4) Evaluate by changing the limits of integration during substitution:

$\int_0^{\pi/3} (\cos^3 x + 1)\sin x\ dx$

$u = \cos x,\ du = -\sin x\ dx$, so

$\sin x\ dx = -du$.

At $x = 0,\ u = \cos 0 = 1$.

At $x = \frac{\pi}{3},\ u = \cos \frac{\pi}{3} = \frac{1}{2}$.

Therefore

$\int_0^{\pi/3} (\cos^3 x + 1)\sin x\ dx$

$= \int_1^{1/2} (u^3 + 1)\ (-du)$

$= -\int_1^{1/2} (u^3 + 1)\ du$

$= \int_{1/2}^1 (u^3 + 1)\ du = \left(\frac{u^4}{4} + u\right)\Big]_{1/2}^1$

$= \left(\frac{1}{4} + 1\right) - \left(\frac{1}{64} + \frac{1}{2}\right) = \frac{47}{64}$.

5) Evaluate $\int x^5(x^3 + 2)^2\ dx$.

The first substitution to try is $u = x^3 + 2$, $du = 3x^2\ dx$, and $x^2\ dx = \frac{du}{3}$. Since we have an x^5 term it appears we may need a different substitution. But

$x^5 = x^3 \cdot x^2 = (u - 2)x^2$ since

$u = x^3 + 2$. So

$\int x^5(x^3 + 2)^2\ dx = \int (u - 2)u^2\ \frac{du}{3}$

$= \frac{1}{3}\int (u^3 - 2u^2)du = \frac{1}{3}\left(\frac{u^4}{4} - \frac{2}{3}u^3\right)$

$= \frac{u^4}{12} - \frac{2u^3}{9} = \frac{(x^3 + 2)^4}{12} - \frac{2(x^3 + 2)^3}{9} + C$.

6) Evaluate $\int \dfrac{x^3}{2 + x^4}\, dx$

Let $u = 2 + x^4$, $du = 4x^3\, dx$

so $x^3\, dx = \dfrac{du}{4}$.

$$\int \frac{x^3}{2 + x^4}\, dx = \int \frac{1}{2 + x^4}\, x^3\, dx$$
$$= \int \frac{1}{u} \frac{du}{4} = \frac{1}{4}\int \frac{1}{u}\, du$$
$$= \frac{1}{4} \ln |u|$$
$$= \frac{1}{4} \ln(2 + x^4)$$
$$= \ln \sqrt[4]{2 + x^4} + C.$$

Note that absolute value signs are not needed because $2 + x^4 > 0$ for all x.

7) $\int 7^{-x}\, dx$

$u = -x$, $du = -dx$, $dx = -du$.
$$\int 7^{-x}\, dx = \int 7^u(-du) = -\int 7^u\, du$$
$$= -\frac{7^u}{\ln 7} = -\frac{7^{-x}}{\ln 7} + C.$$

8) Find the area of the shaded region:

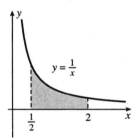

The area is $\displaystyle\int_{1/2}^{2} \frac{1}{x}\, dx$
$$= \ln x \Big]_{\frac{1}{2}}^{2} = \ln 2 - \ln \frac{1}{2}$$
$$= \ln 2 - \ln 2^{-1} = \ln 2 + \ln 2$$
$$= 2 \ln 2 \approx 1.39.$$

9) Evaluate

a) $\int_1^{\sqrt{3}} \frac{1}{1 + x^2}\, dx$

$$\int_1^{\sqrt{3}} \frac{1}{1 + x^2}\, dx = \tan^{-1} x \Big]_1^{\sqrt{3}}$$

$$= \tan^{-1}(\sqrt{3}) - \tan^{-1}(1)$$

$$= \frac{\pi}{3} - \frac{\pi}{4} = \frac{\pi}{12}.$$

b) $\int_0^1 \frac{1}{\sqrt{4 - x^2}}$

This is almost the form needed for $\sin^{-1} x$ except we have $\sqrt{4 - x^2}$ instead of $\sqrt{1 - x^2}$. Make the constant 1:

$$\sqrt{4 - x^2} = \sqrt{4\left(1 - \frac{x^2}{4}\right)} = 2\sqrt{1 - \left(\frac{x}{2}\right)^2}.$$

Then

$$\int_0^1 \frac{1}{2\sqrt{1 - \left(\frac{x}{2}\right)^2}}\, dx$$

$$= \left(u = \frac{x}{2}, du = \frac{dx}{2}\right)$$

$$= \int_0^1 \frac{1}{\sqrt{1 - u^2}}\, du = \sin^{-1} u \Big]_0^1$$

$$= \sin^{-1} \frac{x}{2} \Big]_0^1 = \sin^{-1} \frac{1}{2} - \sin^{-1} 0$$

$$= \frac{\pi}{6} - 0 = \frac{\pi}{6}.$$

c) $\int \frac{2x}{\sqrt{1 - x^2}}\, dx$

Although this may appear to involve $\sin^{-1}$, this is a simple substitution.

$u = 1 - x^2, du = -2x\, dx$

$$\int \frac{2x}{\sqrt{1 - x^2}}\, dx = -\int (2x)(1 - x^2)^{-1/2}\, dx$$

$$= -\int u^{-1/2}\, du = \frac{-u^{1/2}}{1/2} = -2u^{1/2}$$

$$= -2\sqrt{1 - x^2} + C.$$

d) $\int \frac{1}{2x\sqrt{x^2 - 1}}\, dx$

$$\int \frac{1}{2x\sqrt{x^2 - 1}}\, dx = \frac{1}{2}\int \frac{1}{x\sqrt{x^2 - 1}}\, dx$$

$$= \frac{1}{2}\sec^{-1} x + C.$$

10) Evaluate $\int_0^1 \frac{1}{4 - x^2}\, dx.$

$$\int_0^1 \frac{1}{4 - x^2}\, dx = \frac{1}{4}\int_0^1 \frac{1}{1 - \left(\frac{x}{2}\right)^2} = dx$$

$$= \left(u = \frac{x}{2},\ du = \frac{1}{2}\, dx\right)$$

$$= \frac{1}{2}\int_0^1 \frac{1}{1 - \left(\frac{x}{2}\right)^2}\, \frac{1}{2}\, dx$$

$$= \frac{1}{2}\int \frac{1}{1 - u^2}\, du = \frac{1}{2}\tanh^{-1} u$$

$$= \frac{1}{2}\tanh^{-1}\frac{x}{2}\Big]_0^1$$

$$= \frac{1}{2}\tanh^{-1}\frac{1}{2} - \frac{1}{2}\tanh^{-1} 0$$

$$= \frac{1}{2}\left(\frac{1}{2}\ln 3\right) - \frac{1}{2}(0) = \frac{\ln 3}{4}.$$

B. Recall that a function f is *even* if $f(-x) = f(x)$ and *odd* if $f(-x) = -f(x)$.
If f is continuous on $[-a, a]$ then:

Page
415

if f is even, $\int_{-a}^a f(x)\, dx = 2\int_0^a f(x) = dx$

if f is odd, $\int_{-a}^a f(x)\, dx = 0.$

11) Evaluate $\int_{-\pi/2}^{\pi/2} \sin x\, dx.$

Since $f(x) = \sin x$ is odd, this integral is 0.

12) Evaluate $\int_{-3}^3 (x^4 + x^2)\, dx.$

$$\int_{-3}^3 (x^4 + x^2)\, dx = 2\int_0^3 (x^4 + x^2)\, dx$$

$$= 2\left[\frac{x^5}{5} + \frac{x^3}{3}\right]_0^3$$

$$= 2\left[\left(\frac{3^5}{5} + \frac{3^3}{3}\right) - (0)\right]$$

$$= \frac{576}{5}.$$

Section 5.6   The Logarithm Defined as an Integral

This section gives an alternative way to define ln x, and then, in turn, define e^x, a^x, and $\log_a x$. The properties that you have already learned for logarithms and exponentials follow from these definitions.

Concepts to Master

A. Integral definition of ln x; Properties and the derivative of ln x; Definitions of e

B. Inverse definition of e^x; Properties and the derivative of e^x; Definitions of a^x and $\log_a x$

Summary and Focus Questions

Page 418

A. The *natural logarithm function* may be defined as

$$\ln x = \int_1^x \frac{1}{t}\, dt \text{ for } x > 0.$$

The function $y = \ln x$ has these properties:

domain of ln is $(0, \infty)$
range of ln is all reals

$\ln 1 = 0$

$\ln(xy) = \ln x + \ln y$

$\ln\left(\frac{x}{y}\right) = \ln x - \ln y$

$\ln(x^r) = r \ln x$

$\lim\limits_{x \to \infty} \ln x = \infty$

$\lim\limits_{x \to 0^+} \ln x = -\infty$

ln is increasing, concave down, and one-to-one.

By the Fundamental Theorem of Calculus $\frac{d}{dx}(\ln x) = \frac{1}{x}$.

e is the unique number for which $\ln e = 1$.

Here is another way to determine the value of e:

$$\lim_{x \to 0^+} (1 + x)^{1/x} = e.$$

Equivalently, $\lim\limits_{t \to \infty} \left(1 + \frac{1}{t}\right)^t = e.$

1) Find an expression for the shaded area:

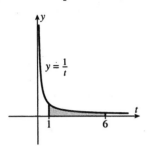

The shaded area is $\int_1^6 \frac{1}{t}\, dt$ which is ln 6.

2) True or False:

  a) ln $e = 1$

True.

  b) $\ln(a - b) = \ln a - \ln b$

False.

  c) $\ln\!\left(\frac{a}{b}\right) = -\ln\!\left(\frac{b}{a}\right)$

True.

3) Write $2 \ln x + 3 \ln y$ as a single logarithm.

$2 \ln x + 3 \ln y = \ln x^2 + \ln y^3 = \ln(x^2 y^3)$.

4) Find the number k on the t-axis such that:

  a) the shaded area is 1.

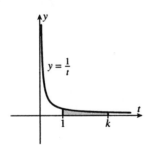

The shaded area is $\int_1^k \frac{1}{t}\, dt = \ln k = 1$,

hence $k = e$.

b) the shaded area is 1.

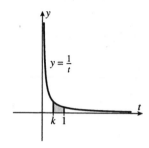

$y = \frac{1}{t}$

5) Find each limit:

a) $\lim\limits_{h \to 0+} (1 + h)^{1/h}$

b) $\lim\limits_{x \to \infty} \left(1 + \frac{1}{x}\right)^{3x}$

c) $\lim\limits_{x \to 1+} x^{1/(x-1)}$

The area $= \int_{k}^{1} \frac{1}{t}\, dt = \ln 1 - \ln k$
$$= -\ln k = 1.$$
Thus $\ln \frac{1}{k} = 1$, so $\frac{1}{k} = e$ or $k = \frac{1}{e}$.

e.

$$\lim\limits_{x \to \infty}\left(1 + \frac{1}{x}\right)^{3x} = \lim\limits_{x \to \infty}\left[\left(1 + \frac{1}{x}\right)^{x}\right]^{3}$$
$$= \left[\lim\limits_{x \to \infty}\left(1 + \frac{1}{x}\right)^{x}\right]^{3} = e^{3}.$$

Let $t = x - 1$, hence $x = 1 + t$.
As $x \to 1^{+}$, $t \to 0^{+}$.
Thus $\lim\limits_{x \to 1^{+}} x^{1/(x-1)} = \lim\limits_{t \to 0^{+}} (1 + t)^{1/t} = e$.

Page 421

B. The *natural exponential function*, $y = \exp(x)$ is defined as the inverse of $y = \ln x$. Thus $\exp(a) = b$ if and only if $\ln b = a$.

Recalling that e is a constant, approximately 2.718, we may write $\exp(x)$ as an exponential:

$$\exp(x) = e^x \text{ for all } x.$$

Thus

$$e^a = b \text{ if and only if } \ln b = a.$$

The function $y = e^x$ has these properties:

domain of e^x is all reals	range of e^x is $(0, \infty)$
$e^{\ln x} = x$ for all $x > 0$.	$\ln e^x = x$ for all x.
$e^0 = 1$	$e^{x+y} = e^x e^y$
$e^{x-y} = \dfrac{e^x}{e^y}$	$(e^x)^r = e^{rx}$
$\lim\limits_{x \to \infty} e^x = \infty$	$\lim\limits_{x \to -\infty} e^x = 0$
$\dfrac{d}{dx}(e^x) = e^x$	

For $a > 0$, a^x is defined as $a^x = e^{x \ln a}$. So, for example, $2^\pi = e^{\pi \ln 2}$.
Properties of exponentials include:

$$a^{x+y} = a^x a^y$$
$$a^{x-y} = \frac{a^x}{a^y}$$
$$a^{xy} = (a^x)^y$$
$$(ab)^x = a^x b^x$$

If $f(x) = a^x$, $f'(x) = a^x \ln a$.
The general logarithmic function $y = \log_a x$ is defined as the inverse of $y = a^x$:

$$\log_a x = y \text{ means } a^y = x.$$

This definition requires $x > 0$, since $a^y > 0$ for all y.
The function $y = \log_a x$ has these properties:

$\log_a(a^x) = x$	$a^{\log_a x} = x$, for $x > 0$
$\log_a(xy) = \log_a x + \log_a y$	$\log_a\!\left(\dfrac{x}{y}\right) = \log_a x - \log_a y$
$\log_a x^c = c \log_a x$	$\lim\limits_{x \to \infty} \log_a x = \infty$
$\lim\limits_{x \to 0+} \log_a x = -\infty$	

$$\log_a x = \frac{\ln x}{\ln a}.$$
If $f(x) = \log_a x$, $f'(x) = \dfrac{1}{x \ln a}$.

6) $\lim\limits_{x\to\infty} e^{-x}$

As $x \to \infty$, $-x \to -\infty$.
Thus $\lim\limits_{x\to\infty} e^{-x} = 0$.

7) True, False:
$e^{\ln x} = x$ for all x.

False. This is only true for $x > 0$.

8) True or False:
   a) $(x + y)^a = x^a + y^a$

False.

   b) $x^{a-b+c} = \dfrac{x^a x^c}{x^b}$

True.

9) By definition $\pi^{\sqrt{2}} = \underline{\quad\quad}$.

$e^{\sqrt{2}\,\ln \pi}$

10) For what a is $(a^x)' = a^x$?

Only for $a = e$.

11) Solve for x:
   a) $e^{x+1} = 10$

By definition, this means $\ln 10 = x + 1$, so
$x = \ln 10 - 1$.

   b) $2^{3x} = 7$

Take $\log_2$ of both sides:
$\log_2 2^{3x} = \log_2 7$
$3x = \log_2 7$
$x = \frac{1}{3}\log_2 7$.

12) $\log_2 10 = \dfrac{\ln \underline{\quad}}{\ln \underline{\quad}}$

$\log_2 10 = \dfrac{\ln 10}{\ln 2}$

Technology Plus for Chapter 5

1) Use a calculator or spreadsheet to estimate $\int_0^5 f(x)\, dx$, where $f(x) = 2^x + x^2$.

a) Use the midpoint rule with 10 intervals.

$\Delta x = \dfrac{5 - 0}{10} = 0.5$.

i	Δx_i	x_i^*	$f(x_i^*)$	$f(x_i^*)\Delta x_i$
1	0.50	0.25	1.252	0.626
2	1.00	0.75	2.244	1.122
3	1.50	1.25	3.941	1.970
4	2.00	1.75	6.426	3.213
5	2.50	2.25	9.819	4.910
6	3.00	2.75	14.290	7.145
7	3.50	3.25	20.076	10.038
8	4.00	3.75	27.517	13.758
9	4.50	4.25	37.090	18.545
10	5.00	4.75	49.471	24.736
				86.063

Thus, $\int_0^5 f(x)\, dx \approx 86.063$.

b) Use the left endpoint for a sample point with 20 equal intervals.

$\Delta x = \dfrac{5 - 0}{20} = 0.25$.

i	Δx_i	x_i^*	$f(x_i^*)$	$f(x_i^*)\Delta x_i$
1	0.25	0.00	1.000	0.250
2	0.50	0.25	1.252	0.313
3	0.75	0.50	1.664	0.416
4	1.00	0.75	2.244	0.561
5	1.25	1.00	3.000	0.750
6	1.50	1.25	3.941	0.985
7	1.75	1.50	5.078	1.270
8	2.00	1.75	6.426	1.607
9	2.25	2.00	8.000	2.000
10	2.50	2.25	9.819	2.455
11	2.75	2.50	11.907	2.977
12	3.00	2.75	14.290	3.572
13	3.25	3.00	17.000	4.250
14	3.50	3.25	20.076	5.019
15	3.75	3.50	23.564	5.891
16	4.00	3.75	27.517	6.879
17	4.25	4.00	32.000	8.000
18	4.50	4.25	37.090	9.272
19	4.75	4.50	42.877	10.719
20	5.00	4.75	49.471	12.368
				79.554

Thus, $\int_0^5 f(x)\, dx \approx 79.554$.

c) We will see later that
$$\int_0^5 f(x)\, dx = \frac{31}{\ln 2} + \frac{125}{3} \approx 86.390.$$
Why is the answer for part a) closer to the actual value than the answer to b) even though b) uses twice as many rectangles?

The function $f(x) = 2^x + x^2$ is increasing, so the left endpoint of each subinterval will give the smallest possible value for $f(x_i^*)$; thus the area of each subrectangle will underestimate the area.

2) Graph $f(x) = x + 2x^2$ and $y = 7x$ for $0 \le x \le 3$ on the same screen and use it to estimate $\int_0^3 f(x)\, dx.$

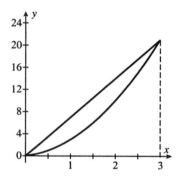

The area under $f(x) = x + 2x^2$ appears to be about $\frac{3}{4}$ of the area under $y = 7x$. That triangular area is $\frac{1}{2}(3)(21) = \frac{63}{2}$.
$$\int_0^3 (x + 2x^2)\, dx \approx \frac{3}{4}\left(\frac{63}{2}\right) = 23.625.$$
We note that
$$\int_0^3 (x + 2x^2)\, dx = \left(\frac{x}{2} + \frac{2x^2}{3}\right)\Big]_0^3$$
$$= \frac{64}{3} = 21.333.$$

Chapter 6 — Applications of Integration

© 1999 by Sidney Harris.

Section 6.1   Areas between Curves

This chapter contains several applications of definite integrals. It will help to remember that each comes from a limit of Riemann sums. In this section we determine the area between two curves. We shall see that sometimes it is better to treat y as a function of x and other times treat x as a function of y.

Concepts to Master

A. Area between $y = f(x)$, $y = g(x)$, $x = a$, $x = b$

B. Area between $x = f(y)$, $x = g(y)$, $y = c$, $y = d$

C. Area enclosed by two curves

Summary and Focus Questions

Page 433

A. The area A of the region bounded by continuous $y = f(x)$ and $y = g(x)$ and $x = a$, $x = b$ with $f(x) \geq g(x)$ on $[a, b]$ may be approximated by a Riemann sum with terms $[f(x_i^*) - g(x_i^*)]\Delta x_i$. Thus the area is

$$A = \lim_{n \to \infty} \sum_{i=1}^{n} [f(x_i^*) - g(x_i^*)]\Delta x_i$$

$$= \int_a^b [f(x) - g(x)]\, dx.$$

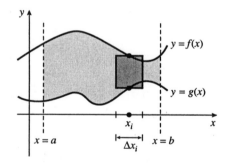

1) Set up a definite integral for the area of the region bounded by $y = x^3$, $y = 3 - x$, $x = 0$, $x = 1$.

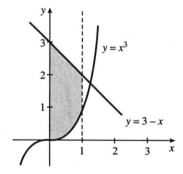

For $0 \leq x \leq 1$, $3 - x \geq x^3$. The area is

$$\int_0^1 [3 - x - x^3]\, dx \left(= \frac{9}{4}\right).$$

Page 437

B. The area of a region bounded by continuous $x = f(y)$, $x = g(y)$, $y = c$, $y = d$ with $f(y) \geq g(y)$ on $[c, d]$ is found in a manner similar to areas in part A. Here, however, x is a function of y and the area is

$$A = \lim_{n \to \infty} \sum_{i=1}^{n} [f(y_i^*) - g(y_i^*)] \Delta y_i$$

$$= \int_c^d [f(y) - g(y)] \, dy.$$

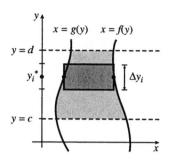

2) Set up a definite integral for the area of the region bounded by $x = 4 - y^2$, $x = y - 2$, $y = 2$, $y = -1$.

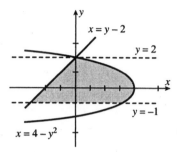

For $-1 \leq y \leq 2$, $4 - y^2 \geq y - 2$.

The area is

$$\int_{-1}^{2} [4 - y^2 - (y - 2)] \, dy$$

$$= \int_{-1}^{2} [6 - y^2 - y] \, dy \left(= \frac{27}{2}\right).$$

Page 436

C. Some regions may be described so that either the $\int \ldots dx$ form or the $\int \ldots dy$ form may be used. You should select the form with the easier integral.

To find the area of regions such as the types sketched below, the x coordinates of the points of intersection a, b, c, ... must be found. This is done by setting $f(x) = g(x)$ and solving for x. Often a sketch of the functions will help or a graphing calculator can be used to find approximate values for the points of intersection.

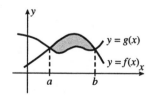

$$\text{Area} = \int_a^b [f(x) - g(x)] \, dx$$

$$\text{Area} = \int_a^b [g(x) - f(x)] dx$$
$$+ \int_b^c [f(x) - g(x)] dx$$

3) Write an expression involving definite integrals for the area of each.

a) the region bounded by $y = (x + 1)^2$ and $y = x + 3$.

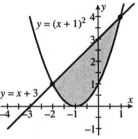

The "top" curve is $y = x + 3$ and "bottom" curve is $y = (x + 1)^2$. First find where they intersect:

$$(x + 1)^2 = x + 3$$
$$x^2 + 2x + 1 = x + 3$$
$$x^2 + x - 2 = 0$$
$$(x - 1)(x + 2) = 0 \text{ at } x = -2, -1.$$

The area is $\displaystyle\int_{-2}^{1} [(x + 3) - (x + 1)^2] \, dx \left(= \frac{9}{2}\right)$.

b) the shaded region.

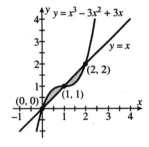

For $0 \le x \le 1$, $x^3 - 3x^2 + 3x \ge x$ while for $1 \le x \le 2$, $x \ge x^3 - 3x^2 + 3x$. Thus the area is the sum of two integrals:

$$\int_0^1 [(x^3 - 3x^2 + 3x) - x] \, dx$$
$$+ \int_1^2 (x - (x^3 - 3x^2 + 3x)) \, dx$$
$$= \int_0^1 (x^3 - 3x^2 + 2x) \, dx$$
$$+ \int_1^2 (-x^3 + 3x^2 - 2x) \, dx$$
$$\left(= \frac{1}{4} + \frac{1}{4} = \frac{1}{2}\right).$$

c) the shaded region

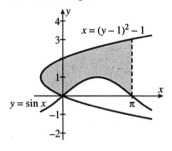

4) Write an expression involving definite integrals for the area of the regions bounded by:

a) $y = 2x$ and $y = 8 - x^2$

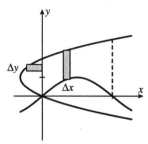

Using the y-axis to divide the region into two parts, the total area is two integrals. The area to the left of the y-axis is found using x as a function of y. The area to the right of the y-axis is found using y as a function of x: solving $x = (y - 1)^2 - 1$ for y gives $y = 1 + \sqrt{x + 1}$.
The total area is

$$\int_0^2 -((y - 1)^2 - 1)\,dy$$
$$+ \int_0^\pi ((1 + \sqrt{x + 1}) - \sin x)\,dx$$
$$= \int_0^2 (1 - (y - 1)^2)\,dy$$
$$+ \int_0^\pi (1 + \sqrt{x + 1} - \sin x)\,dx$$
$$\left(= \frac{2(\pi + 1)^{3/2}}{3} + \pi - \frac{4}{3}\right).$$

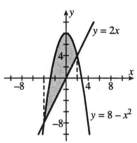

To find the points of intersection set
$$2x = 8 - x^2$$
$$x^2 + 2x - 8 = 0$$
$$(x - 2)(x + 4) = 0, \text{ at } x = 2, -4.$$

The area is $\displaystyle\int_{-4}^2 (8 - x^2 - 2x)\,dx\ (= 36).$

b) The shaded area between $y = \sin x$ and $y = \cos x$.

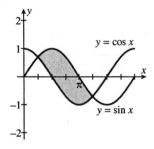

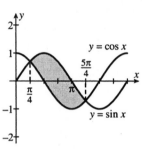

$\sin x = \cos x;\ \dfrac{\sin x}{\cos x} = 1;\ \tan x = 1.$

So $x = \dfrac{\pi}{4}$ and $x = \dfrac{5\pi}{4}.$

$\text{Area} = \displaystyle\int_{\pi/4}^{5\pi/4} (\sin x - \cos x)\,dx\ (= 2\sqrt{2}).$

c) $y = x,\ y = -x,\ y = 2x - 3$

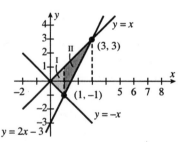

Divide the area into two regions:
Region I is bounded by $x = 0$, $x = 1$, $y = x$, and $y = -x$.
Region II is bounded by $x = 1$, $x = 3$, $y = x$, and $y = 2x - 3$.
The area is

$$\int_0^1 [(x - (-x)]\,dx + \int_1^3 [x - (2x - 3)]\,dx$$

$$= \int_0^1 2x\,dx + \int_1^3 (3 - x)\,dx\ (= 3).$$

Section 6.2   Volumes

This section shows you how to calculate the volume of a solid object if you can determine its cross sectional areas. The types of solids here include irregular ones (such as an orange slice) and solids of revolution (such as a cone) which are obtained by rotating an area about an axis. In all cases it helps in setting up the integral to determine an expression for the area of a typical cross section.

Concepts to Master

A. Determine the volume of a solid by the slicing method
B. Volume of a solid of revolution

Summary and Focus Questions

Page **440**

A. Suppose the cross-sectional area of a solid cut by planes perpendicular to an x-axis is known to be $A(x)$ for each $x \in [a, b]$. The volume of a typical slice is approximately $A(x_i^*)\Delta x$ and thus the total volume obtained by this *method of slicing* is

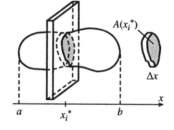

$$V = \lim_{n \to 0} \sum_{i=1}^{n} A(x_i^*)\Delta x = \int_a^b A(x) \ dx.$$

In many problems, the x-axis will have to be chosen carefully so that the function $A(x)$ can be determined.

1) Set up a definite integral for the volume of a right triangular solid that is 1 m, 1 m, and 2 m on its edges (see the figure).

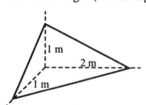

Draw an x-axis along the 2 meter side with the origin in the corner.

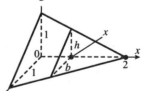

The area of the triangle determined by slicing perpendicular to the x-axis at point x is $A(x) = \frac{1}{2}bh$. By similar triangles, $\frac{2-x}{2} = \frac{h}{1}$, so $h = 1 - \frac{x}{2}$. Likewise, $b = 1 - \frac{x}{2}$, so $A(x) = \frac{1}{2}\left(1 - \frac{x}{2}\right)^2$.
The volume is $\int_0^2 \frac{1}{2}\left(1 - \frac{x}{2}\right)^2 \ dx \left(= \frac{1}{3}\right)$.

B. A solid of revolution is an object obtained by rotating a planar region about a line. The line may be an x-axis, y-axis or some other line. Sometimes the resulting solid will have a cross section in the shape of a disk; other times the cross section will be washer shaped. The volume will be approximated by a sum of volumes of disks or washers which will suggest the definite integral for the exact volume.

Page 442

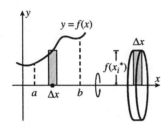

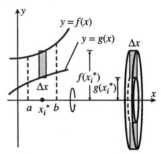

If the region under $y = f(x)$ is rotated about the x-axis, the volume may be approximated by a sum of volumes of disks. The volume of a typical disk is $\pi(\text{radius})^2(\text{thickness}) \approx \pi[f(x_i^*)]^2\Delta x$. Thus the volume is

$$V = \int_a^b \pi[f(x)]^2 \, dx.$$

If the region between $y = f(x)$ and $y = g(x)$ is rotated about the x-axis, the volume may be approximated by a sum of volumes of washers. The volume of a typical washer is $\pi(\text{outer radius})^2(\text{thickness}) - \pi(\text{inner radius})^2(\text{thickness}) = \pi[(f(x_i^*))^2 - g(x_i^*))^2]\Delta x$. Thus the volume

$$V = \int_a^b \pi[(f(x))^2 - (g(x))^2] \, dx.$$

2) Set up a definite integral for the volume of each solid of revolution.

a) Rotate the region $y = 4 - x^2$, $y = 0$, $x = 0$, $x = 2$ about the x-axis.

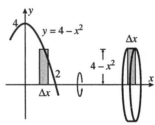

Volume of disk $= \pi(4 - (x_i^*)^2)^2\Delta x$.

Volume of solid $= \int_0^2 \pi(4 - x^2)^2 \, dx$

$$\left(= \frac{256}{15}\pi\right).$$

b) The solid obtained by rotating about the y-axis the region bounded by $y = \sqrt{x}, x = 0, y = 4$.

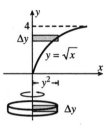

$$\text{Volume of disk} = \pi((y_i^*)^2)^2 \Delta y$$
$$= \pi(y_i^*)^4 \Delta y.$$
$$\text{Volume of solid} = \int_0^4 \pi y^4 \, dy \left(= \frac{1024\pi}{5}\right).$$

c) The solid obtained by rotating about the line $y = 3$ the region bounded by $y = 10 - x^2, x = 0, x = 2, y = 3$.

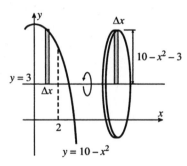

$$\text{Volume of disk} = \pi(10 - (x_i^*)^2 - 3)^2 \Delta x$$
$$\text{Volume of solid} = \int_0^2 \pi(10 - x^2 - 3)^2 \, dx$$
$$= \int_0^2 \pi(7 - x^2)^2 \, dx \left(= \frac{1006}{15}\pi\right).$$

3) Set up a definite integral for the volume of the solid obtained by rotation of the region between $x = 1$, $x = 2$, $y = x^2$, and $y = x^3$ about the x-axis.

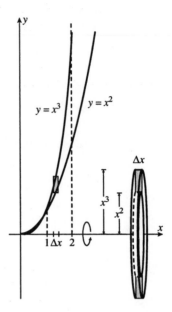

Volume of washer

$$= \pi[((x_i^*)^3)^2 - ((x_i^*)^2)^2]\Delta x.$$

Volume of solid

$$= \int_1^2 \pi[(x^3)^2 - (x^2)^2]\,dx$$
$$= \pi\int_1^2 (x^6 - x^4)\,dx \left(= \frac{418}{35}\pi\right).$$

Section 6.3   Volumes by Cylindrical Shells

This section presents another way to determine the definite integral for a solid of revolution. This time the volume is approximated by a sum of volumes of concentric cylindrical shells. As in the previous section, the key will be to visualize the solid as composed of many shells whose volumes can be determined.

Concepts to Master

Volumes of solids of revolution by the shell method

Summary and Focus Questions

Page 451

The volume of the solid obtained by rotating the region bounded by $y = f(x)$, $y = 0$, $x = a$, $x = b$ about the y-axis may be approximated by a sum of volumes of shells. The volume of a typical shell is

$$2\pi(\text{radius})(\text{height})(\text{thickness})$$
$$= 2\pi x_i{}^* f(x_i{}^*)\Delta x.$$

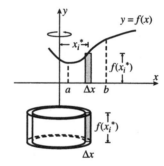

Thus the total volume by this *method of (cylindrical) shells* is

$$V = \int_a^b 2\pi x f(x)\ dx.$$

Volumes of solids of revolution about the x-axis are computed similarly.

1) Set up a definite integral for the volume of the solid obtained by rotating about the y-axis the region bounded by $y = \sqrt{x}$, $y = 0$, $x = 1$, $x = 4$.

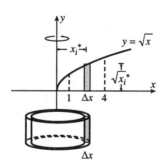

Volume of shell $= 2\pi x_i{}^*(\sqrt{x_i{}^*})\Delta x$
$= 2\pi(x_i{}^*)^{3/2}\Delta x$

Volume of solid $= \int_1^4 2\pi x^{3/2}\ dx \left(= \frac{124\pi}{5}\right).$

2) Set up a definite integral for the volume of the solid obtained by rotating about the x-axis the region bounded by $x = y^2$, $y = 0$, $x = 4$.

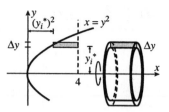

$$\text{Volume of shell} = 2\pi y_i^*(4 - (y_i^*)^2)\Delta y$$
$$= 2\pi(4y_i^* - (y_i^*)^3)\Delta y.$$

$$\text{Volume of solid} = \int_0^2 2\pi(4y - y^3)\,dy \, (= 12\pi).$$

3) A solid "dog dish" is obtained by rotating the region bounded by $y = 0$, $y = x^3$, $x = 0$, $x = 1$ about the y-axis. Set up two integrals for the volume—one by the slicing method and the other using shells.

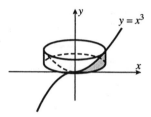

1. Using the slicing method, the slices along the y-axis are washers:

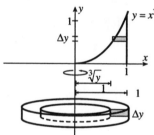

$$\text{Volume of washer} = \pi[1^2 - (\sqrt[3]{y})^2]\Delta y.$$

$$\text{Volume of solid} = \int_0^1 \pi(1 - y^{2/3})\,dy \left(= \tfrac{2\pi}{5}\right).$$

2. Using shells

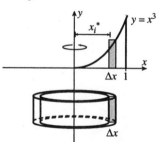

$$\text{Volume of shell} = 2\pi x_i^*(x_i^*)^3\Delta x$$
$$= 2\pi(x_i^*)^4\Delta x.$$

$$\text{Volume of solid} = \int_0^1 2\pi x^4\,dx \left(= \tfrac{2\pi}{5}\right).$$

Section 6.4   Work

Work is a measure of effort expended by a force moving an object from point a to point b. If the force varies then the total amount of work done is determined by a definite integral.

Concepts to Master

Work done by a varying amount of force along an axis

Summary and Focus Questions

Page 456

The work done to move an object a distance d with a constant force F is

$$W = Fd.$$

The units of work are foot-pounds; in the metric system, the units are newton-meters which is called a Joule (J).

If an object moves along an x-axis from a to b by a varying force $F(x)$, the work done is

$$W = \int_a^b F(x)\, dx.$$

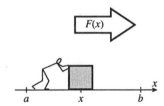

1) How much work is done lifting an 800 pound piano upward 6 feet?

The force is a constant 800 pounds and the distance is 6 feet. The work is
$W = (800)(6) = 4800$ ft-lb.

2) How much work is done moving a particle along an x-axis from 0 to $\frac{\pi}{2}$ feet if at each point x the amount of force is $\cos x$ pounds?

$$W = \int_0^{\pi/2} \cos x\, dx = \sin x \Big]_0^{\pi/2} = 1 \text{ ft-lb.}$$

3) Find an integral for how much work is done in pumping over the edge all the liquid of density ρ out of a 2 foot long trough in the shape of an equilateral triangle with sides of 1 foot.

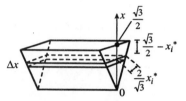

The work done to pump out a slice of the liquid is (density · gravitational constant · volume · height raised). The density is ρ; g is the gravitational constant. The volume of a slice is

$$\left(\frac{2}{\sqrt{3}}x_i^*\right)(2)(\Delta x) = \frac{4}{\sqrt{3}}x_i^*\Delta x.$$

The height raised is $\frac{\sqrt{3}}{2} - x_i^*$.

The work done lifting the slice is

$$\rho g\left(\frac{4}{\sqrt{3}}x_i^*\Delta x\right)\left(\frac{\sqrt{3}}{2} - x_i^*\right).$$

The total work is

$$\lim_{n\to\infty} \sum_{i=1}^{n} \rho g\left(\frac{4}{\sqrt{3}}x_i^*\right)\left(\frac{\sqrt{3}}{2} - x_i^*\right)\Delta x$$

$$= \int_0^{\sqrt{3}/2} \rho g\frac{4}{\sqrt{3}}x\left(\frac{\sqrt{3}}{2} - x\right) dx \left(= \frac{\rho g}{4}\right).$$

Section 6.5   Average Value of a Function

As the graph of a function f varies from $x = a$ to $x = b$, $f(x)$ assumes many values. This section shows how the average of these values may be determined by $\int_a^b f(x)\, dx$. This section also contains a restatement of the Mean Value Theorem—this time in terms of integrals.

Concepts to Master

A. Average value of a function
B. Mean Value Theorem for Integrals

Summary and Focus Questions

Page 460

A. The *average value* of a continuous function $y = f(x)$ over a closed interval $[a, b]$ is

$$f_{\text{ave}} = \frac{1}{b - a} \int_a^b f(x)\, dx$$

One way to think of this is that $\int_a^b f(x)\, dx$ is the area under $y = f(x)$ and so dividing it by the width of the area $(b - a)$ yields the (average) height of $f(x)$.

1) Find the average value of $f(x) = 2x + 6x^2$ over $[1, 4]$.

$$f_{\text{ave}} = \frac{1}{4 - 1} \int_1^4 (2x + 6x^2)\, dx$$
$$= \frac{1}{3}(x^2 + 2x^3)\Big|_1^4 = 47.$$

2) Find the average value of $\cos x$ over $[0, \pi]$.

$$\frac{1}{\pi - 0} \int_0^\pi \cos x\, dx = \frac{1}{\pi} \sin x \Big|_0^\pi$$
$$= \frac{1}{\pi}(0 - 0) = 0.$$

Intuitively this makes sense because the "right half" of the cosine curve (the part between $\frac{\pi}{2}$ and π) is below the x-axis and is symmetric with the other half above the x-axis. The cosine values average zero.

Page 461

B. *The Mean Value Theorem for Integrals:*

If f is continuous on a closed interval $[a, b]$ there exists $c \in [a, b]$ such that
$$\int_a^b f(x)\, dx = f(c)(b - a).$$

For $f(x) \geq 0$, we can think of $f(c)(b - a)$ as the area of a rectangle with height $f(c)$ and width $b - a$. The Mean Value Theorem for Integrals says there is some point $c \in [a, b]$ so that the area of this rectangle equals the area under the curve.

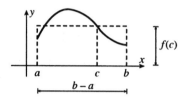

3) Mark on the graph a number c guaranteed by the Mean Value Theorem for $f(x)$:

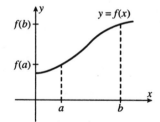

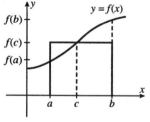

The number c is chosen so that the area under $f(x)$ from a to b equals the area of the rectangle with width $b - a$ and height $f(c)$.

4) Let $F'(x) = f(x)$ be continuous on $[a, b]$. State the conclusion of the Mean Value Theorem for Integrals for f in terms of the function $F(x)$. Use the Fundamental Theorem of Calculus.

The conclusion is "there is a number c such that $\int_a^b f(x)\, dx = f(c)(b - a)$." By the Fundamental Theorem $\int_a^b f(x)\, dx = F(b) - F(a)$. Also $f(c) = F'(c)$. So the conclusion, in terms of $F(x)$, becomes "there is a number $c \in (a, b)$ such that $F(b) - F(a) = F'(c)(b - a)$"— precisely the Mean Value Theorem of Chapter 4!

5) For $f(x) = x^2$ on $[1, 4]$, find a value for c in the Mean Value Theorem for Integrals.

$\int_1^4 x^2\, dx = \left.\dfrac{x^3}{3}\right]_1^4 = 21$. So $21 = f(c)(4 - 1)$.
$f(c) = 7, c^2 = 7, c = \sqrt{7}$.

Technology Plus for Chapter 6

1) Use a graphing calculator to zoom in where the functions $y = 8 - x^2$ and $y = x^2 - 2x + 2$ intersect. Then find the area between the two curves.

Using a $[-3, 3]$ by $[-1, 9]$ window we see

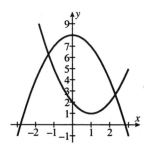

The graphs intersect at approximately $x = -1.30$ and $x = 2.30$.

The area is

$$\int_{-1.30}^{2.30} (8 - x^2) - (x^2 - 2x + 2) \, dx$$

$$= \int_{-1.30}^{2.30} (-2x^2 - 2x + 6) \, dx$$

$$= \frac{-2}{3}x^3 - x^2 + 6x \Big]_{-1.30}^{2.30} \approx 15.624$$

2) Use a spreadsheet or a calculator to find the Riemann sum for estimating the volume of the solid obtained by rotating $y = e^x$, $0 \le x \le 2$, about the x-axis. Use $n = 20$ and the midpoint of each subinterval.

$$\Delta x = \frac{2-0}{20} = -0.10.$$

n	x_i^*	$f(x_i^*)$	$f(x_i^*)\Delta x_i$
1	0.05	3.472	0.374
2	0.15	4.241	0.424
3	0.25	5.180	0.518
4	0.35	6.326	0.633
5	0.45	7.727	0.773
6	0.55	9.438	0.944
7	0.65	11.527	1.153
8	0.75	14.080	1.408
9	0.85	17.197	1.720
10	0.95	21.004	2.100
11	1.05	25.655	2.565
12	1.15	31.335	3.133
13	1.25	38.272	3.827
14	1.35	46.746	4.675
15	1.45	57.096	5.710
16	1.55	69.737	6.974
17	1.65	85.177	8.518
18	1.75	104.035	10.404
19	1.85	127.069	12.707
20	1.95	155.202	15.520
			84.052

$$\int_0^2 \pi(e^x)^2 \, dx \approx 84.052.$$

Chapter 7 — Techniques of Integration

"I THINK YOU SHOULD BE MORE EXPLICIT HERE IN STEP TWO."

© 1999 by Sidney Harris.

Section 7.1   Integration by Parts

The integration by parts formula in this section is simply the product rule for differentiation, $(uv)' = uv' + u'v$, stated in terms of antiderivatives.

Concepts to Master

Evaluate integrals using integration by parts

Summary and Focus Questions

Page 469

The integration by parts formula is

$$\int u\, dv = uv - \int v\, du.$$

Integration by parts is just a rule for replacing the problem of evaluating $\int u\, dv$ with hopefully an easier problem of evaluating $\int v\, du$. The key to successful use is to identify the two terms u and v' in $\int u\, dv$ (remember $u\, dv = uv'\, dx$), and to check that $\int v\, du$ is indeed simpler.

Example: Find $\int x \ln x\, dx$.

Initially, we might choose $u = x$ because $du = dx$ is simpler. However, the companion choice $dv = \ln x\, dx$ would mean v is more difficult. So we choose $u = \ln x$, $dv = x\, dx$. Then $du = \frac{1}{x}\, dx$, $v = \frac{x^2}{2}$ and $v\, du = x\frac{1}{x}\, dx = dx$.

$$\int x \ln x\, dx = (\ln x)\,\frac{x^2}{2} - \int \frac{x^2}{2}\left(\frac{1}{x}\right) dx = \frac{1}{2}x^2 \ln x - \frac{1}{2}\int x\, dx$$
$$= \frac{1}{2}x^2 \ln x - \frac{1}{4}x^2 + C$$

1) Give an appropriate choice for u and dv in each by parts integral and evaluate:

a) $\int x^2 \ln x \, dx$

$u = $ _____

$dv = $ _____

$u = \ln x,\ du = \frac{1}{x}\,dx.$

$dv = x^2\,dx,\ v = \frac{x^3}{3}.$

Thus $v\,du = \frac{x^3}{3} \cdot \frac{1}{x}\,dx = \frac{x^2}{3}\,dx$ and

$\int x^2 \ln x \, dx = (\ln x)\frac{x^3}{3} - \int \frac{x^2}{3}\,dx$

$= \frac{x^3}{3} \ln x - \frac{1}{3}\int x^2\,dx = \frac{x^3}{3} \ln x - \frac{1}{9}x^3 + C.$

b) $\int x \cos 2x \, dx$

$u = $ _____

$dv = $ _____

$u = x,\ du = dx$

$dv = \cos 2x\,dx,\ v = \frac{1}{2}\sin 2x.$

Then $v\,du = \frac{1}{2}\sin 2x\,dx$ and

$\int x \cos x \, dx = \frac{1}{2}x \sin 2x - \frac{1}{2}\int \sin 2x\,dx$

$= \frac{1}{2}x \sin 2x - \frac{1}{2}\left(-\frac{1}{2}\cos 2x\right)$

$= \frac{1}{2}x \sin 2x + \frac{1}{4}\cos 2x + C.$

2) Evaluate $\int x^2 e^{5x}\,dx.$

Let $u = x^2$ and $dv = e^{5x}\,dx.$

Then $du = 2x\,dx,\ v = \frac{1}{5}e^{5x},$ and

$v\,du = \frac{1}{5}e^{5x}\,2x\,dx = \frac{2}{5}xe^{5x}\,dx.$

Thus $\int x^2 e^{5x}\,dx = x^2\left(\frac{1}{5}e^{5x}\right) - \int \frac{2}{5}xe^{5x}\,dx$

$= \frac{1}{5}x^2\,e^{5x} - \frac{2}{5}\int x\,e^{5x}\,dx.$

The last integral is easier than the original so we are on the right track. Apply by parts to it:

$\int x\,e^{5x}\,dx = \begin{pmatrix} u = x,\ du = dx, \\ dv = e^{5x}\,dx,\ v = \frac{1}{5}e^{5x} \end{pmatrix}$

$= x\left(\frac{1}{5}e^{5x}\right) - \int \frac{1}{5}e^{5x}\,dx$

$= \frac{1}{5}xe^{5x} - \frac{1}{25}e^{5x}.$

Thus the original integral is

$\frac{1}{5}x^2 e^{5x} - \frac{2}{5}\left(\frac{1}{5}xe^{5x} - \frac{1}{25}e^{5x}\right)$

$= \frac{1}{5}x^2 e^{5x} - \frac{2}{25}xe^{5x} + \frac{2}{125}e^{5x} + C.$

3) Evaluate $\displaystyle\int_1^e \ln x\, dx$

Let $u = \ln x$ and $dv = 1\, dx$.

Then $du = \dfrac{1}{x}\, dx$, $v = x$, and

$v\, du = x\dfrac{1}{x}\, dx = dx$.

Thus $\displaystyle\int_1^e \ln x\, dx = x \ln x - \int dx$

$= x \ln x - x\Big]_1^e = (e \ln e - e) - (1 \ln 1 - 1)$

$= 1.$

4) Evaluate $\displaystyle\int \sin x \cos x\, dx$ by parts.

$\displaystyle\int \sin x \cos x\, dx \quad \left(\begin{array}{l} u = \sin x,\ du = \cos x\, dx, \\ dv = \cos x\, dx,\ v = \sin x \end{array}\right)$

$= (\sin x)(\sin x) - \displaystyle\int (\sin x)\cos x\, dx.$

Thus

$\displaystyle\int \sin x \cos x\, dx = \sin^2 x - \int \sin x \cos x\, dx.$

Hence $2\displaystyle\int \sin x \cos x\, dx = \sin^2 x,$

so $\displaystyle\int \sin x \cos x\, dx = \dfrac{1}{2} \sin^2 x + C.$

Note: This integral could have been evaluated with the simple substitution $u = \sin x$, $du = \cos x\, dx$ to yield $\int u\, du$. We could also have solved this with the identity $\sin x \cos x = \dfrac{1}{2} \sin 2x$ and the substitution $u = 2x$. We chose the above method to illustrate integration by parts.

Section 7.2   Trigonometric Integrals

Certain combinations of trigonometric functions are easy to integrate. For example, $\int (\sin^4 x \cos x)\, dx$ may be evaluated with the Chain Rule substitution $u = \sin x$, $du = \cos x\, dx$ to produce $\int u^4\, du$. This section will call upon your ability to remember trigonometric identities and use them to rewrite integrals into forms that may be solved using techniques that you have already learned.

Concepts to Master

Evaluate integrals using trigonometric identities

Summary and Focus Questions

Page 476 A summary of the techniques to evaluate integrals with certain combinations of trigonometric integrands is given on the next page. Note that m and n must be integers. If the condition(s) are satisfied, then the original integral may be rewritten as indicated.

Original Integral	Condition(s)	Identities to Use	Technique	Rewritten Form	u
$\int \sin^m x \cos^n x\, dx$	$m = $ odd	$\sin^2 x = 1 - \cos^2 x$	Factor out one $\sin x$ for du	$\int (1 - \cos^2 x)^{(m-1)/2} \cos^n x \sin x\, dx$	$u = \cos x$
	$n = $ odd	$\cos^2 x = 1 - \sin^2 x$	Factor out one $\cos x$ for du	$\int (1 - \sin^2 x)^{(n-1)/2} \sin^m x \cos x\, dx$	$u = \sin x$
	Both m and n odd	Use either of the above			
	Both m and n even	$\sin^2 x = \dfrac{1 - \cos 2x}{2}$, $\cos^2 x = \dfrac{1 + \cos 2x}{2}$	Substitute the identities to get started	$\int \left(\dfrac{1-\cos 2x}{2}\right)^{m/2}\left(\dfrac{1 + \cos 2x}{2}\right)^{n/2} dx$	

Original Integral	Condition(s)	Identities to Use	Technique	Rewritten Form	u
$\int \tan^m x \sec^n x\, dx$	$m = $ odd	$\tan^2 x = \sec^2 x - 1$	Factor out $\sec^2 x$ for du	$\int (\sec^2 x - 1)^{(m-1)/2} \sec^{n-1} x \tan x \sec x\, dx$	$u = \tan x$
	$n = $ even	$\sec^2 x = \tan^2 x + 1$	Factor out $\sec x \tan x$ for du	$\int \tan^m x (\tan^2 x + 1)^{(n-2)/2} \sec^2 x\, dx$	$u = \sec x$
	$m = $ even, $n = $ odd	$\tan^2 x = \sec^2 x - 1$	Substitute the identity to get started	$\int (\sec^2 x - 1)^{m/2} \sec^n x\, dx$	

Integrals containing $\cot^m x \csc^n x$ are handled similarly to $\int \tan^m x \sec^n x\, dx$ using the identity $\cot^2 x = \csc^2 x - 1$.

Original Integral	Identities to Use	Technique	Rewritten Form
$\int \sin mx \sin nx\, dx$	$\sin A \sin B = \dfrac{\cos(A - B) - \cos(A + B)}{2}$	Substitute the identity and simplify	$\int \frac{1}{2}(\cos(m - n)x - \cos(m + n)x)\, dx$
$\int \cos mx \cos nx\, dx$	$\cos A \cos B = \dfrac{\cos(A - B) + \cos(A + B)}{2}$	Substitute the identity and simplify	$\int \frac{1}{2}(\cos(m - n)x + \cos(m + n)x)\, dx$
$\int \sin mx \cos nx\, dx$	$\sin A \cos B = \dfrac{\sin(A - B) + \sin(A + B)}{2}$	Substitute the identity and simplify	$\int \frac{1}{2}(\sin(m - n)x + \sin(m + n)x)\, dx$

1) Evaluate $\int \sin^5 x \, dx$.

This fits the form $\int \sin^m x \cos^n x \, dx$ with $m = 5, n = 0$.

$$\int \sin^5 x \, dx = \int \sin^4 x \sin x \, dx$$
$$= \int (1 - \cos^2 x)^2 \sin x \, dx$$
$$(u = \cos x, \, du = -\sin x \, dx)$$
$$= -\int (1 - u^2)^2 \, du$$
$$= -\int (1 - 2u^2 + u^4) \, du$$
$$= -u + \frac{2}{3}u^3 - \frac{1}{5}u^5$$
$$= -\cos x + \frac{2}{3}\cos^3 x - \frac{1}{5}\cos^5 x + C.$$

2) Evaluate $\int \tan x \sec^4 x \, dx$.

Here $m = 1$ and $n = 4$ so we have our choice of identities. We select the $n =$ even case.

$$\int \tan x \sec^4 x \, dx$$
$$= \int \tan x (\sec^2 x)\sec^2 x \, dx$$
$$= \int \tan x (\tan^2 x + 1)\sec^2 x \, dx$$
$$= \int (\tan^3 x + \tan x)\sec^2 x \, dx$$
$$(u = \tan x, \, du = \sec^2 x \, dx)$$
$$= \int (u^3 + u) \, du = \frac{u^4}{4} + \frac{u^2}{2}$$
$$= \frac{1}{4}\tan^4 x + \frac{1}{2}\tan^2 x + C.$$

3) Evaluate $\int \sin 6x \cos 2x \, dx$.

The form is $\sin A \cos B$ where $A = 6x$ and $B = 2x$.

$$\int \sin 6x \cos 2x \, dx$$
$$= \int \frac{1}{2}(\sin 4x + \sin 8x) \, dx$$
$$= \frac{1}{2}\int \sin 4x \, dx + \frac{1}{2}\int \sin 8x \, dx$$
$$(\text{Two substitutions: } u = 4x \text{ and } u = 8x.)$$
$$= \frac{1}{2}\left(\frac{-\cos 4x}{4}\right) + \frac{1}{2}\left(\frac{-\cos 8x}{8}\right)$$
$$= -\frac{1}{8}\cos 4x - \frac{1}{16}\cos 8x + C.$$

4) Evaluate $\int \cos^4 x \, dx$.

Here $m = 0$ and $n = 4$ are both even.

$\int \cos^4 x \, dx = \int (\cos^2 x)^2 \, dx$

$= \int \left(\frac{1 + \cos 2x}{2} \right)^2 dx$

$= \frac{1}{4} \int (1 + 2 \cos 2x + \cos^2 2x) \, dx$

$= \frac{1}{4} \int \left(1 + 2 \cos 2x + \frac{1 + \cos 4x}{2} \right) dx$

$= \frac{1}{4} \int \left(\frac{3}{2} + 2 \cos 2x + \frac{1}{2} \cos 4x \right) dx$

$= \frac{1}{4} \left[\frac{3}{2} x + \sin 2x + \frac{1}{2} \frac{\sin 4x}{4} \right]$

$= \frac{3}{8} x + \frac{1}{4} \sin 2x + \frac{1}{32} \sin 4x + C.$

5) Evaluate $\int \tan^2 x \sec x \, dx$.

Here $m = 2$ is even and $n = 1$ is odd.

$\int \tan^2 x \sec x \, dx = \int (\sec^2 x - 1) \sec x \, dx$

$= \int (\sec^3 x - \sec x) \, dx$

$= \int \sec^3 x \, dx - \int \sec x \, dx$

$= \frac{1}{2} (\sec x \tan x + \ln|\sec x + \tan x|) - \ln|\sec$

$x + \tan x| + C$

$= \frac{1}{2} (\sec x \tan x - \ln|\sec x + \tan x|) + C.$

($\int \sec^3 x \, dx$ is evaluated by parts)

6) Evaluate $\int \frac{\sin^3 x}{\sqrt{\cos x}} \, dx$

Here $m = 3$ and $n = -\frac{1}{2}$. Although n is not

an integer, the same technique works.

$\int \frac{\sin^3 x}{\sqrt{\cos x}} \, dx = \int (\cos x)^{-1/2} \sin^2 x \sin x \, dx$

$= \int (\cos x)^{-1/2} (1 - \cos^2 x) \sin x \, dx$

$(u = \cos x, \ du = -\sin x \, dx)$

$= - \int u^{-1/2} (1 - u^2) \, du$

$= - \int (u^{-1/2} - u^{3/2}) \, du$

$= - (2u^{1/2} - \frac{2}{5} u^{5/2})$

$= \frac{2}{5} \cos^{5/2} x - 2 \sqrt{\cos x} + C.$

Section 7.3 Trigonometric Substitution

Until now we have been using direct substitutions ($u = \ldots$, $du = \ldots dx$), to simplify an integrand. This section describes a form of "inverse" substitution to evaluate $\int f(x)\, dx$ by replacing x (and dx) by a function of another variable: t, θ, ... The result may be an integral easier to evaluate.

Concepts to Master

Evaluate integrals containing $\sqrt{a^2 - x^2}$, $\sqrt{a^2 + x^2}$, and $\sqrt{x^2 - a^2}$.

Summary and Focus Questions

Page
483

"Trigonometric substitution" uses substitutions of the form

$$x = a \sin \theta,$$
$$x = a \tan \theta, \text{ or}$$
$$x = a \sec \theta.$$

The resulting integrals are then evaluated by the techniques of Section 8.2.

The table below summarizes the three corresponding substitutions. Drawing a mnemonic triangle may help you determine the correct substitutions.

Remember that the triangle is drawn for $0 \le \theta \le \frac{\pi}{2}$ only; relationships should be worked out using the identities $\sin^2 \theta + \cos^2 \theta = 1$ and $\tan^2 \theta = \sec^2 \theta - 1$.

Integrand contains	Mnemonic triangle	Substitution
$\sqrt{a^2 - x^2}$	triangle with hypotenuse a, opposite side x, adjacent side $\sqrt{a^2 - x^2}$, angle θ	$x = a \sin \theta$, where $-\frac{\pi}{2} \le \theta \le \frac{\pi}{2}$ $dx = a \cos \theta\, d\theta$ $\sqrt{a^2 - x^2} = a \cos \theta.$
$\sqrt{a^2 + x^2}$	triangle with hypotenuse $\sqrt{a^2 + x^2}$, opposite side x, adjacent side a, angle θ	$x = a \tan \theta$, where $-\frac{\pi}{2} < \theta < \frac{\pi}{2}$ $dx = a \sec^2 \theta\, d\theta$ $\sqrt{a^2 + x^2} = a \sec \theta.$
$\sqrt{x^2 - a^2}$	triangle with hypotenuse x, opposite side $\sqrt{x^2 - a^2}$, adjacent side a, angle θ	$x = a \sec \theta$, where $0 \le \theta < \frac{\pi}{2}$ or $\pi \le \theta < \frac{3\pi}{2}$ $dx = a \sec \theta \tan \theta\, d\theta$ $\sqrt{x^2 - a^2} = a \tan \theta.$

It may be necessary to algebraically transform an integrand before substitution. One technique is completing the square:

Example: Evaluate $\int \dfrac{dx}{\sqrt{x^2 + 4x + 5}}$.

The integrand $\dfrac{1}{\sqrt{x^2 + 4x + 5}}$ does not apparently meet the form $\dfrac{1}{\sqrt{1 + u^2}}$, but will do so after completing the square:

$$x^2 + 4x + 5 = x^2 + 4x + 4 + 5 - 4 = (x + 2)^2 + 1.$$

Thus $\int \dfrac{dx}{\sqrt{x^2 + 4x + 5}} = \int \dfrac{dx}{\sqrt{1 + (x + 2)^2}}$.

Let $u = x + 2$. Then $du = dx$ and $\int \dfrac{dx}{\sqrt{1 + (x + 2)^2}} = \int \dfrac{1}{\sqrt{1 + u^2}}\, du$.

We now use a trigonometric substitution:

$\dfrac{u}{1} = \tan\theta$

$\dfrac{\sqrt{1 + u^2}}{1} = \sec\theta$

Let $u = \tan\theta$, $du = \sec^2\theta\, d\theta$, $\sqrt{1 + u^2} = \sec\theta$.

Then $\int \dfrac{1}{\sqrt{1 + u^2}}\, du = \int \dfrac{1}{\sec\theta} \sec^2\theta\, d\theta = \int \sec\theta\, d\theta = \ln|\sec\theta + \tan\theta|$
$$= \ln|\sqrt{1 + u^2} + u| = \ln|\sqrt{x^2 + 4x + 5} + x + 2| + C.$$

1) Rewrite each of the following using a trigonometric substitution:

a) $\int x^2 \sqrt{4 - x^2}\, dx$

$\dfrac{x}{2} = \sin\theta$

$\dfrac{\sqrt{4 - x^2}}{2} = \cos\theta$

Let $x = 2 \sin\theta$, $dx = 2 \cos\theta\, d\theta$,

$\sqrt{4 - x^2} = 2 \cos\theta$. Then

$\int x^2 \sqrt{4 - x^2}\, dx$

$= \int (2 \sin\theta)^2 (2 \cos\theta)(2 \cos\theta\, d\theta)$

$= 16 \int \sin^2\theta \cos^2\theta\, d\theta.$

b) $\int x^2 \sqrt{x^2 - 4}\, dx$

$\frac{x}{2} = \sec \theta$

$\frac{\sqrt{x^2 - 4}}{2} = \tan \theta$

Let $x = 2 \sec \theta$, $dx = 2 \sec \theta \tan \theta\, d\theta$,

$\sqrt{x^2 - 4} = 2 \tan \theta$. Then

$\int x^2 \sqrt{x^2 - 4}\, dx$

$= \int (2 \sec \theta)^2 (2 \tan \theta)(2 \sec \theta \tan \theta\, d\theta)$

$= 16 \int \sec^3 \theta \tan^2 \theta\, d\theta.$

c) $\int \frac{x + 1}{\sqrt{4 + x^2}}\, dx$

$\frac{x}{2} = \tan \theta$

$\frac{\sqrt{4 + x^2}}{2} = \sec \theta$

Let $x = 2 \tan \theta$, $dx = 2 \sec^2 \theta\, d\theta$,

$\sqrt{4 + x^2} = 2 \sec \theta$. Then

$\int \frac{x + 1}{\sqrt{4 + x^2}}\, dx$

$= \int \frac{2 \tan \theta + 1}{2 \sec \theta} \cdot 2 \sec^2 \theta\, d\theta$

$= \int (2 \tan \theta + 1) \sec \theta\, d\theta$

$= \int 2 \tan \theta \sec \theta\, d\theta + \int \sec \theta\, d\theta.$

2) $\int$ Evaluate $\frac{x}{1 + x^2}\, dx$

While a trigonometric substitution will lead you to a solution, a direct substitution is all that is needed:

$u = 1 + x^2$, $du = 2x\, dx$

$\int \frac{x}{1 + x^2}\, dx = \frac{1}{2} \int u^{-1}\, du = \frac{1}{2} \ln u$

$= \frac{1}{2} \ln(1 + x^2) = \ln \sqrt{1 + x^2} + C.$

3) Rewrite the following so that a trigonometric substitution may be used.

$$\int \frac{x^2}{\sqrt{13 - 6x + x^2}}\, dx$$

$\sqrt{13 - 6x + x^2}$ does not look friendly until we complete the square:

$$13 - 6x + x^2 = 13 + (x^2 - 6x + 9 - 9)$$
$$= x^2 - 6x + 9 + 4$$
$$= (x - 3)^2 + 4.$$

Let $u = x - 3$, $du = dx$, $x = u + 3$, and $13 - 6x + x^2 = u^2 + 4$.

Then $\int \dfrac{x^2}{\sqrt{13 - 6x + x^2}}\, dx = \int \dfrac{(u + 3)^2}{\sqrt{u^2 + 4}}\, du.$

The integral is now ready for the substitution $u = 2 \tan \theta$, $du = 2 \sec^2 \theta\, d\theta$.

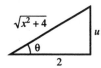

Section 7.4   Integration of Rational Functions by Partial Fractions

This section shows you how to rewrite a rational function so that it may be integrated. There are no new calculus concepts here, only algebraic techniques for rewriting the integrand.

Concepts to Master

A. Evaluate integrals containing rational functions by partial fractions; method of determining coefficients.

B. Evaluate integrals using rationalizing substitutions.

Summary and Focus Questions

Page 490

A. A rational function is a function of the form $\frac{P(x)}{Q(x)}$ where $P(x)$ and $Q(x)$ are polynomials. The *method of partial fractions* for evaluating integrals involves writing a rational $\frac{P(x)}{Q(x)}$ integrand as a sum of rational functions each of which you already know how to integrate. Here are the steps:

Step 1: If the degree of $P(x)$ is greater than or equal to the degree of $Q(x)$, use polynomial long division to write $\frac{P(x)}{Q(x)} = H(x) + \frac{R(x)}{Q(x)}$, where the degree of $R(x)$ is less than the degree of $Q(x)$.

Step 2: Write $Q(x)$ in factored form as a product of powers of distinct linear and irreducible quadratics.

Step 3: Write $\frac{R(x)}{Q(x)} = T_1 + T_2 + \ldots + T_n$

where the T_i are terms corresponding to the distinct factors of $Q(x)$ found in Step 1. The forms of T_i are given in the table on the next page.

Type of $Q(x)$ factor	Corresponding T_i term
$x - a$	$\dfrac{A}{x - a}$
$(x - a)^k$	$\dfrac{A_1}{x - a} + \dfrac{A_2}{(x - a)^2} + \cdots + \dfrac{A_k}{(x - a)^k}$
$ax^2 + bx + c$	$\dfrac{Ax + B}{ax^2 + bx + c}$
$(ax^2 + bx + c)^k$	$\dfrac{A_1x + B_1}{ax^2 + bx + c} + \dfrac{A_2x + B_2}{(ax^2 + bx + c)^2} + \cdots + \dfrac{A_kx + B_k}{(ax^2 + bx + c)^k}$

Step 4: Write $R(x) = Q(x)[T_1 + T_2 + \ldots + T_n]$. Then multiply out the right side, combining like terms.

Step 5: Equate coefficients of like terms in the equation in Step 4. This gives a system of linear equations.

Step 6: Solve the linear system in Step 5. This is frequently done by solving for one variable in one equation and using the result to eliminate that variable in the other equations.

Shortcut: Sometimes some of the unknown coefficients can be found after Step 3 by judicious substitution of values for x. The rest of the coefficients may be found by continuing with Steps 4, 5, and 6.

Here is a complete example: $\displaystyle\int \frac{x^3 + 5x^2 - 4x + 4}{(x^2 - 2x + 1)(x^2 + x + 1)}\, dx.$

Step 1: Because the degree of the numerator is less than the degree of the denominator, we can skip this step.

Step 2: $Q(x) = (x^2 - 2x + 1)(x^2 + x + 1) = (x - 1)^2(x^2 + x + 1)$. Note that $x^2 + x + 1$ is irreducible.

Step 3: $\dfrac{x^3 + 5x^2 - 4x + 4}{(x - 1)^2(x^2 + x + 1)} = \dfrac{A}{(x - 1)} + \dfrac{B}{(x - 1)^2} + \dfrac{Cx + D}{x^2 + x + 1}.$

Step 4: Multiply by the denominator $Q(x)$:
$$x^3 + 5x^2 - 4x + 4$$
$$= A(x - 1)(x^2 + x + 1) + B(x^2 + x + 1) + (Cx + D)(x - 1)^2.$$
Multiply out the right side:
$$x^3 + 5x^2 - 4x + 4$$
$$= Ax^3 - A + Bx^2 + Bx + B + Cx^3 - 2Cx^2 + Cx + Dx^2 - 2Dx + D$$
and collect like terms:
$$x^3 + 5x^2 - 4x + 4$$
$$= (A + C)x^3 + (B - 2C + D)x^2 + (B + C - 2D)x + (-A + B + D).$$

Step 5: Equate coefficients from the equality in Step 4:
$$1 = A + C$$
$$5 = B - 2C + D$$
$$-4 = B + C - 2D$$
$$4 = -A + B + D$$

Step 6: Solve the system.

From the first equation, $A = 1 - C$ and thus the others become:

$$5 = B - 2C + D \qquad\qquad 5 = B - 2C + D$$
$$-4 = B + C - 2D \qquad \text{or} \qquad -4 = B + C - 2D$$
$$4 = -(1 - C) + B + D \qquad\qquad 5 = B + C + D$$

Subtracting the last equation from the first equation yields $0 = -3C$ or $C = 0$. So the system is:

$$5 = B + D$$
$$-4 = B - 2D$$

Again subtracting, $9 = 3D$, $D = 3$. Thus $B = 2$ and $A = 1$.

Therefore $\dfrac{x^3 + 5x^2 - 4x + 4}{(x^2 - 2x + 1)(x^2 + x + 1)} = \dfrac{1}{x - 1} + \dfrac{2}{(x - 1)^2} + \dfrac{3}{x^2 + x + 1}$.

Finally, the original integral may be written as the sum of three integrals, each of which we know how to evaluate:

$$\int \frac{x^3 + 5x^2 - 4x + 4}{(x^2 - 2x + 1)(x^2 + x + 1)}\, dx = \int \frac{1}{x - 1}\, dx + \int \frac{2}{(x - 1)^2}\, dx + \int \frac{3}{x^2 + x + 1}\, dx.$$

$$\int \frac{1}{x - 1}\, dx = \ln|x - 1| \text{ (substitute } u = x - 1).$$

$$\int \frac{2}{(x - 1)^2}\, dx = \frac{-2}{x - 1} \text{ (substitute } u = x - 1).$$

$$\int \frac{3}{x^2 + x + 1}\, dx = \int \frac{3}{(x + 1/2)^2 + 3/4}\, dx = 4\int \frac{1}{\left[\frac{2}{\sqrt{3}}\left(x + \frac{1}{2}\right)\right]^2 + 1}\, dx$$

$$= 2\sqrt{3}\, \tan^{-1}\!\left[\frac{2x + 1}{\sqrt{3}}\right].$$

Thus $\displaystyle\int \frac{x^3 + 5x^2 - 4x + 4}{(x^2 - 2x + 1)(x^2 + x + 1)}\, dx$

$$= \ln|x - 1| - \frac{2}{x - 1} + 2\sqrt{3}\, \tan^{-1}\!\left[\frac{2x + 1}{\sqrt{3}}\right] + C \text{ (Whew!)}$$

(In this example we could have found $B = 2$ by setting $x = 1$ in Step 4. The rest of the procedure would be followed to find A, C, and D.)

1) Write the partial fraction form for each.

a) $\dfrac{2x + 1}{x(x + 8)}$

b) $\dfrac{4x^2 + 11}{(x + 1)^2(x^2 - 3x + 5)}$

c) $\dfrac{3}{x(x^2 + 2x + 2)^2}$

d) $\dfrac{x^3}{x^2 + 2x + 1}$

2) Evaluate $\displaystyle\int \dfrac{6 - x}{x(x + 3)}\, dx$.

$\dfrac{A}{x} + \dfrac{B}{x + 8}$

$\dfrac{A}{x + 1} + \dfrac{B}{(x + 1)^2} + \dfrac{Cx + D}{x^2 - 3x + 5}$

$\dfrac{A}{x} + \dfrac{Bx + C}{x^2 + 2x + 2} + \dfrac{Dx + E}{(x^2 + 2x + 2)^2}$

First write the fraction (using long division):

$$x - 2 + \frac{3x + 2}{x^2 + 2x + 1} = x - 2 + \frac{3x + 2}{(x + 1)^2}.$$

The partial fraction form is

$$x - 2 + \frac{A}{x + 1} + \frac{B}{(x + 1)^2}.$$

(1) Not needed.

(2) $Q(x) = x(x + 3)$

(3) $\dfrac{6 - x}{x(x + 3)} = \dfrac{A}{x} + \dfrac{B}{x + 3}$

(4) Clear fractions:
$$6 - x = A(x + 3) + Bx$$
Combine terms:
$$6 - x = (A + B)x + 3A$$

(5) Equate coefficients:
$$A + B = -1$$
$$3A = 6$$

(6) Solving, we see $A = 2$ from the second equation. Thus
$$2 + B = -1, B = -3.$$
Therefore, $\displaystyle\int \frac{6 - x}{x(x + 3)}\, dx = \int\left(\frac{2}{x} + \frac{-3}{x + 3}\right)dx$
$$= 2 \ln|x| - 3 \ln|x + 3| + C.$$

Note: At Step 4, after clearing fractions to get $6 - x = A(x + 3) + Bx$ we could substitute $x = 0$ to get $6 = A(0 + 3) + 0$, so $A = 2$, and we could substitute $x = -3$ to get
$$6 - (-3) = A(0) + B(-3)$$
$$9 = -3B, B = -3.$$
Sometimes, substituting carefully selected values of x can shorten the work.

**Page
497**

B. In some cases the substitution $u = \sqrt[n]{g(x)}$ in the form $u^n = g(x)$ will transform an integrand involving radicals into a rational function.

Example: Evaluate $\int \dfrac{x}{\sqrt{x-2}}\, dx$.

Let $u = \sqrt{x-2}$, $u^2 = x - 2$, $x = u^2 + 2$, $dx = 2u\, du$.

Then $\int \dfrac{x}{\sqrt{x-2}}\, dx = \int \dfrac{u^2+2}{u}\, 2u\, du = 2\int (u^2 + 2)\, du = \dfrac{2u^3}{3} + 4u$

$\qquad\qquad = \dfrac{2}{3}(x-2)^{3/2} + 4\sqrt{x-2} + C.$

1) Evaluate $\int \dfrac{\sqrt[3]{x}}{\sqrt[3]{x}+1}\, dx$.

Let $u = \sqrt[3]{x}$, $x = u^3$, $dx = 3u^2\, du$.

The integral becomes

$\int \dfrac{u}{u+1}\, 3u^2\, du = 3\int \dfrac{u^3}{u+1}\, du.$

By long division

$\dfrac{u^3}{u+1} = u^2 - u + 1 + \dfrac{-1}{u+1}.$

$\int \dfrac{u^3}{u+1}\, du = \int \left(u^2 - u + 1 - \dfrac{1}{u+1} \right) du$

$\qquad = \dfrac{u^3}{3} - \dfrac{u^2}{2} + u - \ln|u+1|.$

Since $u = \sqrt[3]{x}$, this is

$3\left[\dfrac{x}{3} - \dfrac{x^{2/3}}{2} + \sqrt[3]{x} - \ln\left|\sqrt[3]{x} + 1\right| \right] + C.$

2) What substitution should be made for $\int \dfrac{\sqrt[3]{x}}{1 + \sqrt[4]{x}}\, dx$?

$u = \sqrt[12]{x}$, because then $\sqrt[3]{x}\,(= u^3)$ and $\sqrt[4]{x}\,(= u^4)$ are both integer powers of u. (*Note:* 12 is the least common multiple of 3 and 4.)

Section 7.5   Strategy for Integration

There are no new integration techniques in this section; its purpose is to draw together all the techniques so far. In this section you will also encounter some rather simple looking functions for which all the techniques that you have seen so far will fail.

Concepts to Master

A. Review all the integration techniques
B. Functions that are not readily integrable

Summary and Focus Questions

Page
499

A. The two main techniques for evaluating integrals are: *substitution* and *integration by parts*. Substitution can take many forms.

Method	Sample
Straight substitution	$\int \frac{x^2}{x^3 + 1}\, dx,\ u = x^3 + 1,\ \dots$
Integration by parts	$\int x^3 \ln x\, dx,\ u = \ln x,\ dv = x^3\, dx,\ \dots$
Trigonometric	$\int \sin^3 x \cos^2 x\, dx = \int \sin^2 x \cos^2 x \sin x\, dx$ $= \int (1 - \cos^2 x)\cos^2 x \sin x\, dx,\ u = \cos x \dots$
Trigonometric substitution	$\int \frac{x^2}{\sqrt{4 - x^2}}\, dx,\ x = 2 \sin \theta,\ \dots$
Partial fractions	$\int \frac{5}{(x + 2)(x - 3)}\, dx = \int \left(\frac{-1}{x + 2} + \frac{1}{x - 3}\right) dx = \dots$
Rationalizing substitution	$\int \frac{x}{\sqrt{x} + 1}\, dx,\ u = \sqrt{x},\ u^2 = x,\ \dots$

Remember your algebra and trigonometric identities and know the twenty integrals in the Table of Integration formulas (p. 534) of the text.

1) Indicate what technique should be used to begin solving each:

a) $\int \frac{e^x}{e^x - 1}\, dx$

Straight substitution, $u = e^x - 1$.

b) $\int x^2 \sqrt{16 + x^2}\, dx$

Trigonometric substitution, $x = 4 \tan \theta$.

c) $\int \dfrac{2x+1}{x^3(x+1)}\, dx$

Partial fractions, $\dfrac{A}{x} + \dfrac{B}{x^2} + \dfrac{C}{x^3} + \dfrac{D}{x+1}$.

d) $\int \tan^3 x \sec^4 x\, dx$

Trigonometric integral,

$\int \tan^3 x \sec^2 x (\sec^2 x\, dx)$.

e) $\int x \sec^2 x\, dx$

Integration by parts,
$u = x,\ dv = \sec^2 x\, dx$.

f) $\int \dfrac{\csc^2 x}{1 + \cot^2 x}\, dx$

Since $1 + \cot^2 x = \csc^2 x$, this one reduces
to $\int 1\, dx = x + C$.

**Page
503**

B. All the functions you have encountered so far are "elementary," not because
they are necessarily easy, but because it is possible to express each as a
combination (addition, subtraction, multiplication, division, composition)
of polynomial, rational, exponential, logarithmic, trigonometric, inverse
trigonometric, hyperbolic, and inverse hyperbolic functions.

The *derivative* of an elementary function is elementary.
The *antiderivative* of an elementary function need not be elementary.
Thus there are some elementary functions for which there is no simple form
for its antiderivative. $\int \dfrac{1}{\ln x}\, dx$ is one; we will wait until Chapter 11 for this
one.

2) True or False:
$f(x) = e^{x^{x^x}}$ is elementary.

True.

3) True or False:
If $f(x)$ is not elementary, then $\int f(x)\, dx$ is
not elementary.

True.

4) True or False:
If $f(x)$ is elementary, then $\int f(x)\, dx$ is
elementary.

False.

5) Do any of the techniques we have apply to
$\int \cos x^2\, dx$?

No. This one also must wait until
Chapter 11.

Section 7.6   Using Tables of Integrals and Computer Algebra Systems

This section covers two of the common tools to assist in performing integration. To successfully use a table of integrals or a computer algebra system you do need an understanding of the techniques of the previous sections.

Concepts to Master

A. Evaluate integrals using a table of integrals
B. Evaluate integrals using a computer algebra system

Summary and Focus Questions

Page 505

A. There is a table of 120 integration formulas on the back inside covers of your textbook. Most of these formulas are derived using the techniques you have already studied. The integrals are described with variable u and constants a, b, etc., and are grouped by form of the integrand. Sometimes you must use algebra to adjust the integrand before determining the appropriate form. In a few cases, such as integral formula #58, the formula is a "reduction" formula that replaces the given integral by an expression containing another integral.

1) What integral formula from the Table of Integrals in the text may be used for each?

a) $\int \dfrac{dx}{x^2(4 + 5x)}$

   #50, $a = 4$, $b = 5$.

b) $\int \dfrac{\sqrt{4x^2 - 9}}{x}\, dx$

   #41, $u = 2x$, $a = 3$

c) $\int \dfrac{1}{5x^2 - 3}\, dx$

   #20, $u = \sqrt{5}x$, $a = \sqrt{3}$

d) $\int \dfrac{\sqrt{25 + 4x^2}}{x}\, dx$

   #23, $u = 2x$, $a = 5$

e) $\int 3x \cos^{-1} 2x\, dx$

   #91, $u = 2x$

**Page
507**

B. Computer algebra system (CAS) programs for computers and calculators are a great tool for the symbolic manipulation involved in finding antiderivatives. Here are some cautions to observe when using a CAS:

1) There is usually no constant of integration displayed (answers may appear like x^2 or $x^2 + 5$ rather than $x^2 + C$).

2) An answer you obtain by hand may not look like the one produced by the CAS, yet they are the same. You may need to apply algebra and various identities to transform one to another. Here are a few common differences:

a) Some systems will expand an integrand into a polynomial rather than use a substitution, as in $\int (2x + 5)^4 \, dx$, $u = 2x + 5$, $du = 2 \, dx$.

b) Fractional answers and constants are often grouped differently or factored out. For example, $\frac{1}{2}x + \frac{1}{4x}$ may be expressed as $\frac{2x^2 + 1}{4x}$.

c) Hyperbolic trigonometric functions are usually expressed with logarithms. You will often see $\ln(x + \sqrt{x^2 + 1})$ rather than $\sinh^{-1} x$.

2) Use a CAS to evaluate the integrals in question 1.

Answers depend upon the particular CAS used. Here are answers using a Texas Instruments TI-92:

a) $\dfrac{5x \ln |5x + 4| - 5x \ln |x| - 4}{16x}$

b) $-\left(3 \tan^{-1} \dfrac{\sqrt{4x^2 - 9}}{3} - \sqrt{4x^2 - 9}\right)$

c) $\dfrac{-\sqrt{15}\, \ln\left(\dfrac{|\sqrt{15}x + 3|}{|\sqrt{15}x - 3|}\right)}{30}$

d) $\dfrac{10 \ln(\sqrt{4x^2 + 25} - 5) - 5 \ln x^2 + 2\sqrt{4x^2 + 25}}{2}$

e)

$\dfrac{-3((8x^2 - 1)\sin^{-1}(2x) + 2x(\sqrt{-(4x^2 - 1)} - 2\pi x))}{16}$

Section 7.7   Approximate Integration

We have seen that one way to approximate the value of a definite integral that represents an area is to cover the area with approximating rectangles and compute the corresponding Riemann sum. This section considers three ways to create sums that approximate the value of a given integral.

Concepts to Master

A. Approximate a definite integral using the Midpoint Rule; using the Trapezoidal Rule; using Simpson's Rule

B. Estimate the maximum error

Summary and Focus Questions

Page 512

A. Suppose f is an integrable function on a closed interval $[a, b]$ which has been partitioned $\{a = x_0, x_1, x_2, \ldots, x_{n-1}, x_n = b\}$ into n equal subintervals of length $\Delta x = \dfrac{b - a}{n}$. There are three rules in this section to approximate $\displaystyle\int_a^b f(x)\, dx$. Each rule replaces portions of the graph of $y = f(x)$ with lines segments or curves and sums the areas under these replacements.

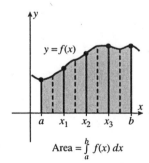

Area $= \displaystyle\int_a^b f(x)\, dx$

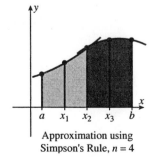

Approximation using
Midpoint Rule, $n = 4$

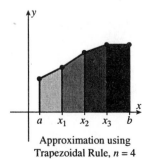

Approximation using
Trapezoidal Rule, $n = 4$

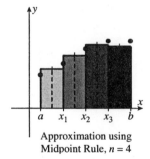

Approximation using
Simpson's Rule, $n = 4$

(1) *Midpoint Rule:* Let $\bar{x}_i = \frac{x_{i-1} + x_i}{2}$, the midpoint of the ith subinterval. An approximation to $\int_a^b f(x)\,dx$ is the Riemann sum:

$$\Delta x[f(\bar{x}_1) + f(\bar{x}_2) + \dots + f(\bar{x}_n)].$$

The Midpoint Rule replaces each portion of the graph by a horizontal line segment through $(\bar{x}_i, f(\bar{x}_i))$ and sums the areas of the resulting rectangles under these segments.

(2) *Trapezoidal Rule:* An approximation to $\int_a^b f(x)\,dx$ is the sum:

$$\frac{\Delta x}{2}[f(x_0) + 2f(x_1) + 2f(x_2) + \dots + 2f(x_{n-1}) + f(x_n)].$$

The Trapezoidal Rule replaces portions of the graph by straight line segments from $(x_{i-1}, f(x_{i-1}))$ to $(x_i, f(x_i))$ and sums the areas of the trapezoidal regions under these segments.

(3) *Simpson's Rule:* An approximation to $\int_a^b f(x)\,dx$ is the sum:

$$\frac{\Delta x}{3}[f(x_0) + 4f(x_1) + 2f(x_2) + 4f(x_3) + \dots + 4f(x_{n-1}) + f(x_n)].$$

Simpson's Rule replaces pairs of portions of the graph by parabolas and sums the areas of the regions under these parabolas.

1) Estimate $\int_{-1}^5 2^x\,dx$ to three decimal places with $n = 6$ subintervals using:

a) the Midpoint Rule

$$\Delta x = \frac{5 - (-1)}{6} = 1.$$

i	$\bar{x}_i$	$f(\bar{x}_i)$
1	−0.5	.707
2	0.5	1.414
3	1.5	2.828
4	2.5	5.656
5	3.5	11.314
6	4.5	22.627

The approximation is
$1(.707 + 1.414 + 2.828 + 5.656$
$\quad + 11.314 + 22.627) = 44.546.$

b) the Trapezoidal Rule

i	x_i	$f(x_i)$	term
0	−1	.5	.5
1	0	1	2(1) = 2
2	1	2	2(2) = 4
3	2	4	2(4) = 8
4	3	8	2(8) = 16
5	4	16	2(16) = 32
6	5	32	32

The approximation is

$\frac{1}{2}(.5 + 2 + 4 + 8 + 16 + 32 + 32)$
$= 47.250.$

c) Simpson's Rule

i	x_i	$f(x_i)$	term
0	−1	.5	.5
1	0	1	4(1) = 4
2	1	2	2(2) = 4
3	2	4	4(4) = 16
4	3	8	2(8) = 16
5	4	16	4(16) = 64
6	5	32	32

The approximation is

$\frac{1}{3}(.5 + 4 + 4 + 16 + 16 + 64 + 32)$
$= 45.500.$

$\left(\text{The actual value of } \int_{-1}^{5} 2^x \, dx \text{ is}\right.$
$\left.\frac{63}{\ln 4} \approx 45.445.\right)$

2) For $\int_{1}^{5} (40 - x) \, dx$ and any n, the Midpoint approximation will be:

a) too large

b) too small

c) exact

c) exact, because $y = 40 - x$ is linear. The area of the midpoint rectangle will equal the area under $y = 40 - x$.

3) For $\int_1^5 (40 - x^2)\, dx$ and any n, the Trapezoidal Rule approximation will be:

   a) too large

   b) too small

   c) exact

b) too small. Since the graph of $y = 40 - x^2$ is concave downward for $1 \le x \le 5$, all inscribed trapezoids will have area less than the area under $y = 40 - x^2$.

4) For $\int_1^5 (40 - x^2)\, dx$ and any even n, Simpson's Rule approximation will be:

   a) too large

   b) too small

   c) exact

c) exact, because Simpson's Rule replaces the function with portions of approximating parabolas and since $y = 40 - x^2$ is a parabola, the approximation is exact.

5) The distances across a small pond were measured at 10 m intervals and are given in the figure. Use the Trapezoidal Rule to estimate the area covered by the pond.

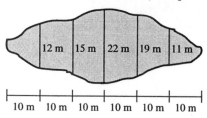

$\Delta x = 10$ m

i	dist	term
0	0	0
1	12	$2(12) = 24$
2	15	$2(15) = 30$
3	22	$2(22) = 44$
4	19	$2(19) = 38$
5	11	$2(11) = 22$
6	0	0

The pond area is approximately
$$\frac{10}{2}(0 + 24 + 30 + 44 + 38 + 22 + 0)$$
$$= 790 \text{ m}^2.$$

Section 7.7 *Approximate Integration* **249**

**Page
515**

B. The *error of estimation* is the difference between the actual value of $\int_a^b f(x)\, dx$ and the estimate. Here is a table that describes, under the conditions given, an upper bound on the error of estimation for the three methods in this sections.

Rule	Condition for all $x \in [a, b]$	Maximum Error of Estimation
Midpoint	$|f''(x)| \le M$	$\dfrac{M(b-a)^3}{24n^2}$
Trapezoidal	$|f''(x)| \le M$	$\dfrac{M(b-a)^3}{12n^2}$
Simpson's	$|f^{(4)}(x)| \le M$	$\dfrac{M(b-a)^5}{180n^4}$

6) Find the maximum error in estimating $\int_{-1}^{3} x^3\, dx$ with $n = 8$ using the Trapezoidal Rule.

The maximum is $\dfrac{M(3-(-1))^3}{12(8)^2} = \dfrac{M}{12}$.

To find a value for M we note $f''(x) = 6x$.

For $-1 \le x \le 3, |f''(x)| = |6x| \le 6 \cdot 3 = 18$.

We choose $M = 18$. Thus the maximum error of estimation is $\dfrac{M}{12} = \dfrac{18}{12} = 1.5$.

7) Find the maximum error in estimating $\int_{1}^{3} \sin 2x\, dx$ with $n = 4$ using Simpson's Rule.

$\dfrac{M(3-1)^3}{180(4)^4} = \dfrac{M}{1440}$.

For $f(x) = \sin 2x, f^{(4)}(x) = 16 \sin 2x$. For $1 \le x \le 3, |\sin 2x| \le 1$, so $|f^{(4)}(x)| \le 16$.

We choose $M = 16$. Thus the maximum error is $\dfrac{M}{1440} = \dfrac{16}{1440} = \dfrac{1}{90}$.

Section 7.8   Improper Integrals

We will see in this section that it is possible for a region to be "infinitely long" and still have a finite area. This section covers improper integrals $\int_a^b f(x)\,dx$ where either a or b is infinity (or $-\infty$) or the integrand grows infinitely large somewhere between a and b. These are evaluated as limits of ordinary definite integrals.

Concepts to Master

A. Definition of improper integrals; Type I and Type II integrals; convergence; divergence

B. Comparison Test for improper integrals

Summary and Focus Questions

Page 523

A. Improper integrals may take several forms. They may be of Type I where there is an infinite limit of intergration, or they may be of Type II and have a "trouble point" where the function is not defined and grows infinitely large or small. Improper integrals are defined and evaluated as limits. The various Type I and Type II improper integrals are defined in this table:

Definition	Necessary Condition	Example
Type 1		
$\int_a^\infty f(x)\,dx = \lim_{t\to\infty}\int_a^t f(x)\,dx$	$\int_a^t f(x)\,dx$ exists for all $t \geq a$	$\int_1^\infty e^{-x}\,dx$
$\int_{-\infty}^b f(x)\,dx = \lim_{t\to-\infty}\int_t^b f(x)\,dx$	$\int_t^b f(x)\,dx$ exists for all $t \leq b$	$\int_{-\infty}^0 \sin x\,dx$
$\int_{-\infty}^\infty f(x)\,dx = \int_{-\infty}^a f(x)\,dx + \int_a^\infty f(x)\,dx$	Both integrals on the right exist	$\int_{-\infty}^\infty x^2\,dx$
Type 2		
$\int_a^b f(x)\,dx = \lim_{t\to a+}\int_t^b f(x)\,dx$	f is continuous on $(a, b]$ and not continuous at a; $\int_t^b f(x)\,dx$ exists for $t \in (a, b]$	$\int_0^1 \ln x\,dx$
$\int_a^b f(x)\,dx = \lim_{t\to b-}\int_a^t f(x)\,dx$	f is continuous on $[a, b)$ and not continuous at b; $\int_a^t f(x)\,dx$ exists for all $t \in [a, b)$	$\int_0^1 \frac{1}{1-x}\,dx$
$\int_a^b f(x)\,dx = \int_a^c f(x)\,dx + \int_c^b f(x)\,dx$	f is continuous on $[a, c)$ and $(c, b]$ and not continuous at c; Both improper integrals on the right exist	$\int_{-1}^1 \frac{1}{x^2}\,dx$

If an improper integral exists, it is said to *converge;* otherwise it *diverges.*
Improper integrals are evaluated as limits according to their definitions.

1) Give a definition of each:

a) $\displaystyle\int_0^\infty \sin 2x \, dx$

$\displaystyle\lim_{t\to\infty} \int_0^t \sin 2x \, dx$

b) $\displaystyle\int_0^1 \frac{1}{\sqrt{x}} \, dx$

$\displaystyle\lim_{t\to 0^+} \int_t^1 \frac{1}{\sqrt{x}} \, dx$

c) $\displaystyle\int_{-\infty}^\infty e^{-x^2} \, dx$

$\displaystyle\int_{-\infty}^0 e^{-x^2} \, dx + \int_0^\infty e^{-x^2} \, dx$ (The choice of zero was arbitrary; any real number would do.)

d) $\displaystyle\int_{-2}^2 \frac{1}{(x-1)^2} \, dx$

$\displaystyle\int_{-2}^1 \frac{1}{(x-1)^2} \, dx + \int_1^2 \frac{1}{(x-1)^2} \, dx$

e) $\displaystyle\int_1^\infty \frac{1}{1-x} \, dx$

This one is both types and should be rewritten as

$$\int_1^2 \frac{1}{1-x} \, dx + \int_2^\infty \frac{1}{1-x} \, dx$$

$$= \lim_{t\to 1^+} \int_t^2 \frac{1}{1-x} \, dx + \lim_{t\to\infty} \int_2^t \frac{1}{1-x} \, dx.$$

2) Make a table of values for $\displaystyle\int_t^1 \frac{1}{\sqrt{x}} \, dx$ for
$t = .5, .05, .005, .0005$ and $.00005$. Does it
appear that $\displaystyle\int_0^1 \frac{1}{\sqrt{x}} \, dx$ converges?

t	$\displaystyle\int_t^1 \frac{1}{\sqrt{x}}\, dx$
.5	.586
.05	1.553
.005	1.859
.0005	1.955
.00005	1.986

$\displaystyle\int_0^1 \frac{1}{\sqrt{x}} \, dx$ appears to converge to 2. (It does, because $\displaystyle\int_t^1 \frac{1}{\sqrt{x}} \, dx = 2 - 2\sqrt{t}$. As $t\to 0$, this approaches 2.)

3) Evaluate $\int_0^{\pi/2} \sec^2 x\, dx$.

$$\int_0^{\pi/2} \sec^2 x\, dx = \lim_{t \to \pi/2^-} \int_0^t \sec^2 x\, dx$$
$$= \lim_{t \to \pi/2^-} \tan x\Big|_0^t = \lim_{t \to \pi/2^-} \tan t$$

which does not exist (the limit is ∞). This integral diverges.

4) Evaluate $\int_1^{\infty} xe^{-x^2} dx$.

$$\int_1^{\infty} xe^{-x^2}\, dx = \lim_{t \to \infty} \int_1^t xe^{-x^2}\, dx$$

$$(u = -x^2,\ du = -2x\, dx)$$

$$= \lim_{t \to \infty}\left(-\tfrac{1}{2}e^{-x^2}\Big|_1^t\right)$$

$$= \lim_{t \to \infty}\left(-\tfrac{1}{2}(e^{-t^2} - e^1)\right)$$

$$= -\tfrac{1}{2}(0 - e) = \tfrac{e}{2}.$$

5) a) Find the area of the shaded region.

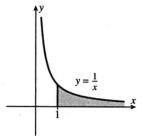

$$\int_1^{\infty} \tfrac{1}{x}\, dx = \lim_{t \to \infty} \int_1^t \tfrac{1}{x}\, dx = \lim_{t \to \infty} \ln t\Big|_1^t$$
$$= \lim_{t \to \infty} \ln t = \infty.$$

The improper integral does not exist; the area is infinite.

b) Evaluate $\int_1^{\infty} \dfrac{\pi}{x^2}\, dx$.

$$\int_1^{\infty} \tfrac{\pi}{x^2}\, dx = \lim_{t \to \infty} \int_1^t \tfrac{\pi}{x^2}\, dx = \lim_{t \to \infty} \tfrac{-\pi}{x}\Big|_1^t$$
$$= \lim_{t \to \infty}\left(\tfrac{-\pi}{t} + \tfrac{\pi}{1}\right) = \pi.$$

c) (For fun) The region in part a) has an infinite area. How can it be painted with a finite amount of paint?

Step 1: Take the region, $0 \le y \le \tfrac{1}{x}, 1 \le x \le \infty$, and revolve it about the x-axis.

Step 2: The volume of the resulting solid of revolution is given by $\int_1^{\infty} \tfrac{\pi}{x^2}\, dx$, which by part b) is π.

Step 3: Fill the solid of revolution with π gallons of paint.

Step 4: Dip the region in part a) in the solid of revolution. The region is then painted with at most π gallons of paint!

Page
529

B. Sometimes improper integrals can be shown to converge without actually determining the exact value to which it converges. (It is useful information to know an improper integral exists before it will be approximated, for instance.) One such method to determine existence is the *Comparison Test for Improper Integrals:*

Suppose $f(x)$ and $g(x)$ are continuous and $f(x) \geq g(x) \geq 0$ for $x \geq a$. If $\int_a^\infty f(x)\, dx$ converges then $\int_a^\infty g(x)\, dx$ converges.

Similar statements can be made for the other forms of improper integrals.

6) True or False:

If $f(x) \geq g(x) \geq 0$ for $x \geq a$ and f and g are continuous, then if $\int_a^b g(x)\, dx$ diverges, then $\int_a^\infty f(x)\, dx$ diverges.

True. This statement is just the contrapositive of the Comparison Test for Improper Integrals.

7) Show that $\int_1^\infty \dfrac{|\sin x|}{x^2}\, dx$ converges.

For $1 \leq x \leq \infty, 0 \leq |\sin x| \leq 1$. Thus

$0 \leq \dfrac{|\sin x|}{x^2} \leq \dfrac{1}{x^2}$. $\int_1^\infty \dfrac{1}{x^2}\, dx$ converges (to 1).

Thus $\int_1^\infty \dfrac{|\sin x|}{x^2}\, dx$ converges, (but we don't

know to what value, except that it will be

between 0 and 1.)

Technology Plus for Chapter 7

1) Use a CAS to find the partial fraction decomposition of
$$\frac{12x^3}{x^6 + x^5 + x^4 - x^2 - x - 1}.$$

2) Use a CAS to find each.

a) $\displaystyle\int \frac{\sqrt{4x^2 - 1}}{x^2}\,dx$

b) $\displaystyle\int \frac{e^x}{x}\,dx$

3) Use a spreadsheet or calculator to estimate $\displaystyle\int_0^5 (25x - x^3)\,dx$ by

a) the Midpoint Rule with $n = 20$.

Using a Texas Instruments TI-92, the expansion is

$$\frac{-4x - 8}{x^2 + x + 1} + \frac{6}{x^2 + 1} + \frac{3}{x + 1} + \frac{1}{x - 1}.$$

$$\frac{2x\ln\left(\sqrt{4x^2 - 1} + 2x\right) - \sqrt{4x^2 - 1}}{x} + C$$

The CAS will probably return something like "$\int \frac{e^x}{x}\,dx$," which means the integral is not elementary.

$$\Delta x = \frac{5 - 0}{20} = 0.25$$

i	$\overline{x}_i$	$f(\overline{x}_i)$
1	0.125	3.123
2	0.375	9.322
3	0.625	15.381
4	0.875	21.205
5	1.125	26.701
6	1.375	31.775
7	1.625	36.334
8	1.875	40.283
9	2.125	43.529
10	2.375	45.979
11	2.625	47.537
12	2.875	48.111
13	3.125	47.607
14	3.375	45.979
15	3.625	42.990
16	3.875	38.689
17	4.125	32.936
18	4.375	25.635
19	4.625	16.693
20	4.875	6.018
		625.781

$$\sum_{i=1}^{20} f(\overline{x}_i) \approx 625.781$$

The integral is approximately
$(0.25)(625.781) = 156.445.$

b) the Trapezoidal Rule with $n = 20$.

$$\Delta x = \frac{5 - 0}{20} = 0.25$$

i	x_i	$f(x_i)$	term
0	0.00	0.000	0.000
1	0.25	6.234	12.469
2	0.50	12.375	24.750
3	0.75	18.328	36.656
4	1.00	24.000	48.000
5	1.25	29.297	58.594
6	1.50	34.125	68.250
7	1.75	38.391	76.781
8	2.00	42.000	84.000
9	2.25	44.859	89.719
10	2.50	46.875	93.750
11	2.75	47.953	95.906
12	3.00	48.000	96.000
13	3.25	46.922	93.844
14	3.50	44.625	89.250
15	3.75	41.016	82.031
16	4.00	36.000	72.000
17	4.25	29.484	58.969
18	4.50	21.375	42.750
19	4.75	11.578	23.156
20	5.00	0.000	0.000
			1246.875

The integral is approximately
$\frac{1}{2}(0.25)(1246.875) = 155.859$.

c) Simpson's Rule with $n = 20$.

$$\Delta x = \frac{5 - 0}{20} = 0.25$$

i	x_i	$f(x_i)$	term
0	0.00	0.000	0.000
1	0.25	6.234	12.469
2	0.50	12.375	49.500
3	0.75	18.328	36.656
4	1.00	24.000	96.000
5	1.25	29.297	58.594
6	1.50	34.125	136.500
7	1.75	38.391	76.781
8	2.00	42.000	168.000
9	2.25	44.859	89.719
10	2.50	46.875	187.500
11	2.75	47.953	95.906
12	3.00	48.000	192.000
13	3.25	46.922	93.844
14	3.50	44.625	178.500
15	3.75	41.016	82.031
16	4.00	36.000	144.000
17	4.25	29.484	58.969
18	4.50	21.375	85.500
19	4.75	11.578	23.156
20	5.00	0.000	0.000
			1865.625

The integral is approximately
$\left(\frac{1}{3}\right)(0.25)(1865.625) = 155.469$.

Chapter 8 — Further Applications of Integration

"THAT WRAPS IT UP — THE MASS OF THE UNIVERSE."

© 1999 by Sidney Harris.

Section 8.1   Arc Length

The section uses integrals to calculate the length of a curve (called arc length). As with other integral applications, arc length will be the limit of the sums of approximations of the lengths of small pieces of the curve.

Concepts to Master

A. Length of a curve
B. Arc length function and its derivative

Summary and Focus Questions

**Page
541**

A. If f' is continuous on $[a, b]$, the *length of the curve $y = f(x)$, $a \le x \le b$* (arc length) is
$$L = \int_a^b \sqrt{1 + (f'(x))^2}\, dx.$$
For curves described as $x = g(y)$ for $c \le y \le d$, the arc length is
$$\int_c^d \sqrt{1 + (g'(y))^2}\, dy.$$

1) Find a definite integral for the length of
$y = \frac{1}{x}$, $1 \le x \le 3$.

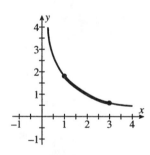

$y' = -\frac{1}{x^2}$, so
$$L = \int_1^3 \sqrt{1 + \left(\frac{-1}{x^2}\right)^2}\, dx = \int_1^3 \frac{\sqrt{x^4 + 1}}{x^2}\, dx.$$

2) Show that the circumference of a circle of radius r is $2\pi r$.

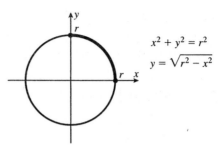

$$x^2 + y^2 = r^2$$
$$y = \sqrt{r^2 - x^2}$$

The circumference is 4 times the arc length of $y = \sqrt{r^2 - x^2}$ in the first quadrant, $0 \le x \le r$.

$y' = -\dfrac{x}{\sqrt{r^2 - x^2}}$, so the arc length is

$$\int_0^r \sqrt{1 + \left(\frac{-x}{\sqrt{r^2 - x^2}}\right)^2}\, dx$$

$$= \int_0^r \sqrt{1 + \frac{x^2}{r^2 - x^2}}\, dx = \int_0^r \frac{r}{\sqrt{r^2 - x^2}}\, dx.$$

We use the substitution $x = r \sin\theta$, $dx = r \cos\theta\, d\theta$,

$$\int \frac{r}{r \cos\theta} \cdot r \cos\theta\, d\theta = \int r\, d\theta$$

$$= r\theta = r \sin^{-1}\!\left(\frac{x}{r}\right)$$

Thus, $\displaystyle\int_0^r \frac{r}{\sqrt{r^2 - x^2}}\, dx = r \sin^{-1}\!\left(\frac{x}{r}\right)\Big]_0^r$

$$= r(\sin^{-1}(1) - \sin^{-1}(0)) = r\!\left(\frac{\pi}{2} - 0\right)$$

$$= \frac{r\pi}{2}.$$

The entire circle is $4\!\left(\dfrac{r\pi}{2}\right) = 2\pi r.$

3) What is the length of the arc given by
$x = 1 + y^{3/2}, 0 \leq y \leq 4$?

$$x' = \frac{3}{2}y^{1/2}$$

$$L = \int_0^4 \sqrt{1 + \left(\frac{3}{2}y^{1/2}\right)^2}\, dy$$

$$= \int_0^4 \sqrt{1 + \frac{9}{4}y}\, dy$$

$$\left(u = 1 + \frac{9}{4}y,\, du = \frac{9}{4}\, dy\right)$$

$$= \frac{4}{9}\int \sqrt{u}\, du = \frac{4}{9}\frac{u^{3/2}}{3/2} = \frac{8}{27}u^{3/2}$$

$$= \frac{8}{27}\left(1 + \frac{9}{4}y\right)^{3/2}\Big]_0^4$$

$$= \frac{8}{27}(10\sqrt{10} - 1).$$

4) Write an expression involving definite integrals for the length of the curve $y = \sqrt[3]{x}$ from $x = -1$ to $x = 8$.

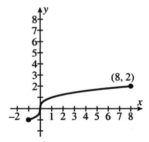

Since a vertical tangent exists at $x = 0$, the arc length expression $\int_{-1}^{8} \ldots\, dx$ would result in an improper integral. Therefore, we choose to treat x as a function of y.
From $y = \sqrt[3]{x}$, $x = y^3$ and $x' = 3y^2$.
The arc length is

$$\int_{-1}^{2} \sqrt{1 + (3y^2)^2}\, dy = \int_{-1}^{2} \sqrt{1 + 9y^4}\, dy.$$

B. For a smooth function $y = f(x)$, $a \le x \le b$, let $s(x)$ be the distance travelled along the curve from $(a, f(a))$ to $(x, f(x))$. The function $s(x)$ may be written as

$$s(x) = \int_a^x \sqrt{1 + [f'(t)]^2}\, dt.$$

The differential $ds = \sqrt{1 + \left(\frac{dy}{dx}\right)^2}\, dx$ may be used to express arc length:

$$L = \int ds.$$

This form is handy because, from $ds = \sqrt{1 + \left(\frac{dy}{dx}\right)^2}$ we can write

$$(ds)^2 = (dx)^2 + (dy)^2.$$

This last form can be used to remember *both* arc length formulas:

$$ds = \sqrt{1 + \left(\frac{dy}{dx}\right)^2}\, dx \text{ and } ds = \sqrt{1 + \left(\frac{dx}{dy}\right)^2}\, dy.$$

5) Find the arc length function for a particle moving along the curve $y^2 = 4x^3$ and starting at $(0, 0)$.

$$y^2 = 4x^3, y = 2x^{3/2}$$

$$\frac{dy}{dx} = 3x^{1/2}$$

$$s(x) = \int_0^x \sqrt{1 + (3t^{1/2})^2}\, dt$$

$$= \int_0^x \sqrt{1 + 9t}\, dt$$

$$(u = 1 + 9t, du = 9\, dt)$$

$$= \frac{1}{9}\frac{(1 + 9t)^{3/2}}{3/2} = \frac{2}{27}(1 + 9t)^{3/2}\Big|_0^x$$

$$= \frac{2}{27}(1 + 9x)^{3/2} - \frac{2}{27}.$$

6)

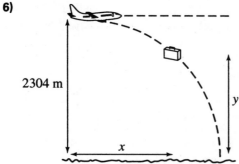

2304 m

y

x

A plane is flying at 2304 m when a suitcase falls from the cargo area. Because of air resistance, the suitcase trajectory is $y = 2304 - \frac{x^{3/2}}{6}$. Find the distance the suitcase travels along its path to the ground.

When the suitcase travels x meters horizontally, its height is $y = 2304 - \frac{x^{3/2}}{6}$.

$$y' = -\frac{3}{2}\frac{x^{1/2}}{6} = -\frac{\sqrt{x}}{4}$$

$$s(x) = \int_0^x \sqrt{1 + \left(-\frac{\sqrt{t}}{4}\right)^2}\, dt$$

$$= \frac{1}{4}\int_0^x \sqrt{16 + t}\, dt$$

$$(u = 16 + t,\ du = dt)$$

$$= \frac{1}{4}\int u^{1/2}\, du = \frac{1}{6}u^{3/2}$$

$$= \frac{1}{6}(16 + t)^{3/2}\Big|_0^x$$

$$= \frac{(16 + x)^{3/2} - 64}{6}.$$

When $y = 0$,

$$2304 - \frac{x^{3/2}}{6} = 0$$

$$x^{3/2} = 13824$$

$$x = 576$$

$$s(576) = \frac{(16 + 576)^{3/2} - 64}{6} \approx 2389.997 \text{ m}.$$

Section 8.2   Area of a Surface of Revolution

Objects such as a football or a glass bowl, which are the surfaces of solids of revolution, may have their surface areas calculated by a definite integral. As with other area applications, it is helpful to draw a representative figure for a typical term in the Riemann sum.

Concepts to Master

Calculate the surface area of a solid of revolution.

Summary and Focus Questions

Page 548

If $f(x) \geq 0$ and $f'(x)$ is continuous on $[a, b]$, the *surface area* of the solid of revolution obtained by rotating $f(x)$ about the x-axis is

$$S = \lim_{n \to \infty} \sum_{i=1}^{n} 2\pi f(x_i^*) \sqrt{1 + (f'(x_i^*))^2} \, \Delta x$$

$$= \int_a^b 2\pi f(x) \sqrt{1 + (f'(x))^2} \, dx.$$

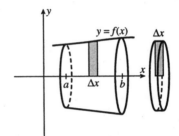

For curves described by $x = g(y)$, $c \leq y \leq d$, and rotated about the x-axis,

$$S = \int_c^d 2\pi y \sqrt{1 + \left(\frac{dx}{dy}\right)^2} \, dy.$$

Recalling that $(ds)^2 = (dx)^2 + (dy)^2$, more compact notation may be used which represents both the above cases of rotation about the x-axis.

$$S = \int 2\pi y \, ds, \text{ where } ds = \sqrt{1 + \left(\frac{dy}{dx}\right)^2} \, dx \text{ or } ds = \sqrt{1 + \left(\frac{dx}{dy}\right)^2} \, dy$$

For rotation about the y-axis, $S = \int 2\pi x \, ds$.

1) Find the surface area of the solid obtained by revolving $y = \sin 2x$, $0 \le x \le \frac{\pi}{2}$, about the x-axis.

$$S = \int_0^{\pi/2} 2\pi \sin 2x \sqrt{1 + (2\cos 2x)^2}\, dx$$

$$\left(\begin{array}{l} u = 2\cos 2x, \; du = -4\sin 2x \\ \text{At } x = 0, \, u = 2; \, x = \frac{\pi}{2}, \, u = -2 \end{array} \right)$$

$$= -\frac{\pi}{2} \int_2^{-2} \sqrt{1 + u^2}\, du = \frac{\pi}{2} \int_{-2}^{2} \sqrt{1 + u^2}\, du$$

(Table of Integrals, #21)

$$= \frac{\pi}{2}\left(\frac{u}{2}\sqrt{1 + u^2} + \frac{1}{2}\ln\left|u + \sqrt{1 + u^2}\right| \right)\Big]_{-2}^{2}$$

$$= \frac{\pi}{2}\left(\frac{2}{2}\sqrt{5} + \frac{1}{2}\ln\left|2 + \sqrt{5}\right| \right.$$
$$\left. -\left(\frac{-2}{2}\sqrt{5} + \frac{1}{2}\ln\left| -2 + \sqrt{5}\right| \right) \right)$$

$$= \frac{\pi}{2}[2\sqrt{5} + \frac{1}{2}\ln(2 + \sqrt{5})$$
$$- \frac{1}{2}\ln(-2 + \sqrt{5})].$$

2) Set up two equivalent definite integrals for the surface area of the solid obtained by revolving $y = x^2$, $0 \le x \le 2$ about the y-axis.

In both solutions $s = \int 2\pi x\, ds$.

I. $ds = \sqrt{1 + \left(\frac{dy}{dx}\right)^2}\, dx$.

$y = x^2$ so $\frac{dy}{dx} = 2x$.

$$S = \int_0^2 2\pi x\, ds = \int_0^2 2\pi x \sqrt{1 + (2x)^2}\, dx$$

$$= \int_0^2 2\pi x \sqrt{1 + 4x^2}\, dx$$

II. $ds = \sqrt{1 + \left(\frac{dy}{dx}\right)^2}\, dy$.

Since $y = x^2$, $0 \le x \le 2$,

$x = \sqrt{y}$, $0 \le y \le 4$ and $\frac{dx}{dy} = \frac{1}{2\sqrt{y}}$.

$$S = \int_0^4 2\pi x\, ds$$

$$= \int_0^4 2\pi \sqrt{y} \sqrt{1 + \left(\frac{1}{2\sqrt{y}}\right)^2}\, dy$$

$$= \int_0^4 2\pi \sqrt{y} \sqrt{\frac{1 + 4y}{4y}}\, dy$$

$$= \int_0^4 \pi \sqrt{1 + 4y}\, dy.$$

The integral in part II is equal to the integral in part I since $y = x^2$, $dy = 2x\, dx$, and for $0 \le x \le 2$, we have $0 \le y \le 4$.

Section 8.3   Applications to Physics and Engineering

This section gives additional applications of definite integrals, including hydrostatic pressure and the total force on an object immersed in a liquid to a given depth and centers of mass (balance points) of two dimensional objects.

Concepts to Master

A. Calculate the force on a vertical plane due to hydrostatic pressure; the pressure at a given depth

B. Moments on inertia about the x-axis and y-axis; centroid (center of mass)

C. Theorem of Pappus

Summary and Focus Questions

Page 555

A. The pressure on a plate suspended horizontally in a liquid with density ρ at a depth d is

$$P = \rho g d,$$

where g is the gravitational constant. P is in units of newtons per meter^2 (= pascals) or in pounds per ft^2.

Suppose a plate is suspended vertically in a liquid with mass density ρ and the total depth of the liquid is H. Suppose the surface of the plate can be described as bounded by $x = f(y)$, $x = k(y)$, $y = c$, $y = d$ with $f(y) \geq k(y)$ for all $y \in [c, d]$. Then the force due to liquid pressure on a section of the plate is approximately

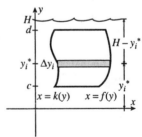

(density)(gravitational constant)(depth)(area of section)
$= \rho g(H - y_i^*)[f(y_i^*) - k(y_i^*)]\Delta y_i$

Thus the *total hydrostatic force on the surface is* $\displaystyle\int_c^d \rho g(H - y)(f(y) - k(y))\, dy$.

1) A circular plate of radius 3 cm is suspended horizontally at a depth of 12 m in an oil having density 1500 kg/m^3. What is the pressure on the plate?

Since the plate is horizontal, the pressure is uniform at all points.
$P = \rho g d = (1500 \text{ kg/m}^3)(9.8 \text{ m/s}^2)(12 \text{ m})$
$= 176{,}400 \text{ Pa (pascals)} = 176.4 \text{ kPa}.$

2) A flat isosceles triangle is suspended in water as in the figure. Water density is 1000 kg/m³. Set up a definite integral for the total hydrostatic force.

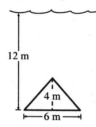

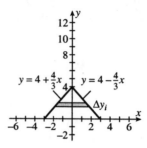

Setting up coordinates as shown, the sides of the triangle are

$$y = 4 - \tfrac{4}{3}x, \text{ so } x = -\tfrac{3}{4}y + 3, \text{ and}$$
$$y = 4 + \tfrac{4}{3}x, \text{ so } x = \tfrac{3}{4}y - 3.$$

The force on the triangle is

$$\int_0^4 1000(9.8)(12 - y)\left[\left(-\tfrac{3}{4}y + 3\right) - \left(\tfrac{3}{4}y - 3\right)\right] dy$$
$$= \int_0^4 9800(12 - y)\left(6 - \tfrac{3}{2}y\right) dy.$$

Page 557

B. A lamina is a thin flat sheet of material determined by a region R, such as in the figure, whose density is uniformly ρ, a constant.

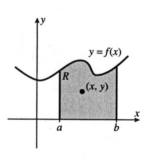

Let $A = \int_a^b f(x)\, dx$ be the area of R. The *centroid* (or *center of mass*) of R is the point $(\bar{x}, \bar{y})$ where

$$\bar{x} = \tfrac{1}{A}\int_a^b x f(x)\, dx \text{ and}$$

$$\bar{y} = \tfrac{1}{A}\int_a^b \tfrac{1}{2}[f(x)]^2\, dx.$$

If the lamina was suspended in the air by a string attached at the centroid $(\bar{x}, \bar{y})$, then it would balance.

The *moment of R about the x-axis* is $M_x = \rho\int_a^b \tfrac{1}{2}[f(x)]^2\, dx$ and the *moment about the y-axis* is $M_y = \rho\int_a^b x f(x)\, dx$. These are measures of the tendency of the lamina to rotate about the x-axis and y-axis, respectively. Note that $\bar{x} = \dfrac{M_y}{\rho A}$ and $\bar{y} = \dfrac{M_x}{\rho A}$.

3) Find M_x, M_y, and the center of mass of the region bounded by $y = 4 - x^2$, $x = 0$, $y = 0$, with density ρ.

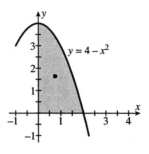

The area of the region is
$$A = \int_0^2 (4 - x^2)\, dx = 4x - \frac{x^3}{3}\Big|_0^2 = \frac{16}{3}.$$
For M_y and $\bar{x}$:

$$\int_a^b x f(x)\, dx = \int_0^2 x(4 - x^2)\, dx$$
$$= \int_0^2 (4x - x^3)\, dx$$
$$= \left(2x^2 - \frac{x^4}{4}\right)\Big|_0^2 = 4.$$

Therefore, $M_y = 4\rho$ and
$$\bar{x} = \frac{1}{16/3\rho}(4\rho) = \frac{3}{4}.$$
For M_x and $\bar{y}$:

$$\int_a^b \frac{1}{2}[f(x)]^2\, dx = \frac{1}{2}\int_0^2 (4 - x^2)^2\, dx$$
$$= \frac{1}{2}\int_0^2 (16 - 8x^2 + x^4)\, dx$$
$$= \frac{1}{2}\left(16x - \frac{8}{3}x^3 + \frac{x^5}{5}\right)\Big|_0^2$$
$$= \frac{128}{15}.$$

Therefore $M_x = \frac{128}{15}\rho$ and
$$\bar{y} = \frac{1}{16/3\rho}\left(\frac{128}{15}\rho\right) = \frac{8}{5}.$$

**Page
562**

C. *The Theorem of Pappus:* If a region R is revolved about the line L that does not intersect R, the resulting solid has volume $V = 2\pi dA$, where A is the area of R and d is the perpendicular distance between the centroid of R and the line L. (Notice that $2\pi d$ is the distance traveled by the centroid.)

Be careful that you do not interpret dA as a differential; in this case dA means the product of the distance and the area.

4) Find the volume of the solid of revolution obtained by rotating the region in question 1 about:

a) the x-axis.

$\bar{y} = \frac{8}{5}$ is the distance from the center of mass to the axis of revolution (the x-axis). The area is $A = \frac{16}{3}$. Thus

$$V = 2\pi\left(\frac{8}{3}\right)\left(\frac{16}{3}\right) = \frac{256\pi}{9}.$$

b) the y-axis

$\bar{x} = \frac{3}{4}$, so $V = 2\pi\left(\frac{3}{4}\right)\left(\frac{16}{3}\right) = 8\pi.$

Section 8.4   Applications to Economics and Biology

This section, like the previous ones, gives additional applications of definite integrals, including consumer surplus (the total amount of money that consumers save if they buy at a certain price) and various models of blood flow.

Concepts to Master

Determining the total amount or total change of a quantity in economic and biological applications

Summary and Focus Questions

Page 564

Applications of definite integrals in this section and elsewhere all involve calculating the total amount of a quantity (that can be represented by a continuous $y = f(x)$) within a certain range ($a \le x \le b$). The idea of summing the values (Riemann sums) is extended to the continuous case (limit) to give the definite integral ($\int_a^b f(x)\, dx$). Here are two examples, one from economics and one from biology.

1. A demand function $p(x)$ is the price necessary to sell x items. Let X be the amount actually available and P be the price (constant) for that amount, $P = p(X)$. The *consumer's surplus* is

$$\int_0^X [p(x) - P]\, dx.$$

   This is the total amount of money that could be saved by consumers if they buy when the price is P dollars.

2. The *cardiac output* of the heart is the rate of flow of blood through the heart per unit time. We calculate this by injecting dye into the heart and measuring the concentration levels $c(t)$ over some time interval $[0, T]$. If A is the amount of dye injected then the ratio of the dye to the total of the concentrations is the cardiac output:

$$\frac{A}{\int_0^T c(t)\, dt}$$

1) Find a definite integral for the consumer surplus determined by 1000 units available and a demand function of $p(x) = 85 - 0.03x$.

At $X = 1000$,
$P = 85 - 0.03(1000) = 55$.
The consumer surplus is

$$\int_0^{1000} [85 - 0.03x - 55]\,dx$$

$$= \int_0^{1000} (30 - 0.03x)\ dx.$$

2) The number of black fly hatchlings as measured on the Marquette Bay during the month of June is increasing at an estimated rate of $1000 + e^{1.1t}$ per day (t is measured in days). How much does the fly population increase between June 5 and June 12?

If we integrate the rate of change of the population from day 5 to day 12, that will calculate the total change in the population.

$$\int_5^{12} (1000 + e^{1.1t})\ dt$$

$$= 1000t + \frac{e^{1.1t}}{1.1}\Big]_5^{12} \approx 498{,}018 \text{ flys.}$$

Section 8.5   Probability

This section shows how calculus is used to define and calculate some basic probability concepts. The probability of an event is a number between 0 and 1 that represents the likelihood that the event will occur. We can represent probability as an area under a portion of a curve where the area under the entire curve is one. Areas, of course, are determined by definite integrals.

Concepts to Master

A. Continuous random variable, probability density function; the probability of an event

B. The mean (average value) of a random variable

C. Normal distributions

Summary and Focus Questions

Page 569

A. A *continuous random variable* is a variable whose values range over an entire interval. For example, the distances that an Olympic athlete can throw a javelin may be somewhere between 60 and 90 meters.

A *probability density function* for a random variable X is a non-negative function $y = f(x)$, with total area under the curve equal to one.
The *probability* that X realizes a value somewhere between a and b is

$$P(a \le X \le b) = \int_a^b f(x)\ dx.$$

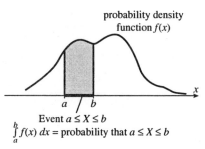

probability density function $f(x)$

Event $a \le X \le b$
$\int_a^b f(x)\ dx$ = probability that $a \le X \le b$

1) For any probability density function, what is $\int_{-\infty}^{\infty} f(x)\ dx$?

1; this is the area under the entire curve.

2) The time it takes before a drug begins being absorbed into the bloodstream may be modeled with the exponential probability density function

$$f(x) = \begin{cases} 0, & \text{for } x \leq 0 \\ 2e^{-2x}, & \text{for } x > 0. \end{cases}$$

a) What does $\int_{-\infty}^{10} f(x)\, dx$ represent?

The probability that it takes up to 10 minutes for the blood to begin absorbing the drug.

b) Find the probability that the drug takes at least 1 minute before starting to be absorbed.

$$P(1 \leq X < \infty) = \int_{1}^{\infty} 2e^{-2x}\, dx$$
$$= \lim_{t \to \infty} \int_{1}^{t} 2e^{-2x}\, dx$$
$$= \lim_{t \to \infty} (-e^{-2x}) \Big]_{1}^{t}$$
$$= \lim_{t \to \infty} (-e^{-2t} + e^{-2}) = e^{-2}.$$

B. The *mean* of a random variable X with probability density $y = f(x)$ is

$$\mu = \int_{-\infty}^{\infty} x f(x)\, dx.$$

μ is the average value of X in the long run.

Page
571

3) Show that the graph below is a probability density function and calculate the mean.

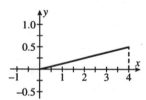

The function is $y = \frac{1}{8}x$ for $0 \leq x \leq 4$. The area under the curve is a triangle with value $= \frac{1}{2}\left(\frac{1}{2}\right)(4) = 1$. Thus this is a probability density function. The mean is

$$\int_{0}^{4} x\left(\frac{1}{8}x\right) dx = \frac{1}{8}\int_{0}^{4} x^2\, dx = \frac{x^3}{24}\Big]_{0}^{4} = \frac{8}{3}.$$

C. A *normal* probability distribution with mean μ and standard deviation σ for a random variable X has a probability density function of the form

$$f(x) = \frac{1}{\sigma\sqrt{2\pi}}\, e^{-(x-\mu)^2/(2\sigma^2)}$$

The graph is a bell shaped curve, symmetric about the line $y = \mu$. The standard deviation σ determines how spread out the shape of the bell curve will be. The larger the value of σ, the flatter the bell curve will be.

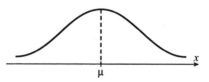

Example: If it has been determined that the length of life of an Acme 100-watt bulb is normally distributed with mean 160 hours and standard deviation 20 hours, what is the probability that an Acme 100-watt bulb will last at least 200 hours?

$$P(200 \le X \le \infty) = \int_{200}^{\infty} \frac{1}{20\sqrt{2\pi}} e^{-(x-160)^2/(2(20)^2)}\, dx$$

$$= \frac{1}{20\sqrt{2\pi}} \int_{200}^{\infty} e^{-(x-160)^2/2(20)^2)}\, dx.$$

This integral will need to be approximated using a calculator, computer or one of the approximation techniques.

$$\frac{1}{20\sqrt{2\pi}} \int_{200}^{\infty} e^{-(x-160)^2/(2(20)^2)}\, dx \approx \frac{1}{20\sqrt{2\pi}}(2.506) \approx 0.028.$$

About 2.8% of all Acme bulbs last at least 200 hours.

4) Suppose that the length of time it takes to complete an income tax form is normally distributed with mean 50 minutes and standard deviation 20 minutes. Write definite integral expressions for:

a) the probability that the form can be completed in 10 minutes or less.

$$\int_{0}^{10} \frac{1}{20\sqrt{2\pi}} e^{-((x-50)^2/2(20)^2)}\, dx.$$

(We may use 0 instead of $-\infty$ because completion times are non-negative.)

b) the probability that the form is completed in between 30 and 60 minutes.

$$\int_{30}^{60} \frac{1}{20\sqrt{2\pi}} e^{-((x-50)^2/2(20)^2)}\, dx.$$

Technology Plus for Chapter 8

1) Use a CAS to evaluate the integral for the arc length of $y = x - \cos x$, $-1 \leq x \leq 1$.

$y' = 1 + \sin x$

The length of the curve is

$$\int_{-1}^{1} \sqrt{1 + (1 + \sin x)^2} \, dx \approx 2.932.$$

2) Find the surface area of the solid obtained by rotating $y = \tan x$ about the x-axis, $0 \leq x \leq \frac{\pi}{3}$. Use a CAS to evaluate the integral.

The surface area is

$$\int_{0}^{\pi/3} 2\pi \, (\tan x) \, \sqrt{1 + (\sec^2 x)^2} \, dx \approx 10.502.$$

3) The "Empirical Rule" says that for a normal distribution, the probability that the random variable lies within one standard deviation of the mean is about 68%.

a) State the Empirical Rule in terms of a definite integral.

$$\int_{-\sigma}^{\sigma} \frac{1}{\sigma\sqrt{2\pi}} \, e^{-(x-\mu)^2/2\sigma^2} \, dx \approx 68\%.$$

b) Use a CAS to verify this for $\mu = 0$ and $\sigma = 1$.

$$\int_{-1}^{1} \frac{1}{\sqrt{2\pi}} \, e^{-x^2/2} \, dx \approx 0.68269.$$

Chapter 9 — Differential Equations

"DOES THIS APPLY ALWAYS, SOMETIMES, OR NEVER?"

© 1999 by Sidney Harris.

Section 9.1 Modeling with Differential Equations

A differential equation is an equation that relates an unknown function and one or more of its derivatives. This section defines the basic terminology for differential equations and looks at several phenomena that are modeled by differential equations. Right now we do not know how to solve very many kinds of differential equations; that will come in later sections.

Concepts to Master

A. Differential equation; order; degree; family of solutions to a differential equation

B. Initial condition(s); solution to an initial value problem

Summary and Focus Questions

Page 581

A. A *differential equation* is an equation involving x, y, y', y'', ..., $y^{(n)}$. The n in the highest $y^{(n)}$ is the *order* of the equation and the *degree* is the exponent that $y^{(n)}$ has. For example, $2(y''')^4 + 5xy' + 7x = 0$ has degree 4 and order 3.

We shall see that differential equations can be used to model various population changes and predator-prey relationships. For example, two models for population growth/decay are

$\dfrac{dP}{dt} = kP$ (exponential growth—grows ever faster with time), and

$\dfrac{dP}{dt} = kP\left(1 - \dfrac{P}{K}\right)$ (logistic growth—over time approaches a constant population K.)

A *particular solution* to a differential equation is a function $y = f(x)$ that satisfies the differential equation. A general solution is an expression with arbitrary constants that represents the family of all particular solutions.

We know how to solve differential equations of the form $y' = f(x)$ by integration, $y = \int f(x)\, dx$. In future sections we will solve several other types.

1) The equation $x^2(y'')^3 + 4xy' - 2y + x = 0$
has degree _____ and order _____.

degree 3, order 2

2) Is $y = e^{-2t}$ a solution to $y'' - 2y' - 8y = 0$?

Yes. $y = e^{-2t}$, $y' = -2e^{-2t}$, $y'' = 4e^{-2t}$.
Thus $y'' - 2y' - 8y$
$$= 4e^{-2t} - 2(-2e^{-2t}) - 8(e^{-2t}) = 0.$$

3) Guess a solution to $y'' = -9y$.

From your knowledge of trigonometric
functions, you should see that one solution
is $y = \sin 3x$. (There are many others.)

B. *Initial boundary conditions* are given values that are used to determine a
particular solution from the general. For example, the equation $y' = 10x$ has
general solution $y = 5x^2 + C$. If we specify an initial condition of $y(1) = 12$,
then $12 = 5 + C$ implies $C = 7$ so the particular solution for the condition
$y(1) = 12$ is $y = 5x^2 + 7$.

**Page
585**

4) The general solution to $y' = 2xy$ is
$y = Ce^{x^2}$ where $C > 0$. Find a solution to
the initial-value problem $y' = 2xy$,
$y(1) = 3$.

$y = Ce^{x^2}$. At $x = 1$, $3 = Ce^{1^2} = Ce$. Thus
$C = \frac{3}{e}$; the particular solution is
$y = \frac{3}{e}e^{x^2} = 3e^{x^2-1}$.

Section 9.2 Direction Fields and Euler's Method

Often it is impossible to find an explicit formula for the solution to a differential equation. Nevertheless, graphical and numeric approaches may be used to approximate solutions. In this section we obtain the general shapes of solutions using directional fields and estimate particular solutions using Euler's method.

Concepts to Master

A. Solution curve; directional fields

B. Euler's Method

Summary and Focus Questions

Page 586

A. The graph of a particular solution to a differential equation is called a *solution curve*. The general solution to a first order differential equation of the form

$$y' = F(x, y)$$

is a family of functions whose graphs are related. One way to visualize the general shape of solution curves is to draw a *direction field*—short line segments at various points (x, y) whose slope is $F(x, y)$. These line segments indicate the direction in which the curves proceed at each point.

For example, $y' = x + y - 1$ has this table of values for y' and direction field:

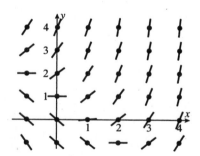

y'		y					
		−1	0	1	2	3	4
	−1	−3	−2	−1	0	1	2
	0	−2	−1	0	1	2	3
x	1	−1	0	1	2	3	4
	2	0	1	2	3	4	5
	3	1	2	3	4	5	6
	4	2	3	4	5	6	7

The solution curves are obtained by "connecting the dots" following the directions indicated by the slopes. (Later we will see that the solution is $y' = -x + Ce^x$.)

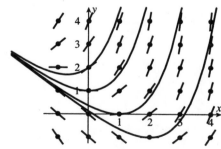

1) Find the directional field and sketch the general solution to $y' = xy + y$.

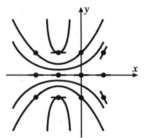

	y		
y'	-1	0	1
-2	1	0	-1
-1	0	0	0
x 0	-1	0	1
1	-2	0	2
3	-4	0	4

Later we will see the equation has general solution $y = Ce^{(x^2/2)+x}$.

B. Euler's method is a way to find approximations to initial value problems based on following tangent lines in direction fields.

Page
590

For a differential equation $\dfrac{dy}{dx} = F(x, y)$ with initial condition (x_0, y_0), Euler's

method constructs a table of (x, y) values that estimate the solution. The x_i values are evenly spaced out (a *step size* of h between successive x_i). Each y_{i+1} is calculated from the direction field at (x_i, y_i):

x	y
x_0	y_0 (given)
$x_1 = x_0 + h$	$y_1 = y_0 + hF(x_0, y_0)$
$x_2 = x_1 + h$	$y_2 = y_1 + hF(x_1, y_1)$
$x_3 = x_2 + h$	$y_3 = y_2 + hF(x_2, y_2)$
$\vdots$	
x_{n+1}	$y_{n+1} = y_n + hF(x_n, y_n)$

As you step from x_0 to x_1, to x_2, and so on, the y_i become less accurate. This can be partially offset by choosing h to be smaller.

2) Approximate five solution values to
$\dfrac{dy}{dx} = 2xy$, $y(1) = 3$ with step size 0.2.

$$h = 0.02$$

x	**y**
$x_0 = 1$	$y_0 = 3$
$x_1 = 1.2$	$y_1 = 3 + (0.2)(2(1)(3)) = 4.2$
$x_2 = 1.4$	$y_2 = 4.2 + (0.2)(2(1.2)(4.2)) = 6.216$
$x_3 = 1.6$	$y_3 = 9.696$
$x_4 = 1.8$	$y_4 = 15.903$
$x_5 = 2.0$	$y_5 = 27.353$

3) Repeat question 2), finding ten solutions with step size 0.1.

$$h = 0.1$$

x	**y**
$x_0 = 1$	$y_0 = 3.000$
$x_1 = 1.1$	$y_1 = 3.600$
$x_2 = 1.2$	$y_2 = 4.392$
$x_3 = 1.3$	$y_3 = 5.446$
$x_4 = 1.4$	$y_4 = 6.892$
$x_5 = 1.5$	$y_5 = 8.783$
$x_6 = 1.6$	$y_6 = 11.418$
$x_7 = 1.7$	$y_7 = 15.072$
$x_8 = 1.8$	$y_8 = 20.197$
$x_9 = 1.9$	$y_9 = 27.468$
$x_{10} = 2$	$y_{10} = 37.906$

4) The solution to $\dfrac{dy}{dx} = 2xy$, $y(1) = 3$ is
$y = 3e^{x^2-1}$. Which value for $x = 2$, in question 2) or 3), is the better approximation?

The exact value is $y = 3e^{2^2-1} \approx 60.257$. The value of 37.906 in question 3) is better than the estimate of 27.353 in question 2). Neither is very close because h is rather large and we have used several steps.

Section 9.3   Separable Equations

Some differential equations can be solved explicitly. This section shows you how to solve one particular first-order type (separable equations) where y' may be written as an expression involving only the variable x divided by an expression involving only the variable y.

Concepts to Master

Separable first-order differential equation; solution of a separable equation

Summary and Focus Questions

Page 595

A *separable first-order differential* equation has the form $\dfrac{dy}{dx} = \dfrac{g(x)}{h(y)}$.

To solve a separable equation:

1. Rewrite the equation as $h(y)\,dy = g(x)\,dx$.

2. Integrate to get $\int h(y)\,dy = \int g(x)\,dx$.

3. Solve, if possible, the results of step 2 for y.

Example:

The separable equation $y' = \dfrac{1}{3y^2}$ has general solution

$$\int 3y^2\,dy = \int 1\,dx$$
$$y^3 = x + C$$
$$y = \sqrt[3]{x + C}.$$

Some particular solutions are $y = \sqrt[3]{x + 1}$, $y = \sqrt[3]{x - 4}$, and so on. The family of solutions is pictured below:

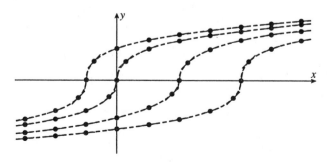

1) Is $y' = xy + x$ separable?

Yes. $\dfrac{dy}{dx} = x(y + 1) = \dfrac{x}{(y + 1)^{-1}}$.

2) Is $\dfrac{dy}{dx} = 2xy$ separable?

Yes. $\dfrac{dy}{dx} = 2xy$ may be rewritten as $\dfrac{dy}{y} = 2x\,dx$.

3) Sketch a direction field for $\dfrac{dy}{dx} = 2xy$.

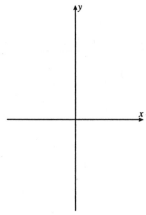

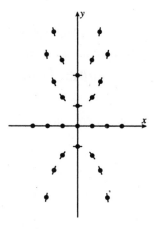

4) Solve $\dfrac{dy}{dx} = 2xy$ and draw three solution curves. (Compare your curves to the direction field in question 3.)

$\dfrac{dy}{dx} = 2xy$, so $\dfrac{dy}{y} = 2x\,dx$

$\displaystyle\int \dfrac{dy}{y} = \int 2x\,dx$.

$\ln |y| = x^2 + C$, so $|y| = e^{x^2 + C} = e^{x^2}e^C$.

$y = Ke^{x^2}$, K a real number.

Solution curves for $K = 1, 2$, and -1 are given in the following figure.

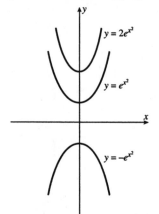

5) Solve the equation $y' = \dfrac{\sin x}{\cos y}$.

The equation is separable.

$\dfrac{dy}{dx} = \dfrac{\sin x}{\cos y}$

$\cos y \, dy = \sin x \, dx$

$\displaystyle\int \cos y \, dy = \int \sin x \, dx$

$\sin y = -\cos x + C$

$y = \sin^{-1}(-\cos x + C).$

6) Find the particular solution for the initial condition $y(0) = 2$ for the equation $e^{x^2}yy' + x = 0$.

First separate: $e^{x^2}y\dfrac{dy}{dx} = -x$

$y\dfrac{dy}{dx} = -xe^{-x^2}$

Integrate: $\displaystyle\int y \, dy = \int -xe^{-x^2} \, dx$

$\dfrac{y^2}{2} = \dfrac{1}{2}e^{-x^2} + C$

$y^2 = e^{-x^2} + C$

$y = (e^{-x^2} + C)^{1/2} = \sqrt{e^{-x^2} + C}$

Find the particular solution:

At $x = 0$, $y = 2$, so $\sqrt{e^{-0} + C} = 2$.

Thus $1 + C = 4$, $C = 3$.

The particular solution is $y = \sqrt{e^{-x^2} + 3}$.

7) Solve $x\dfrac{dy}{dx} = -y^2$ $(x > 0)$ with the initial condition of $x_0 = 1$, $y_0 = \dfrac{1}{3}$.

Separate variables: $-y^{-2} \, dy = x^{-1} \, dx$.

Integrate: $\displaystyle\int -y^{-2} \, dy = \int x^{-1} \, dx$.

$\dfrac{1}{y} = \ln|x| + C = \ln x + C.$

$y = \dfrac{1}{\ln x + C}$ is the general solution.

For $x_0 = 1$, $y_0 = \dfrac{1}{3}$, $\dfrac{1}{3} = \dfrac{1}{\ln 1 + C} = \dfrac{1}{C}$, so

$C = 3$. The particular solution is

$y = \dfrac{1}{\ln x + 3}.$

Section 9.4 Exponential Growth and Decay

This section is devoted to one particular differential equation, $\frac{dy}{dx} = ky$, where the rate of change of the quantity y is proportional to the size of y itself. So, for $k > 0$, the larger the value of y becomes, the greater the rate of change of y will be; in other words, exponential growth.

Concepts to Master

A. Solution of $\frac{dy}{dx} = ky$; applications to growth and decay

B. Continuously compounded interest

Summary and Focus Questions

Page 604

A. The solution to the differential equation $\frac{dy}{dt} = ky$ is $y(t) = y(0) \cdot e^{kt}$.

This solution may be used in problems where a quantity (y) of a substance varies with time in such a manner that the rate of change (y') of the quantity is proportional to y, that is, $y' = ky$, for some constant k. If $k > 0$ this is *natural growth; if $k < 0$, this is natural decay. k* is called the *relative growth rate.* Many growth and decay problems involving $y(t) = y(0)e^{kt}$ ask "given $y = y_1$ at time t_1 and $y = y_2$ at time t_2, find y at some third time t_3." The procedure for solving $y' = ky$ depends on whether or not one of the times given is $t = 0$:

If $t_1 = 0$, then we are given $y(0)$. Just solve $y_2 = y(0) \cdot e^{kt_2}$ for k.

If neither t_1 nor t_2 are zero, solve the equations $y_1 = y(0) \cdot e^{kt_1}$ and $y_2 = y(0) \cdot e^{kt_2}$ simultaneously for $y(0)$ and k. You first divide one equation by the other to eliminate the $y(0)$ term, and then solve for k.

Example: The half-life of radium is approximately 1600 years. What is the mass of a 10 mg sample of radium after 50 years?

At $t = 0$, $y = 10$ and at $t = 1600$, $y = 5$ (half the mass).

Thus $\quad y(t) = 10e^{kt}$

$$5 = 10e^{k(1600)}$$

$$\frac{1}{2} = e^{k(1600)}$$

$$k(1600) = \ln \frac{1}{2}$$

$$k = \frac{\ln 1/2}{1600}.$$

Thus $y(t) = 10e^{\frac{\ln 1/2}{1600} t} = 10(e^{\ln 1/2})^{t/1600} = 10\left(\frac{1}{2}\right)^{t/1600}$.

Finally, at $t = 50$, $y(50) = 10\left(\frac{1}{2}\right)^{50/1600} \approx 9.79$ mg.

1) A village had a population of 1000 in 1980 and 1200 in 1990. Assuming the population is experiencing natural growth, what will be the population in 2010?

Let 1980 be $t = 0$. Then $y(0) = 1000$. At $t_1 = 10$ (year 1990), $y = 1200$
so $1200 = 1000e^{k(10)}$
$1.2 = e^{10k}$
$10k = \ln 1.2$
$k = \frac{1}{10} \ln 1.2$
Thus $y = 1000e^{(1/10 \ln 1.2)t} = 1000(e^{\ln 1.2})^{t/10}$
$= 1000(1.2)^{t/10}$.
In the year 2010, $t = 30$, so
$y = 1000(1.2)^{30/10} = 1000(1.2)^3 \approx 1728$.

2) One hour after a bacteria culture was started there were 300 organisms and after 2 hours from the start there were 900. How many organisms will there be 4 hours after the start?

At $t = 1$, $y = 300$; $t = 2$, $y = 900$.
Thus $300 = y(0)e^{k \cdot 1}$
$900 = y(0)e^{k \cdot 2}$
Dividing the second equation by the first
$\frac{900}{300} = \frac{y(0)e^{2k}}{y(0)e^{k}}$; $3 = e^k$; $k = \ln 3$.
Now use $300 = y(0)e^{k \cdot 1}$ to solve for $y(0)$:
$300 = y(0)e^{\ln 3}$; $300 = y(0)3$; $y(0) = 100$.
The model is
$y = 100e^{\ln 3 t} = 100(e^{\ln 3})^t = 100 \cdot 3^t$.
So $y = 100 \cdot 3^t$.
At $t = 4$, $y = 100 \cdot 3^4 = 8100$.

3) A sample of an isotope intially weighs 90 mg. After 100 days 60 mg remain. After how many days will the sample be down to 20 mg in mass?

At $t_1 = 0$, $y(0) = 90$.
At $t_2 = 100$, $y = 60$, so $60 = 90e^{k(100)}$.
$\frac{2}{3} = e^{100k}$; $100k = \ln \frac{2}{3}$; $k = \frac{1}{100} \ln \frac{2}{3}$
Thus $y = 90e^{(1/100 \ln 2/3)t} = 90(e^{\ln 2/3})^{t/100}$
$= 90\left(\frac{2}{3}\right)^{t/100}$.
When will $y = 20$?
$20 = 90\left(\frac{2}{3}\right)^{t/100}$
$\frac{2}{9} = \left(\frac{2}{3}\right)^{t/100}$
$\frac{2}{9} = \left(\frac{2}{3}\right)^{t/100}$
$\ln\left(\frac{2}{9}\right) = \ln\left(\frac{2}{3}\right)^{t/100} = \frac{t}{100} \ln\left(\frac{2}{3}\right)$
$t = \dfrac{100 \ln \frac{2}{9}}{\ln \frac{2}{3}} \approx 371$ days.

B. If an initial amount A_0 is invested at an annual rate of r% compounded n times per year, then the amount accumulated after t years is

$$A(t) = A_0\left(1 + \frac{r}{n}\right)^{nt}.$$

If, instead, interest is *compounded continuously* (take the limit as $n \to \infty$) then

$$A(t) = A_0 e^{rt}.$$

4) Suppose $1000 is invested at 4%.

   a) Find the value of the investment after 5 years if interest is compounded quarterly.

$$A(5) = 1000\left(1 + \frac{.04}{4}\right)^{4(5)} \approx \$1220.19.$$

   b) Find the value after 5 years if interest is compounded continuously.

$$A(5) = 1000 e^{.04(5)} \approx \$1221.40.$$

   c) Write a differential equation that models compounding continuously at 4%; include an initial amount of $1000.

$\frac{dA}{dt} = .04A$ with initial condition
$A(0) = 1000.$

Section 9.5   The Logistic Equation

This section presents an alternative to the natural growth model of the previous section. It takes into account that often a population increases exponentially in its early stages and then levels off at some carrying capacity.

Concepts to Master

Solution of $\frac{dP}{dt} = kP\left(1 - \frac{P}{K}\right)$; applications to growth and decay

Summary and Focus Questions

Page 613

The *logistic differential equation* is $\frac{dP}{dt} = kP\left(1 - \frac{P}{K}\right)$.

The equation is separable and has solution

$$P(t) = \frac{K}{1 + Ae^{-kt}}, \text{ where } A = \frac{K - P_0}{P_0}.$$

K is called the *carrying capacity* for the equation because $\lim_{t \to \infty} P(t) = K$.

1) a) Find the solution to the initial value problem
$$\frac{dP}{dt} = 0.10P\left(1 - \frac{P}{500}\right), P(1980) = 200,$$
where t is given in years since 1980.

$P_0 = 200$ (in 1980).
$A = \frac{K - P_0}{P_0} = \frac{500 - 200}{200} = 1.5.$
$P(t) = \frac{500}{1 + 1.5e^{-0.1t}}.$

b) What is the population in 1985?

$P(5) = \frac{500}{1 + 1.5e^{-0.1(5)}} \approx 262.$

c) When does the population reach 450?

$P(t) = \frac{500}{1 + 1.5e^{-0.1t}} = 450$
$1 + 1.5e^{-0.1t} = \frac{10}{9}$
$1.5e^{-0.1t} = \frac{1}{9}$
$e^{-0.1t} = \frac{2}{27}$
$-0.1t = \ln\frac{2}{27}$
$t = \frac{\ln\frac{2}{27}}{-0.1} = -10\ln\frac{2}{27} = 10\ln\frac{27}{2} \approx 26 \text{ years}$
(The year 2006).

2) A biologist stocks a shrimp farm pond with 1000 shrimp. The number of shrimp double in one year and the pond has a carrying capacity of 10,000. How long does it take the shrimp population to reach 99% of the pond's capacity?

$P_0 = 1000. \ A = \dfrac{10000 - 1000}{1000} = 9.$

So $P(t) = \dfrac{10000}{1 + 9e^{-kt}}.$

From $P(1) = 2000$, we have

$2000 = \dfrac{10000}{1 + 9e^{-k(1)}} = \dfrac{10000}{1 + 9e^{-k}}$

$1 + 9e^{-k} = 5$

$9e^{-k} = 4$

$e^{-k} = \dfrac{4}{9}$

$-k = \ln \dfrac{4}{9},$ so $k = \ln \dfrac{9}{4}$

$P(t) = \dfrac{10000}{1 + 9e^{-(\ln 9/4)t}} = \dfrac{10000}{1 + 9\left(\dfrac{4}{9}\right)^t}.$

To reach 99% of capacity

$\dfrac{10000}{1 + 9\left(\dfrac{4}{9}\right)^t} = 0.99(10000)$

$1 + 9\left(\dfrac{4}{9}\right)^t = \dfrac{100}{99}$

$9\left(\dfrac{4}{9}\right)^t = \dfrac{1}{99}$

$\left(\dfrac{4}{9}\right)^t = \dfrac{1}{891}$

$t \ln \left(\dfrac{4}{9}\right) = \ln \dfrac{1}{891}$

$t = \dfrac{\ln \dfrac{1}{891}}{\ln \dfrac{4}{9}} \approx 8.376 \text{ years.}$

Section 9.6   Linear Equations

This section considers another type of first-order differential equation—the linear differential equation. Like separable equations, linear differential equations also have an explicit form for their solutions and have many applications modeling physical phenomena.

Concepts to Master

Linear first-order differential equations; solutions by integrating factors

Summary and Focus Questions

Page 622

A first-order differential equation is *linear* if it can be written in the form

$$\frac{dy}{dx} + P(x)y = Q(x)$$

where $P(x)$ and $Q(x)$ are continuous functions on a given interval. To solve a linear equation:

1. Multiply both sides by the integrating factor $I(x) = e^{\int P(x)\,dx}$.

2. Solve $I(x)y = \int I(x)Q(x)\,dx$ for y: $y = \dfrac{\int I(x)Q(x)\,dx}{I(x)}$.

1) Is $y' + \dfrac{x}{y} = x^2 + x$ linear?

No, because of the term $\dfrac{x}{y}$.

2) What integrating factor should be used for $y' + 7y = x^2 + x$?

$P(x) = 7$.
$I(x) = e^{\int 7\,dx} = e^{7x}$.

3) Find the general solution to $xy' + y = e^{-x}$, $x > 0$.

Put in standard form: $y' + \dfrac{1}{x}y = \dfrac{e^{-x}}{x}$.
$I(x) = e^{\int 1/x\,dx} = e^{\ln x} = x$.
$xy = \int x\dfrac{e^{-x}}{x}\,dx = \int e^{-x}\,dx = -e^{-x} + C$.
Thus $y = \dfrac{-e^{-x} + C}{x}$.

4) Solve the initial-value problem
$y' + y = xe^x, y(0) = 1.$

The equation is in standard form with

$P(x) = 1$ and $Q(x) = xe^x$.

$I(x) = e^{\int 1 \, dx} = e^x$.

$e^x y = \int e^x (xe^x) \, dx = \int xe^{2x} \, dx$

(by parts or by integral #96)

$= \frac{1}{2}xe^{2x} - \frac{1}{4}e^{2x} + C$

Thus $y = \dfrac{\frac{1}{2}xe^{2x} - \frac{1}{4}e^{2x} + C}{e^x}$,

or $y = \frac{1}{2}xe^x - \frac{1}{4}e^x + C_1$.

At $x = 0$, $y = 1$. Therefore

$1 = \frac{1}{2}0e^0 - \frac{1}{4}e^0 + C_1$

$C_1 = \frac{5}{4}$.

Thus $y = \frac{1}{2}xe^x - \frac{1}{4}e^x + \frac{5}{4}$ is the particular

solution.

Section 9.7 Predator-Prey Systems

All population models in previous sections involved a single species. This section discusses a model of a two-species predator-prey situation that uses two linked differential equations. The equations are linked because the changes in each population are related to the number of species in both populations. (If there are a lot of rabbits, then the wolf population will increase because there is plenty to eat and wolves to breed.)

Concepts to Master

Model predator-prey populations using two linear differential equations.

Summary and Focus Questions

Page 628

Let $R(t)$ be the number of prey at time t (R stands for rabbit) and let $W(t)$ be the number of predators at time t (W stands for wolf). Several reasonable assumptions are necessary in a predator-prey model:

1. the main cause of death among the prey is that they are eaten by the predators

2. the birth and survival rate of predators depends on the amount of food available; that is, the number of prey

3. the two species encounter each other at a rate proportional to the product of their populations (the more of either species, the more likely they will encounter each other)

The populations $R(t)$ and $W(t)$ are modeled by a pair of *predator-prey equations* (also called the *Lotka-Volterra* equations):

$$\frac{dR}{dt} = kR - aRW \text{ and } \frac{dW}{dt} = -rW + bRW,$$

where k, r, a, and b are positive constants. It is usually impossible to find explicit solutions for R and W. This system of equations is usually analyzed by graphical methods.

An *equilibrium solution* is a solution to the system of equations consisting of constants, where it may be interpreted that a constant number of prey is just the right amount to support a constant number of predators.

The differential equation $\dfrac{dW}{dR} = \dfrac{\frac{dW}{dt}}{\frac{dR}{dt}}$ models the relationship between the populations R and W as time passes. Its solutions are graphed in the R-W plane, called the *phase plane*. Solutions to the predator-prey equations often are closed curves with the equilibrium solution point(s) inside all the closed curves.

1) A part of the Caribbean Sea contains populations of sharks and mullets governed by a system of predator-prey equations with constants
$k = 0.04$, $r = 0.04$, $a = 0.008$, and $b = 0.00001$.

   a) Find the constant solutions for the populations of sharks and mullets described by the equations.

At equilibrium $\dfrac{dR}{dt} = 0$ and $\dfrac{dW}{dt} = 0$.

$\dfrac{dR}{dt} = 0.04R - 0.008RW = 0$

$R(0.04 - 0.008W) = 0$

$0.04 - 0.008W = 0$

$W = 5$ sharks.

$\dfrac{dW}{dt} = -0.04W + 0.00001RW = 0$

$W(-0.04 + 0.00001R) = 0$

$-0.04 + 0.00001R = 0$

$R = 4000$ mullets.

   b) Find $\dfrac{dW}{dR}$ and draw the direction field for these values:

R

	1000	3000	5000	7000
3.5				
W 4.5				
5.5				
6.5				

$\dfrac{dW}{dR} = \dfrac{\frac{dW}{dt}}{\frac{dR}{dt}} = \dfrac{-0.04W + 0.00001RW}{0.04R - 0.008RW}.$

R

	1000	3000	5000	7000
3.5	−0.0088	−0.0010	0.0006	0.0013
W 4.5	−0.0338	−0.0038	0.0023	0.0048
5.5	0.0413	0.0046	−0.0028	−0.0059
6.5	0.0163	0.0018	−0.0011	−0.0023

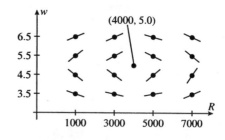

Technology Plus for Chapter 9

1) The solution to the equation $x^2y' + xy = 1$
is $y = \dfrac{\ln x + C}{x}$.
Graph, on one screen, the solutions for
$C = -1, 0, 1, 2$.
Use a $[0, 5]$ by $[-5, 5]$ window.

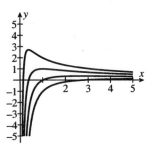

2) An environment has a carrying capacity of
10,000 insects. The population follows a
logistic model with $k = 0.10$. Graph, on the
same screen, the solution curves for initial
populations of 1000, 2000 and 3000.

For an initial population of $P_0 = 1000$,
$$A = \frac{10000 - 1000}{1000} = 9.$$
$$P(t) = \frac{10000}{1 + 9\,e^{-0.02t}}.$$
For $P_0 = 2000$,
$$A = \frac{10000 - 2000}{2000} = 4 \text{ and}$$
$$P(t) = \frac{10000}{1 + 4\,e^{-0.02t}}.$$
For $P_0 = 3000$, $A = \dfrac{10000 - 3000}{3000} = \dfrac{7}{3}$
and $P(t) = \dfrac{10000}{1 + \frac{7}{3}\,e^{-0.02t}}$.

For the window $[0, 500]$ by $[0, 10500]$ the
screen looks like this.

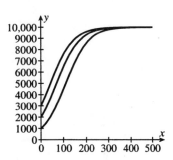

3) Use a spreadsheet or write a program that will approximate $y(2)$, where $y(x)$ is the solution to the initial value problem
$$\frac{dy}{dx} = \frac{1 - xy}{x^2}, \; y(1) = 3.$$
a) Use $h = 0.1$.

$h = 0.100$

i	x_i	y_i
0	1.00	3.000
1	1.10	2.800
2	1.20	2.628
3	1.30	2.479
4	1.40	2.347
5	1.50	2.230
6	1.60	2.126
7	1.70	2.032
8	1.80	1.947
9	1.90	1.870
10	2.00	1.799

b) Use $h = 0.05$.

$h = 0.050$

i	x_i	y_i
0	1.00	3.000
1	1.05	2.900
2	1.10	2.807
3	1.15	2.721
4	1.20	2.640
5	1.25	2.565
6	1.30	2.495
7	1.35	2.428
8	1.40	2.366
9	1.45	2.307
10	1.50	2.251
11	1.55	2.198
12	1.60	2.148
13	1.65	2.100
14	1.70	2.055
15	1.75	2.012
16	1.80	1.971
17	1.85	1.932
18	1.90	1.894
19	1.95	1.858
20	2.00	1.823

c) The exact value is $\frac{\ln 2 + 3}{2} \approx 1.847$. For the smaller value of h, is the approximation more accurate?

Yes.

Chapter 10 — Parametric Equations and Polar Coordinates

© 1999 by Sidney Harris.

Section 10.1   Curves Defined by Parametric Equations

As a point (x, y) moves along a curve over a time interval, each coordinate x and y may be described as a function of a time variable t. The notion of describing a curve with a pair of parametric equations is covered in this section. We will see that graphs of ordinary functions $(y = f(x))$ are one of many types of curves that may be defined parametrically.

Concepts to Master

Parameter; Parametric equations, graphs of curves defined parametrically; Elimination of the parameter

Summary and Focus Questions

Page
641

A set of *parametric equations* has the form

$$x = f(t)$$
$$y = g(t),$$

where f and g are functions of a third variable t, called a *parameter*.
Each value of t determines a point (x, y) in the plane. The collection of all such points is a *curve*.
A curve may be described by several different pairs of equations (each version called a "parametric curve"). Here are two sets of parametric equations for the quarter of the unit circle in the first quadrant:

$$x = t,$$
$$y = \sqrt{1 - t^2}$$
for $t \in [0, 1]$

$$x = \cos t,$$
$$y = \sin t$$
for $t \in \left[0, \frac{\pi}{2}\right]$

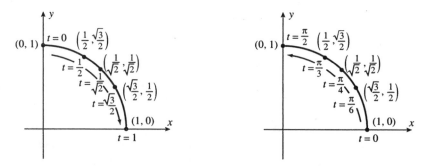

Each has the same graph (set of points). The one on the left is traversed clockwise as t increases; the one on the right is traversed counterclockwise.

Some parametric equations may be combined algebraically to a single equation not involving the parameter t; this process is called *eliminating the parameter.* The method to do so depends greatly on the nature of the parametric equations involved. Sometimes identities need to be employed, especially if trigonometric functions are involved; sometimes you can solve for t in one parametric equation and substitute the result in the other equation.

Example: Eliminate the parameter in $x = 2 + 3t, y = t^2 - 2t + 1.$

Solve $x = 2 + 3t$ for t to get $t = \frac{1}{3}(x - 2)$. Substitute this in the other equation:

$$y = \left[\tfrac{1}{3}(x - 2)\right]^2 - 2\left[\tfrac{1}{3}(x - 2)\right] + 1.$$

$$y = \tfrac{1}{9}x^2 - \tfrac{10}{9}x + \tfrac{25}{9}.$$

Example: Eliminate the parameter in $x = \sin t, y = \cot^2 t.$

$y = \cot^2 t = \dfrac{\cos^2 t}{\sin^2 t} = \dfrac{1 - \sin^2 t}{\sin^2 t}$. Thus $y = \dfrac{1 - x^2}{x^2}$.

We note from $x = \sin t, -1 \le x \le 1$ and $x \ne 0$ (since cot t is undefined when $\sin t = 0$.)

1) Sketch a graph of the curve given by
$x = \sqrt{t}, y = t + 2.$

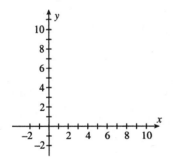

Compute some values and plot points:

t	0	1	4	9
x	0	1	2	3
y	2	3	6	11

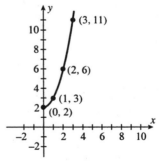

We see this is half a parabola since $x = \sqrt{t}$ implies $x^2 = t.$
Thus $y = x^2 + 2, x \ge 0.$

2) Eliminate the parameter in each

  a) $x = e^t, y = t^2$

Solve for t in terms of y:
$y = t^2, \sqrt{y} = t.$
Thus $x = e^{\sqrt{y}}, x > 0.$
A solution may also be obtained by solving
for t in terms of x:
$x = e^t, t = \ln x, y = (\ln x)^2, x > 0.$

  b) $x = 1 + \cos t, y = \sin^2 t$

$x = 1 + \cos t, x - 1 = \cos t$
$\cos^2 t = (x - 1)^2$
Since $\sin^2 t = y$, and $\cos^2 t + \sin^2 t = 1$
we have $(x - 1)^2 + y = 1.$
Thus $y = 1 - (x - 1)^2.$
Since $x = 1 + \cos t, 0 \le x \le 2.$

  c) $x = 2 \sec t, y = 3 \tan t$

$\sec t = \frac{x}{2}$ and $\tan t = \frac{y}{3}.$
Hence the identity $\tan^2 t + 1 = \sec^2 t$
becomes $\left(\frac{y}{3}\right)^2 + 1 = \left(\frac{x}{2}\right)^2$ or $\frac{x^2}{4} - \frac{y^2}{9} = 1.$

3) Describe the graph of $x = a + bt$,
$y = c + dt$ where $a, b, c,$ and d are
constants.

If both b and d are zero, this is the single
point (a, c). If $b = 0$ and $d \ne 0$ this is a
vertical line through (a, c). If $b \ne 0$, then
$t = \frac{x - a}{b}$ and $y = c + d\frac{(x - a)}{b}$. Thus
$y = \frac{d}{b}x + c - \frac{da}{b}$, so the graph is a line
through (a, c) with slope $\frac{d}{b}$.

4) Does the pair

$$x = e^t$$
$$y = e^{2t}, \; -\infty < t < \infty$$

represent the parabola $y = x^2$?

Since $x = e^t$, $x > 0$. Thus the pair represent that portion of $y = x^2$ to the right of the y-axis.

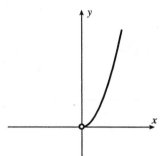

5) Use a graphing calculator to sketch

$$x = t^2 + 3t$$
$$y = 2 - t^3$$

t	x	y
−4	4	66
−3	0	29
−2	−2	10
−1	−2	3
0	0	2
1	4	1
2	10	−6
3	18	−25
4	28	−62

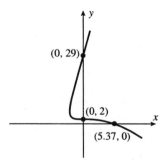

Section 10.2   Tangents and Areas

This section includes methods for finding slopes of tangents and areas involving curves defined parametrically.

Concepts to Master

A. First and second derivative of a function defined by a pair of parametric equations; applications to graphs of curves defined parametrically

B. Area of a region enclosed by curves defined parametrically

Summary and Focus Questions

Page 649

A. The slope of the line tangent to a curve described by $x = f(t)$, $y = g(t)$ is the first derivative of y with respect to x and is given by

$$\frac{dy}{dx} = \frac{\frac{dy}{dt}}{\frac{dx}{dt}} \text{ if } \frac{dx}{dt} \neq 0.$$

The curve will have a horizontal tangent when $\frac{dy}{dt} = 0$ and will have a vertical tangent when $\frac{dx}{dt} = 0$ and $\frac{dy}{dt} \neq 0$.

1) For $x = t^3$, $y = t^2 - 2t$

a) Find $\frac{dy}{dx}$ and $\frac{d^2y}{dx^2}$.

$\frac{dx}{dt} = 3t^2$ and $\frac{dy}{dt} = 2t - 2$, so $\frac{dy}{dx} = \frac{2t - 2}{3t^2}$.

$\frac{d}{dt}\left(\frac{dy}{dx}\right) = \frac{3t^2(2) - (2t - 2)(6t)}{(3t^2)^2} = \frac{4 - 2t}{3t^3}$.

Thus $\frac{d^2y}{dx^2} = \frac{\frac{4 - 2t}{3t^3}}{3t^2} = \frac{4 - 2t}{9t^5}$.

b) Find the horizontal and vertical tangents for the curve.

$\frac{dy}{dx} = 0$ at $t = 1$ ($x = 1$, $y = -1$).

A horizontal tangent is at $(1, -1)$.

$\frac{dy}{dx}$ does not exist at $t = 0$ ($x = 0$, $y = 0$).

A vertical tangent is at $(0, 0)$.

c) Discuss the concavity of the curve.

$\dfrac{d^2y}{dx^2} > 0$ for $0 < t < 2$ $(0 < x < 8)$ and negative for $t < 0$ $(x < 0)$ and $t > 2$ $(x > 8)$. The curve is concave upward for $x \in (0, 8)$ and concave downward for $x \in (-\infty, 0)$ and $x \in (8, \infty)$.

d) Sketch the curve using the information above.

$\dfrac{d^2y}{dx^2} = \dfrac{4 - 2t}{9t^5} = 0$ at $t = 2$ $(x = 8, y = 0)$.
$(8, 0)$ and $(0, 0)$ are inflection points.

t	x	y
-4	-64	24
-3	-27	15
-2	-8	8
-1	-1	3
0	0	0
1	1	-1
2	8	0
3	27	3
4	64	8

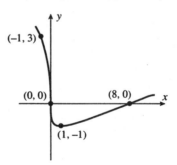

2) Find the equation of the tangent line to the curve $x = 4t^2 + 2t + 1$, $y = 7t + 2t^2$ at the point corresponding to $t = -1$.

At $t = -1$, $x = 3$ and $y = -5$.

$\dfrac{dx}{dt} = 8t + 2 = -6$ at $t = -1$.

$\dfrac{dy}{dt} = 7 + 4t = 3$ at $t = -1$.

Thus $\dfrac{dy}{dx} = \dfrac{3}{-6} = -\dfrac{1}{2}$. The tangent line is

$y + 5 = -\dfrac{1}{2}(x - 3)$.

**Page
651**

B. Suppose the parametric equations $x = f(t)$, $y = g(t)$, $t \in [\alpha, \beta]$ define an integrable function $y = F(x) \geq 0$ over the interval $[a, b]$, where $a = f(\alpha)$, $b = f(\beta)$. The area under $y = F(x)$ is $\displaystyle\int_a^b y\, dx =$

$$\int_\alpha^\beta g(t)\, f'(t)\, dt \text{ if } (f(\alpha), g(\alpha)) \text{ is the left endpoint}$$

or

$$\int_\beta^\alpha g(t)\, f'(t)\, dt \text{ if } (f(\beta), g(\beta)) \text{ is the left endpoint.}$$

3) Find a definite integral for the area under the curve $x = t^2 + 1$, $y = e^t$, $0 \leq t \leq 1$.

The area is $\displaystyle\int_0^1 e^t\,(2t)\,dt$

$=$ (by parts: $u = 2t$ and $dv = e^t\, dt$;

$\qquad du = 2\, dt$, $v = e^t$)

$= 2te^t - \displaystyle\int e^t 2\, dt = 2te^t - 2e^t\Big|_0^1$

$= (2e - 2e) - (0 - 2) = 2$.

4) Find the area inside the loop of the curve given by
$x = 9 - t^2$
$y = t^3 - 3t$.

$$\frac{dy}{dx} = \frac{\frac{dy}{dt}}{\frac{dx}{dt}} = \frac{3t^2 - 3}{-2t}.$$

$\dfrac{dy}{dx} = 0$ at $t = 1$, $t = -1$. $\dfrac{dy}{dx}$ is not defined at $t = 0$. There are horizontal tangents at $(8, 2)$ (where $t = -1$) and $(8, -2)$ (where $t = 1$). There is a vertical tangent at $(9, 0)$ (where $t = 0$).

t	x	y
−3	0	−18
−2	5	−2
−1	8	2
0	9	0
1	8	−2
2	5	2
3	0	18

The graph is

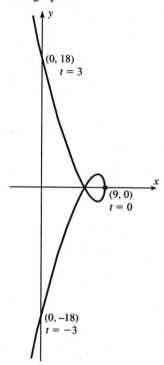

The curve crosses itself (to form a loop) at
$y = 0$.
$t^3 - 3t = 0$
$t(t^2 - 3) = 0$
$t = 0$, $t = \sqrt{3}$, $t = -\sqrt{3}$.
The area of the loop is twice the area under
the top portion which corresponds to
$t \in [-\sqrt{3}, 0]$.

The area is $2 \displaystyle\int_{-\sqrt{3}}^{0} (t^3 - 3t)(-2t)\, dt$

$= 2 \displaystyle\int_{-\sqrt{3}}^{0} (-2t^4 + 6t^2)\, dt$

$= 2 \left(-\tfrac{2}{5}t^5 + 2t^3 \right) \Big]_{-\sqrt{3}}^{0}$

$= \dfrac{24\sqrt{3}}{5}.$

Section 10.3   Arc Length and Surface Area

This section generalizes the formulas in chapter 9 on arc length of curves and surface areas obtained by revolving a curve about a line. In that chapter curves were defined by functions of the form $y = f(x)$; here they are defined parametrically. We must be careful to distinguish arc length of a curve from distance traversed along the curve because a curve defined parametrically may wrap upon itself. The integral for the arc length will compute the total distance traversed along the curve which could be greater the length of the curve.

Concepts to Master

A. Length of a curve defined parametrically
B. Area of a surface of revolution with curve defined parametrically

Summary and Focus Questions

Page 656

A. The length of a curve defined by $x = f(t)$, $y = g(t)$, $t \in [\alpha, \beta]$ with f', g' continuous and the curve traversed only once as t increases from α to β is

$$\int_\alpha^\beta \sqrt{\left(\frac{dx}{dt}\right)^2 + \left(\frac{dy}{dt}\right)^2}\, dt.$$

Since $(ds)^2 = (dx)^2 + (dy)^2$ (see section 9.1) this is $\int ds$.

1) Write a definite integral for the arc length of each:

a) the curve given by $x = \ln t$, $y = t^2$, for $1 \le t \le 2$.

$x'(t) = \frac{1}{t}$ and $y'(t) = 2t$. The arc length is

$$\int_1^2 \sqrt{\left(\frac{1}{t}\right)^2 + (2t)^2}\, dt.$$

b) an ellipse given by $\frac{x^2}{a^2} + \frac{y^2}{b^2} = 1$, where $a, b > 0$.

The ellipse may be defined parametrically by $x = a \cos t$, $y = b \sin t$, $t \in [0, 2\pi]$. The length is

$$\int_0^{2\pi} \sqrt{(-a \sin t)^2 + (b \cos t)^2}\, dt$$

$$= \int_0^{2\pi} \sqrt{a^2 \sin^2 t + b^2 \cos^2 t}\, dt$$

c) The curve given by
$$x = \cos 2t$$
$$y = \sin t$$
$$t \in [0, 2\pi]$$

t	x	y
0	1	0
$\frac{\pi}{4}$	0	$\frac{\sqrt{2}}{2}$
$\frac{\pi}{2}$	−1	1
$\frac{3\pi}{4}$	0	$\frac{\sqrt{2}}{2}$
π	1	0
$\frac{5\pi}{4}$	0	$-\frac{\sqrt{2}}{2}$
$\frac{3\pi}{2}$	−1	−1
$\frac{7\pi}{4}$	0	$-\frac{\sqrt{2}}{2}$
2π	1	0

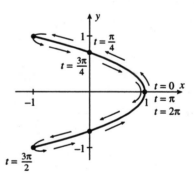

The curve is traversed once for $t \in \left[\frac{\pi}{2}, \frac{3\pi}{2}\right]$ and has length

$$\int_{\frac{\pi}{2}}^{\frac{3\pi}{2}} \sqrt{4 \sin^2 2t + \cos^2 t}\, dt.$$

Page 658

B. If a curve given by $x = f(t)$, $y = g(t)$, $t \in [\alpha, \beta]$, with f', g' continuous and $g(t) \geq 0$ is rotated about the x-axis, the area of the resulting surface of revolution is

$$S = \int_\alpha^\beta 2\pi y \sqrt{\left(\frac{dx}{dt}\right)^2 + \left(\frac{dy}{dt}\right)^2} \, dt.$$

Using the ds notation this is $S = \int 2\pi y \, ds$, the same formula as in section 8.3

2) Find the definite integral for the area of the surface of revolution about the x-axis for each curve:

a) $x = 2t + 1, y = t^3, t \in [0, 2]$

$$S = \int_0^2 2\pi (t^3) \sqrt{(2)^2 + (3t^2)^2} \, dt$$

$$= \int_0^2 2\pi t^3 \sqrt{4 + 9t^4} \, dt.$$

b) a "football" obtained from rotating the ellipse $\dfrac{x^2}{a^2} + \dfrac{y^2}{b^2} = 1, y \geq 0$ about the x-axis.

The top half of the ellipse is $x = a \cos t$, $y = b \sin t, t \in [0, \pi]$.

$$ds = \sqrt{(dx)^2 + (dy)^2}$$

$$= \sqrt{(-a \sin t)^2 + (b \cos t)^2}$$

$$= \sqrt{a^2 \sin^2 t + b^2 \cos^2 t}.$$

Thus the area

$$S = \int_0^\pi 2\pi (b \sin t) \sqrt{a^2 \sin^2 t + b^2 \cos^2 t} \, dt.$$

Section 10.4   Polar Coordinates

Thus far all our graphs have been in a rectangular coordinate system where the two coordinates are distances from perpendicular axes. In this section on polar coordinates the pair of numbers that determines a point are an angle through which to rotate from the *x*-axis and a distance from the origin. You will learn how to convert from one coordinate system to another and see that certain curves are much easier to express in polar coordinates than rectangular coordinates.

Concepts to Master

A. Points in polar coordinates; conversion to and from rectangular to polar coordinates

B. Graphs of equations in polar coordinates

C. Tangents to polar curves

Summary and Focus Questions

Page 660

A. To construct a *polar coordinate system* start with a point called the *pole* and a ray from the pole called the *polar axis.*

Pole

P

A point *P* has polar coordinates (r, θ) if

$|r|$ = the distance form the pole to *P*, and

θ = the measure of a directed angel with initial side the polar axis and terminal side the line through the pole and *P*.

To plot a point *P* with polar coordinates $P(r, \theta)$:

(1) if $r = 0$, *P* is the pole (for any value of θ).

(2) if $r > 0$, rotate the polar axis by the angle θ (counterclockwise for $\theta > 0$, clockwise for $\theta < 0$) and locate *P* on this ray at a distance *r* units from the pole.

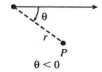

(3) if $r < 0$, rotate the polar axis by the angle θ and then reflect about the pole. P is located on this reflected ray at a distance $-r$ units from the pole.

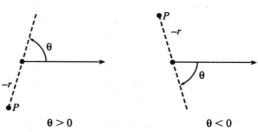

$$\theta > 0 \qquad\qquad\qquad \theta < 0$$

If a rectangular coordinate system is placed upon the polar coordinate system as in the figure, then to change from polar to rectangular:

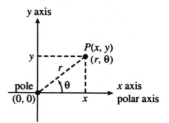

$$x = r \cos \theta$$
$$y = r \sin \theta.$$

To change from rectangular to polar:

θ is a solution to $\tan \theta = \dfrac{y}{x}$ (for $x \neq 0$).

r is a solution to $r^2 = x^2 + y^2$ where $r > 0$ if the terminal side of the angle θ is in the same quadrant as P; if not, then $r \leq 0$.

1) Plot these points in polar coordinates.

A: $\left(4, \frac{\pi}{3}\right)$ B: $\left(-3, \frac{\pi}{4}\right)$

C: $(0, 3\pi)$ D: $\left(2, -\frac{\pi}{4}\right)$

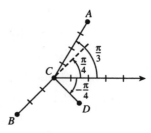

2) Find two other polar coordinates for point with polar coordinates $\left(8, \frac{\pi}{3}\right)$.

There are infinitely many answers including
$\left(8, \frac{7\pi}{3}\right), \left(-8, \frac{4\pi}{3}\right), \left(8, -\frac{5\pi}{3}\right), \ldots$

3) Sometimes, Always, or Never:

   a) The polar coordinates of a point are unique.

   Never

   b) $(r, \theta) = (r, \theta + 2\pi)$.

   Always

   c) $(r, \theta) = (-r, \theta + \pi)$.

   Always

   d) $(r, \theta) = (-r, \theta)$.

   Sometimes. (True when $r = 0$).

4) Find polar coordinates for the point P with rectangular coordinates $P: (-3, 3\sqrt{3})$.

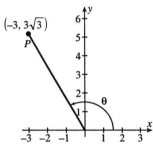

$\tan \theta = \dfrac{y}{x} = \dfrac{3\sqrt{3}}{-3} = -\sqrt{3}.$

A solution is $\theta = \dfrac{2\pi}{3}$.

$r^2 = (-3)^2 + (3\sqrt{3})^2 = 9 + 27 = 36.$

Because the terminal side of $\theta = \dfrac{2\pi}{3}$ lies in the same quadrant as P, $r > 0$. Therefore, $P: \left(6, \dfrac{2\pi}{3}\right)$. *Note:* $\theta = -\dfrac{\pi}{3}$ is another solution to $\tan \theta = -\sqrt{3}$, which results in $r = -6$ and coordinates $\left(-6, -\dfrac{\pi}{3}\right)$.

5) Find the rectangular coordinates for the point with polar coordinates $\left(-4, \dfrac{5\pi}{6}\right)$.

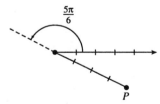

$x = -4 \cos \dfrac{5\pi}{6} = (-4)\left(\dfrac{-\sqrt{3}}{2}\right) = 2\sqrt{3}.$

$y = -4 \sin \dfrac{5\pi}{6} = (-4)\left(\dfrac{1}{2}\right) = -2.$

Thus $P: (2\sqrt{3}, -2)$.

Page 662

B. The procedure for graphing a polar equation is often the same as that you used when you first graphed functions—compute values and plot points. In some cases the graph of a polar equation is easily identified when the equation is transformed into rectangular coordinates. In other cases you may be able to take advantage of symmetry.

For example, $r = 6 \sin \theta$ becomes $r^2 = 6r \sin \theta$ or $x^2 + y^2 = 6y$. This is $x^2 + y^2 - 6y = 0$ or $x^2 + (y-3)^2 = 9$, the circle of radius 3 centered at $(0, 3)$.

6) Sketch a graph of $r = \sin \theta - \cos \theta$.

Compute several points for values of θ.

A: $(-1, 0)$

B: $\left(\frac{1}{2} - \frac{\sqrt{3}}{2}, \frac{\pi}{6}\right)$

C: $\left(0, \frac{\pi}{4}\right)$

D: $\left(1, \frac{\pi}{2}\right)$

E: $\left(\sqrt{2}, \frac{3\pi}{4}\right)$

F: $(1, \pi)$ (same as A)

G: $\left(\frac{-1}{2} + \frac{\sqrt{3}}{2}, \frac{7\pi}{6}\right)$ (same as B)

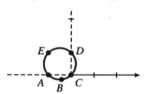

We can verify that the figure is a circle by switching to rectangular coordinates:

Multiply $r = \sin \theta - \cos \theta$ by r.

$r^2 = r \sin \theta - r \cos \theta$

$x^2 + y^2 = y - x$

$x^2 + x + y^2 - y = 0$

Completing the square gives

$$\left(x + \frac{1}{2}\right)^2 + \left(y - \frac{1}{2}\right)^2 = \frac{1}{2},$$

the circle with center $\left(-\frac{1}{2}, \frac{1}{2}\right)$ and radius $\frac{1}{\sqrt{2}}$.

7) Sketch the graph of $r = \cos 3\theta$

Since $\cos 3\theta = \cos(-3\theta)$, the curve is symmetric about the polar axis. Also $\cos 3\theta$ repeats every $\frac{2\pi}{3}$ units.

θ	r
0	1
$\frac{\pi}{12}$	$\frac{\sqrt{2}}{2}$
$\frac{\pi}{6}$	0
$\frac{\pi}{4}$	$-\frac{\sqrt{2}}{2}$
$\frac{\pi}{3}$	-1
$\frac{5\pi}{12}$	$-\frac{\sqrt{2}}{2}$
$\frac{\pi}{2}$	0
$\frac{7\pi}{12}$	$\frac{\sqrt{2}}{2}$
$\frac{2\pi}{3}$	1

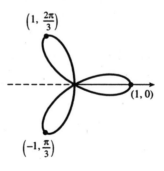

$\left(1, \frac{2\pi}{3}\right)$

$(1, 0)$

$\left(-1, \frac{\pi}{3}\right)$

C. For polar curve $r = f(\theta)$, we can switch to rectangular coordinates $x = f(\theta) \cos \theta,\ y = f(\theta) \sin \theta$. Treating θ as a parameter,

Page 665

$$\frac{dy}{dx} = \frac{\frac{dy}{d\theta}}{\frac{dx}{d\theta}} = \frac{\frac{dr}{d\theta} \sin \theta + r \cos \theta}{\frac{dr}{d\theta} \cos \theta - r \sin \theta}$$

This is the slope of the tangent line to $r = f(\theta)$ at (r, θ).

8) Find the slope of the tangent line to $r = e^{\theta}$ at $\theta = \frac{\pi}{2}$.

$x = r \cos \theta = e^{\theta} \cos \theta$, so

$\frac{dx}{d\theta} = e^{\theta}(-\sin \theta) + e^{\theta}(\cos \theta)$.

$y = r \sin \theta = e^{\theta} \sin \theta$, so

$\frac{dy}{d\theta} = e^{\theta} \cos \theta + e^{\theta} \sin \theta$.

At $\theta = \frac{\pi}{2}$,

$\frac{dx}{d\theta} = -e^{\pi/2}$ and $\frac{dy}{d\theta} = e^{\pi/2}$.

Therefore, $\frac{dy}{dx} = \dfrac{\frac{dy}{d\theta}}{\frac{dx}{d\theta}} = \dfrac{e^{\pi/2}}{-e^{\pi/2}} = -1$.

9) Find the points on $r = \sin \theta - \cos \theta$ where the tangent line is horizontal or vertical.

$\frac{dy}{dx} = \dfrac{\frac{dy}{d\theta}}{\frac{dx}{d\theta}}$ is not defined when $\frac{dx}{d\theta} = 0$

and zero when $\frac{dy}{d\theta} = 0$.

$x = (\sin \theta - \cos \theta) \cos \theta$

$\frac{dx}{d\theta} = (\sin \theta - \cos \theta)(-\sin \theta)$

$\qquad + \cos \theta (\cos \theta + \sin \theta)$

$\qquad = 2 \sin \theta \cos \theta + (\cos^2 \theta - \sin^2 \theta)$

$\qquad = \sin 2\theta + \cos 2\theta$.

$\frac{dx}{d\theta} = 0$ when

$\sin 2\theta = -\cos 2\theta$

$\tan 2\theta = -1$

$2\theta = \frac{3\pi}{4}$ and $\frac{7\pi}{4}$

$\theta = \frac{3\pi}{8}$ and $\frac{7\pi}{8}$.

At $\theta = \frac{3\pi}{8}$, $r = \sin \frac{3\pi}{8} - \cos \frac{3\pi}{8}$

$\qquad = \dfrac{\sqrt{\sqrt{2} + 2}}{2} - \dfrac{\sqrt{2 - \sqrt{2}}}{2} \approx .54$.

At $\theta = \frac{7\pi}{8}$, $r = \sin \frac{7\pi}{8} - \cos \frac{7\pi}{8}$

$\qquad = \dfrac{\sqrt{2 - \sqrt{2}}}{2} - \dfrac{-\sqrt{\sqrt{2} + 2}}{2} \approx 1.31$.

There are vertical tangents at $\left(.54, \frac{3\pi}{8}\right)$ and $\left(1.31, \frac{7\pi}{8}\right)$.

$$y = (\sin\theta - \cos\theta)\sin\theta$$
$$\frac{dy}{d\theta} = (\sin\theta - \cos\theta)\cos\theta$$
$$+ \sin\theta\,(\cos\theta + \sin\theta)$$
$$= 2\sin\theta\cos\theta - (\cos^2\theta - \sin^2\theta)$$
$$= \sin 2\theta - \cos 2\theta = 0.$$

$$\sin 2\theta = \cos 2\theta$$

$$\tan 2\theta = 1$$

$$2\theta = \frac{\pi}{4} \text{ and } \frac{5\pi}{4}$$

$$\theta = \frac{\pi}{8} \text{ and } \frac{5\pi}{8}.$$

At $\theta = \frac{\pi}{8}$, $r = \sin\frac{\pi}{8} - \cos\frac{\pi}{8}$

$$= \frac{\sqrt{2 - \sqrt{2}}}{2} - \frac{\sqrt{2 + \sqrt{2}}}{2} \approx -.54.$$

At $\theta = \frac{5\pi}{8}$, $r = \sin\frac{5\pi}{8} - \cos\frac{5\pi}{8}$

$$= \frac{\sqrt{2 + \sqrt{2}}}{2} - \frac{-\sqrt{2 - \sqrt{2}}}{2} \approx 1.31.$$

There are horizontal tangents at $\left(-.54, \frac{\pi}{8}\right)$ and $\left(1.31, \frac{5\pi}{8}\right)$.

The graph is

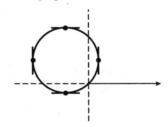

The graph is a circle.

$$r = \sin\theta - \cos\theta$$
$$r^2 = r\sin\theta - r\cos\theta$$
$$x^2 + y^2 = -x + y$$
$$x^2 + x + \frac{1}{4} + y^2 - y + \frac{1}{4} = \frac{1}{2}$$
$$\left(x + \frac{1}{2}\right)^2 + \left(y - \frac{1}{2}\right)^2 = \left(\frac{1}{\sqrt{2}}\right)^2.$$

In rectangular coordinates the center is $\left(-\frac{1}{2}, \frac{1}{2}\right)$ and the radius is $\frac{1}{\sqrt{2}}$.

Section 10.5   Areas and Lengths in Polar Coordinates

This section develops a formula for the area of a region bounded by equations given in polar form and the arc length of a curve given by a polar equation.

Concepts to Master

A. Area of a region described by polar equations

B. Length of a curve described by a polar equation

Summary and Focus Questions

Page 671

A. The area of a region bounded by $\theta = a, \theta = b, r = f(\theta)$ where f is continuous and positive and $0 \le b - a \le 2\pi$ is $\int_a^b \frac{1}{2} [f(\theta)]^2 \, d\theta$.

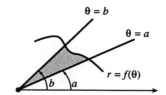

1) Find a definite integral for each region:

   a) the shaded area

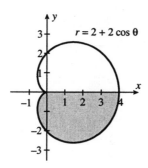

The region is bound by $\theta = 0, \theta = \pi$, $r = 2 + 2\cos\theta$. The area is

$$\int_0^\pi \frac{1}{2}(2 + 2\cos\theta)^2 \, d\theta$$

$$= 2\int_0^\pi (1 + \cos\theta)^2 \, d\theta.$$

b) The region bounded by $\theta = \frac{\pi}{3}$, $\theta = \frac{\pi}{2}$, $r = e^{\theta}$.

The area is $\int_{\pi/3}^{\pi/2} \frac{1}{2} e^{2\theta} \, d\theta$.

c) the shaded area

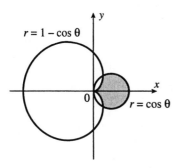

$r = 1 - \cos \theta$

$r = \cos \theta$

The area is between 2 curves so we must first determine where the curves intersect:

$1 - \cos \theta = \cos \theta$, $1 = 2 \cos \theta$, $\cos \theta = \frac{1}{2}$.

Therefore $\theta = \frac{\pi}{3}$, $-\frac{\pi}{3}$.

For $-\frac{\pi}{3} \le \theta \le \frac{\pi}{3}$,

$\cos \theta \ge 1 - \cos \theta$ so the area is

$$= \int_{-\frac{\pi}{3}}^{\frac{\pi}{3}} \frac{1}{2} [\cos \theta]^2 \, d\theta$$

$$- \int_{-\frac{\pi}{3}}^{\frac{\pi}{3}} \frac{1}{2} (1 - \cos \theta)^2 \, d\theta$$

$$= \int_{-\frac{\pi}{3}}^{\frac{\pi}{3}} \left(-\frac{1}{2} + \cos \theta \right) d\theta.$$

2) Find the area above the line $r = \csc \theta$ and inside the circle $r = 2$.

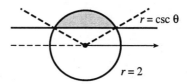

$r = \csc \theta$

$r = 2$

The curves intersect when $\csc \theta = 2$.

$\frac{1}{\sin \theta} = 2$; $\sin \theta = \frac{1}{2}$. $\theta = \frac{\pi}{6}$ and $\frac{5\pi}{6}$.

The area is $\int_{\frac{\pi}{6}}^{\frac{5\pi}{6}} \left(\frac{1}{2}(2)^2 - \frac{1}{2} \csc^2 \theta \right) d\theta$

$$= \frac{1}{2} \int_{\frac{\pi}{6}}^{\frac{5\pi}{6}} (4 - \csc^2 \theta) \, d\theta$$

$$= \frac{1}{2}(4\theta + \cot \theta) \Big|_{\frac{\pi}{6}}^{\frac{5\pi}{6}} = \frac{4\pi}{3} - \sqrt{3}.$$

Page
673

B. A curve given in polar coordinates by $r = f(\theta)$ for $a \leq \theta \leq b$ has arc length

$$\int_a^b \sqrt{[f(\theta)]^2 + [f'(\theta)]^2} \; d\theta.$$

2) Set up a definite integral for the length of each curve.

a) the curve $r = e^{2\theta}$ for $0 \leq \theta \leq 1$.

$$\int_0^1 \sqrt{(e^{2\theta})^2 + (2e^{2\theta})^2} \; d\theta = \sqrt{5} \int_0^1 e^{2\theta} \; d\theta.$$

b) the curve sketched below.

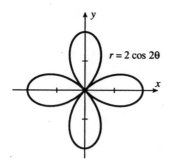

$r = 2 \cos 2\theta$

Since the curve is quite symmetric, the arc length is 8 times the length of half of one leaf:

$$8 \int_0^{\frac{\pi}{4}} \sqrt{[2 \cos 2\theta]^2 + [-4 \sin 2\theta]^2} \; d\theta$$

$$= 8 \int_0^{\frac{\pi}{4}} \sqrt{4 \cos^2 2\theta + 16 \sin^2 2\theta} \; d\theta$$

$$= 16 \int_0^{\frac{\pi}{4}} \sqrt{1 + 3 \sin^2 2\theta} \; d\theta$$

Section 10.6   Conic Sections

Conic sections are the various curves resulting from the intersection of a plane with a cone. This section gives geometric definitions of the conic sections and their equations.

Concepts to Master

Focus-directrix definition of parabolas, ellipses and hyperbolas; vertices; standard form of the equations of conics

Summary and Focus Questions

Page 675

Conic sections are curves obtained by intersecting a plane and a cone in various ways. The resulting curve is either a *parabola, ellipse* or *hyperbola*.
These curves may also be obtained as a certain set of points satisfying a given geometric condition:

Parabola: Given a line (a *directrix*) and a point (the *focus, F*), a point P is on the parabola if the distance from P to the directrix is the same as the distance from P to F.

Ellipse: Given two points (the *foci, F_1* and F_2), a point P is on the ellipse if the sum of the distances from P to F_1 and from P to F_2 is a constant.

Hyperbola: Given two points (the *foci, F_1* and F_2), a point P is on the hyperbola if the difference of the distances form P to F_1 and from P to F_2 is a constant.

The equations of the conics and their graphs in rectangular coordinates are given below.

Conic Section	Equation	Properties	Graphs
Parabola	$x^2 = 4py$	focus: $(0, p)$ vertex: $(0, 0)$ directrix: $y = -p$	$p > 0$
		focus: $(0, p)$ vertex: $(0, 0)$ directrix: $y = -p$	$p < 0$

Parabola $y^2 = 4px$ focus: $(p, 0)$
                                      vertex: $(0, 0)$
                                      directrix: $x = -p$

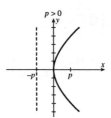

                                      focus: $(p, 0)$
                                      vertex: $(0, 0)$
                                      directrix: $x = -p$

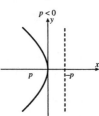

Ellipse $\dfrac{x^2}{a^2} + \dfrac{y^2}{b^2} = 1$ foci: $(c, 0)$, $(-c, 0)$
                                      vertices: $(a, 0)$, $(-a, 0)$
                                      constant sum = $2a$
                                      center: $(0, 0)$
                                      $c^2 = a^2 - b^2$,
                                      $a \geq b > 0$

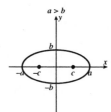

Ellipse $\dfrac{x^2}{b^2} + \dfrac{y^2}{a^2} = 1$ foci: $(0, c)$, $(0, -c)$
                                      vertices: $(0, a)$, $(0, -a)$
                                      constant sum = $2a$
                                      center: $(0, 0)$
                                      $c^2 = a^2 - b^2$,
                                      $a \geq b > 0$

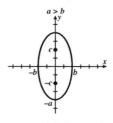

Hyperbola $\dfrac{x^2}{a^2} - \dfrac{y^2}{b^2} = 1$ foci: $(c, 0)$, $(-c, 0)$
                                      vertices: $(a, 0)$, $(-a, 0)$
                                      constant difference = $2a$
                                      center $(0, 0)$
                                      $c^2 = a^2 + b^2$
                                      asymptotes: $y = \pm\dfrac{b}{a}x$

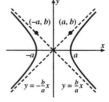

Hyperbola $\dfrac{y^2}{a^2} - \dfrac{x^2}{b^2} = 1$ foci: $(0, c)$, $(0, -c)$
                                      vertices: $(0, a)$, $(0, -a)$
                                      constant difference = $2a$
                                      center: $(0, 0)$
                                      $c^2 = a^2 + b^2$
                                      asymptotes: $y = \pm\dfrac{a}{b}x$

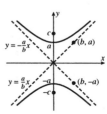

A second degree equation in x and y (with no xy term) represents a conic section whose center or vertex may be shifted from $(0, 0)$. To determine the type of conic, complete the square as in this example.

Example: What type of conic is given by the equation $x^2 + 6x + 4y^2 - 8y = 3$?
Complete the square:
$$x^2 + 6x + 9 + 4(y^2 - 2y + 1) = 3 + 9 + 4(1).$$
$$(x + 3)^2 + 4(y - 1)^2 = 16$$
$$\frac{(x + 3)^2}{16} + \frac{(y - 1)^2}{4} = 1.$$

This is an ellipse shifted 3 units left and one unit upward. Its center is $(-3, 1)$, $a = 4$ and $b = 2$.

1) Find the vertices, foci, and directrix (if a parabola) and sketch the graph of each:

a) $\dfrac{x^2}{144} = 1 + \dfrac{y^2}{25}$

$$\frac{x^2}{144} - \frac{y^2}{25} = 1.$$

This is a hyperbola with $a = 12$, $b = 5$.
$c^2 = 12^2 + 5^2 = 169$, $c = 13$.
Foci: $(13, 0)$, $(-13, 0)$
Vertices: $(12, 0)$, $(-12, 0)$
Asymptotes: $y = \dfrac{5}{12}x$, $y = -\dfrac{5}{12}x$.

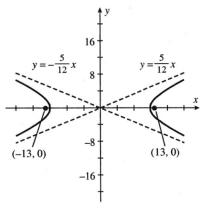

b) $x^2 = 4x + 8y - 4$

$x^2 = 4x + 8y - 4$
$x^2 - 4x + 4 = 8y$
$(x - 2)^2 = 4(2)y$

This is a parabola, shifted two units to the right. The vertex is $(2, 0)$.

$p = 2$; the directrix is $y = -2$.

The focus is $(2, 2)$, also shifted two units.

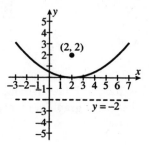

2) What is the conic given by
$2x(6 - x) = y(8 + y)$?

$2x(6 - x) = y(8 + y)$
$2x^2 + 12x = y^2 + 8y$
$2x^2 - 12x + y^2 + 8y = 0$
$2(x^2 - 6x + 9) + y^2 + 8y + 16 = 18 + 16$
$2(x - 3)^2 + (y + 4)^2 = 34$

$$\frac{(x - 3)^2}{17} + \frac{(y + 4)^2}{34} = 1$$

This is an ellipse , $a = \sqrt{34}$, $b = \sqrt{17}$.

Section 10.7   Conic Sections in Polar Coordinates

This section gives alternate definitions for each of the three conic sections that use just one focus and one directrix. This approach leads to a simple form for the equations of parabolas, ellipses and hyperbolas in polar coordinates.

Concepts to Master

Eccentricity; Eccentric definition of conic sections; Equations of conic in polar form

Summary and Focus Questions

Page 682

Another way to define conic sections is to specify a fixed point F (the focus), a fixed line l (the directrix), and a positive constant e called the *eccentricity**. Then the set of all points P such that

$$\frac{\text{distance from } P \text{ to } F}{\text{distance from } P \text{ to } l} = e$$

is a conic section. The value of e determines whether the conic is a parabola, ellipse or hyperbola:

Eccentricity	Type	Graph	
$e = 1$	Parabola		$\dfrac{|PF|}{|Pl|} = 1$
$e < 1$	Ellipse		$\dfrac{|PF|}{|Pl|} = e < 1$
$e > 1$	Hyperbola		$\dfrac{|PF|}{|Pl|} = e > 1$

---

* This use of e for eccentricity is not to be confused with the use of e as the symbol for the base of natural logarithms.

Suppose a conic with eccentricity $e > 0$ is drawn in a rectangular coordinate system with focus F at $(0, 0)$, and one of the lines $x = d$, $x = -d$, $y = d$, or $y = -d$, where $d > 0$, is the directrix l. By superimposing a polar coordinate system the polar equation of the conic has one of these forms:

Conic		Polar Form of Equation		
Equation:	$r = \dfrac{ed}{1 + e\cos\theta}$	$r = \dfrac{ed}{1 - e\cos\theta}$	$r = \dfrac{ed}{1 + e\sin\theta}$	$r = \dfrac{ed}{1 - e\sin\theta}$
Directrix:	$x = d$	$x = -d$	$y = d$	$y = -d$
Parabola ($e = 1$)	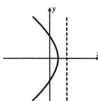	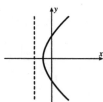	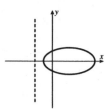	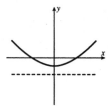
Ellipse ($0 < e < 1$)	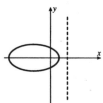	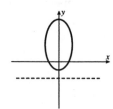	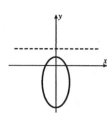	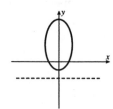
Hyperbola ($e > 1$)	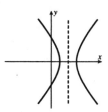	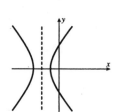	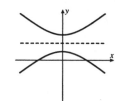	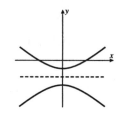

1) The sketch below shows one point P on a conic and its distance from the focus and directrix. What type of conic is it?

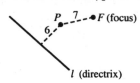

$e = \dfrac{|PF|}{|Pl|} = \dfrac{7}{6} > 1$, so the conic is a hyperbola.

2) What is the polar equation of the conic with:

   a) eccentricity 2, directrix $y = 3$.

$e = 2$ and $d = 3$, so $r = \dfrac{6}{1 + 2 \sin \theta}$.

   b) eccentricity $\frac{1}{2}$, directrix $x = -3$.

$e = \frac{1}{2}$ and $d = 3$, so $r = \dfrac{\frac{3}{2}}{1 - \frac{1}{2} \cos \theta}$,

$r = \dfrac{3}{2 - \cos \theta}$.

   c) directrix $x = 4$ and is a parabola.

$e = 1$ and $d = 4$, so $r = \dfrac{4}{1 + \cos \theta}$.

3) What polar form does the equation of the graph below have?

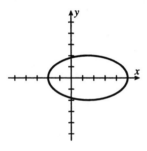

The graph is that of an ellipse. The directrix has the form $x = -d$ so the equation is

$r = \dfrac{ed}{1 - e \cos \theta}$.

4) Find the eccentricity and directrix and identify the conic given by $r = \dfrac{3}{4 - 5 \cos \theta}$.

Divide numerator and denominator by 4.

$r = \dfrac{\frac{3}{4}}{1 - \frac{5}{4} \cos \theta}$, $e = \frac{5}{4}$. Since $ed = \frac{3}{4}$,

$d = \frac{3}{4} \cdot \frac{1}{e} = \frac{3}{4} \cdot \frac{4}{5} = \frac{3}{5}$.

The trigonometric term is $-\cos \theta$ so the directrix is $x = -\frac{3}{5}$. Since $e > 1$, the conic is a hyperbola.

Technology Plus for Chapter 10

1) Use a graphing calculator to sketch a graph of $x = t + 2 \sin 2t$, $y = t + \sin 4t$ for $-9 \le t \le 9$.

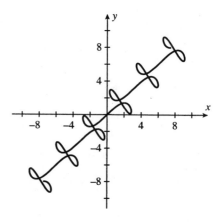

2) Use a CAS to evaluate the definite integral for the arc length of
$x = t + \cos t$
$y = t + \sin t$
for $0 \le t \le 2\pi$.

$\frac{dx}{dt} = 1 - \sin t$, $\frac{dy}{dt} = 1 + \cos t$

The arc length is

$$\int_0^{2\pi} \sqrt{(1 - \sin t)^2 + (1 + \cos t)^2}\, dt$$

≈ 10.037.

3) Sketch the graphs of $r = \cos(2k - 1)\theta$,

$0 \le \theta \le 2\pi$, for $k = 1, 2, 3$ and 4.

How does the graph change as k increases?

$k = 1$

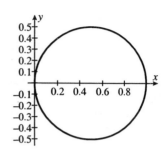

$k = 2$

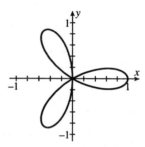

$k = 3$

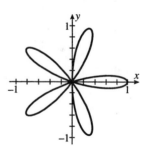

$k = 4$

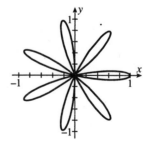

The number of leaves is $2k + 1$.

4) Sketch a graph of $r = \dfrac{1}{1 - \frac{1}{2}\cos\theta}$ on a

graphing calculator. What type of conic is it?

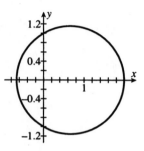

The graph is an ellipse $(e = \frac{1}{2})$.

Chapter 11 —Infinite Sequences and Series

"I'M BEGINNING TO UNDERSTAND ETERNITY, BUT INFINITY IS STILL BEYOND ME."

© 1999 by Sidney Harris.

Section 11.1   Sequences

The main topic of this chapter is the representation of function as a series. A series (introduced in section 12.2) is a sum of an infinite list of numbers—that is, the sum of a sequence of numbers. Thus, this first section describes the basic concepts for sequences.

Concepts to Master

A. Sequences; Limit of a sequence, convergence and divergence

B. Monotone sequence (increasing, decreasing); Bounded sequence; Monotonic Sequence Theorem

C. Sequences defined recursively

Summary and Focus Questions

Page
693

A. A *sequence* is an infinite list of numbers given in a specific order:

$$\{a_n\} = a_1, a_2, a_3, \ldots, a_n, a_{n+1}, \ldots.$$

a_n is called the nth *term* of the sequence.

The sequence $\{a_n\}$ *converges* to a number L, written $\lim\limits_{n \to \infty} a_n = L$, means that the values of a_n get closer and closer to L as n grows larger. The formal definition is:

$\lim\limits_{n \to \infty} a_n = L$ means for all $\epsilon > 0$ there exists a positive integer N such that $|a_n - L| < \epsilon$ for all $n > N$.

If $\lim\limits_{n \to \infty} a_n$ does not exist, we say $\{a_n\}$ *diverges*.

$\lim\limits_{n \to \infty} a_n = \infty$ means the a_n terms grow without bound.

One way to evaluate $\lim\limits_{n \to \infty} a_n$ is to find a real function $f(x)$ such that $f(n) = a_n$ for all n. If $\lim\limits_{n \to \infty} f(x) = L$, then $\lim\limits_{n \to \infty} a_n = L$. The converse is false.

All limit laws for limits of functions at infinity are valid for convergent sequences. A version of the Squeeze Theorem also holds:

If $a_n \leq b_n \leq c_n$ for all $n \geq N$ and $\lim\limits_{n \to \infty} a_n = \lim\limits_{n \to \infty} c_n = L$, then $\lim\limits_{n \to \infty} b_n = L$.

1) Find the fourth term of the sequence

$$a_n = \frac{(-1)^n}{n^2}.$$

$$a_4 = \frac{(-1)^4}{4^2} = \frac{1}{16}.$$

2) $\lim_{n \to \infty} x_n = K$ means
for all _____ there
exists _____ such
that _____ for all
_____.

$\epsilon > 0$
positive integer N
$|x_n - K| < \epsilon$
$n > N.$

3) Determine whether each converges.

a) $a_n = \frac{n}{n^2 + 1}.$

$$\lim_{n \to \infty} \frac{n}{n^2 + 1} = \text{(divide by } n^2)$$

$$\lim_{n \to \infty} \frac{\frac{1}{n}}{1 + \frac{1}{n^2}} = \frac{0}{1 + 0} = 0.$$

Thus a_n converges to 0.

b) $b_n = \frac{n^2 + 1}{2n}.$

b_n grows without bound so $\{b_n\}$ diverges. In this case we may write

$$\lim_{n \to \infty} \frac{n^2 + 1}{2n} = \infty.$$

c) $c_n = \frac{(-1)^n n}{n + 1}.$

The sequence $\frac{n}{n+1}$ converges to 1 so $c_n = \frac{(-1)^n n}{n+1}$ is alternating values near 1 and -1. Thus $\{c_n\}$ diverges.

d) $d_n = \frac{n}{n + 1} + \frac{2}{n^2}.$

$$\lim_{n \to \infty} d_n = \lim_{n \to \infty} \frac{n}{n + 1} + \lim_{n \to \infty} \frac{2}{n^2} = 1 + 0 = 1.$$

4) Find $\lim_{n \to \infty} \frac{\cos n}{n}.$

Since $-1 \le \cos n \le 1$ for all n,

$$-\frac{1}{n} \le \frac{\cos n}{n} \le \frac{1}{n} \text{ for all } n.$$

Because both $\left\{-\frac{1}{n}\right\}$ and $\left\{\frac{1}{n}\right\}$ converge to 0,

$$\lim_{n \to \infty} \frac{\cos n}{n} = 0.$$

5) Find $\lim_{n \to \infty} \frac{\sin x}{n}$.

Since $\sin x$ is a constant as far as n is concerned, $\lim_{n \to \infty} \frac{\sin x}{n} = (\sin x) \lim_{n \to \infty} \frac{1}{n}$
$= (\sin x)\,(0) = 0$.

Silliness: Don't conclude the limit is 6 with this "computation":

$$\lim_{n \to \infty} \frac{\sin x}{n} = \lim_{n \to \infty} \frac{si\!\!\!/n\ x}{n\!\!\!/}$$
$$= \lim_{n \to \infty} six = 6.$$

6) Find $\lim_{n \to \infty} \frac{\ln n}{n}$.

Let $f(x) = \frac{\ln x}{x}$.

$\lim_{x \to \infty} \frac{\ln x}{x} \left(\frac{\infty}{\infty} \text{ form, L'Hospital's Rule} \right)$

$= \lim_{x \to \infty} \frac{\frac{1}{x}}{1} = 0$.

Thus $\lim_{n \to \infty} \frac{\ln n}{n} = 0$.

7) True, False:

a) If $\{a_n\}$ and $\{b_n\}$ converge, then $\{a_n + b_n\}$ converges.

True.

b) If $\{a_n\}$ and $\{b_n\}$ diverge, then $\{a_n + b_n\}$ diverges.

False. For example, $a_n = n^2$ and $b_n = -n^2$ diverge.

8) If $\lim_{n \to \infty} s_n = 4$ and $\lim_{n \to \infty} t_n = 2$, then

a) $\lim_{n \to \infty} (8s_n - 2t_n) = \underline{\qquad}$.

$8(4) - 2(2) = 28$.

b) $\lim_{n \to \infty} \frac{3s_n}{t_n} = \underline{\qquad}$.

$\frac{3(4)}{2} = 6$.

Page
699

B. $\{a_n\}$ is *increasing* means $a_{n+1} \geq a_n$ for all n.
$\{a_n\}$ is *decreasing* means $a_{n+1} \leq a_n$ for all n.
$\{a_n\}$ is *monotonic* if it is either increasing or decreasing.
One way to show that $\{a_n\}$ is increasing is to show that $\dfrac{da_n}{dn} \geq 0$ (treating n
as a real number and a_n as a function of n).
$\{a_n\}$ is *bounded above* means $a_n \leq M$ for some M and all n.
$\{a_n\}$ is *bounded below* means $a_n \geq m$ for some m and all n.
$\{a_n\}$ is *bounded* if it is both bounded above and bounded below.

Monotonic Sequence Theorem: If $\{a_n\}$ is bounded and monotonic, then $\{a_n\}$
converges.

For example, $a_n = \dfrac{1}{e^n + 1}$ is monotonic (decreasing) and bounded ($a_n > 0$

for all n). Therefore $\lim\limits_{n \to \infty} \dfrac{1}{e^n + 1}$ exists (It is 0.)

The converse of the Monotonic Sequence Theorem is "partially true":

   If $\{a_n\}$ converges, then $\{a_n\}$ is bounded.

   However, a convergent sequence need not be monotonic.

9) Is $c_n = \dfrac{1}{3n}$ bounded above? bounded below?

$\left\{\dfrac{1}{3n}\right\}$ is bounded above by $\dfrac{1}{3}$ and bounded
below by 0.

10) Is $s_n = \dfrac{n}{n+1}$ an increasing sequence?

Yes, because
$$\dfrac{ds_n}{dn} = \dfrac{(n+1) - n}{(n+1)^2} = \dfrac{1}{(n+1)^2} > 0.$$

11) True or False:

   a) If $\{a_n\}$ is not bounded below then $\{a_n\}$
       diverges.

True.

   b) If $\{a_n\}$ is decreasing and $a_n \geq 0$ for all
       n then $\lim\limits_{n \to \infty} a_n$ exists.

True.

   c) If $\{a_n\}$ is bounded, then $\{a_n\}$ converges.

False. For example, $a_n = (-1)^n$.

Page
701

C. A sequence $\{a_n\}$ is *defined by a recurrence relation* (defined *recursively*) means:

1. a_1 is defined.

2. a_{n+1}, for $n = 1,2,3, \ldots$, is defined in terms of previous a_i. (Often a_{n+1} is defined using only a_n.)

For example, the sequence $\{a_n\}$ given by

$$a_1 = \frac{1}{2}$$
$$a_{n+1} = \frac{a_n}{2}, n = 1, 2, 3, \ldots$$

is the sequence $\frac{1}{2}, \frac{1}{4}, \frac{1}{8}, \frac{1}{16}, \ldots$. This is a recursive definition of $a_n = \frac{1}{2^n}$.

---

12) Define the sequence $\frac{1}{2}, \frac{3}{4}, \frac{7}{8}, \frac{15}{16}, \frac{31}{32}, \ldots$ with a recurrence relation.

$$a_1 = \frac{1}{2}$$
$$a_{n+1} = a_n + \frac{1}{2^{n+1}}, \text{ for } n = 1, 2, 3, \ldots$$
(Note: a non-recursive answer is
$$a_n = \frac{2^n - 1}{2^n}.)$$

13) Let $a_1 = 2$ and $a_{n+1} = \frac{2a_n}{3}$. Find $\lim\limits_{n \to \infty} a_n$.

$$a_1 = 2, a_2 = \frac{2(2)}{3} = \frac{4}{3},$$

$$a_3 = \frac{2\left(\frac{4}{3}\right)}{3} = \frac{8}{9}, a_4 = \frac{1\left(\frac{8}{9}\right)}{3} = \frac{16}{27},$$

$$a_5 = \frac{2\left(\frac{16}{27}\right)}{3} = \frac{32}{81}.$$

It seems that $\lim\limits_{n \to \infty} a_n = 0$.
This is so since we may write $a_n = 2\left(\frac{2}{3}\right)^{n-1}$.

14) Suppose $\{a_n\}$ is defined as $a_1 = 2$, $a_2 = 1$ and $a_{n+2} = a_{n+1} - a_n$ for $n = 1, 2, 3, \ldots$. Find $\lim\limits_{n \to \infty} a_n$.

Calculate a few terms to understand the pattern.
$a_1 = 2, a_2 = 1$
$a_3 = 1 - 2 = -1$
$a_4 = -1 - 1 = -2$
$a_5 = -2 - (-1) = -1$
$a_6 = -1 - (-2) = 1$
$a_7 = 1 - (-1) = 2$
$a_8 = 2 - 1 = 1$
$a_9 = 1 - 2 = -1$
$\vdots$

The terms cycle through $1, -1, \ldots$ and eventually return to $1, -1$. $\lim\limits_{n \to \infty} a_n$ does not exist.

Section 11.2 Series

A series is the sum of all the terms of an infinite sequence. To determine whether (and if so, to what) a series sums, we build a second sequence: the first term, the sum of the first two terms, the sum of the first three terms, the sum of the first four terms, etc. The sum exists (that is, the series converges) if the limit of this second sequence exists. In this section we define this concept precisely and look at some specific series which converge and others which do not.

Concepts to Master

A. Infinite series, Partial sums; Convergent and divergent series; Convergent series laws

B. Geometric series; Value of a converging geometric series; Harmonic series

Summary and Focus Questions

Page 704

A. Adding up all the terms of a sequence $\{a_n\}$ is an *(infinite) series*:

$$\sum_{k=1}^{\infty} a_k = a_1 + a_2 + a_3 + \dots + a_n + \dots$$

If we stop adding after n terms, we have the *nth partial sum* of the series:

$$s_n = \sum_{k=1}^{n} a_k = a_1 + a_2 + a_3 + \dots + a_n$$

A series *converges* if the limit of its sequence of partial sums exists. In other words $\sum_{n=1}^{\infty} a_n$ *converges* (to s) means $\lim_{n \to \infty} s_n$ exists (and is s).

If $\lim_{n \to \infty} s_n$ does not exist, then $\sum_{n=1}^{\infty} a_n$ *diverges*.

Test for Divergence:

If $\lim_{n \to \infty} a_n$ does not exist or $\lim_{n \to \infty} a_n \neq 0$, $\sum_{n=1}^{\infty} a_n$ diverges.

Thus, for example, the series $\sum_{n=1}^{\infty} 2^n$ diverges, since $\lim_{n \to \infty} 2^n \neq 0$.

However, just because the terms a_n get small is not enough to conclude that $\sum_{n=1}^{\infty} a_n$ converges.

If $\sum_{n=1}^{\infty} a_n$ converges (to L) and $\sum_{n=1}^{\infty} b_n$ converges (to M) then:

$$\sum_{n=1}^{\infty} (a_n \pm b_n) \text{ converges (to } L \pm M).$$

$$\sum_{n=1}^{\infty} ca_n \text{ converges (to } cL) \text{ for } c \text{ any constant.}$$

1) A series converges if the _____ of the sequence of _____ exists.

limit, partial sums

2) Find the first four partial sums of $\displaystyle\sum_{n=1}^{\infty} \frac{1}{n^2}$.

$s_1 = \frac{1}{1^2} = 1.$

$s_2 = \frac{1}{1^2} + \frac{1}{2^2} = 1 + \frac{1}{4} = \frac{5}{4}.$

$s_3 = \frac{1}{1^2} + \frac{1}{2^2} + \frac{1}{3^2} = \frac{5}{4} + \frac{1}{9} = \frac{49}{36}.$

$s_4 = \frac{1}{1^2} + \frac{1}{2^2} + \frac{1}{3^2} + \frac{1}{4^2} = \frac{49}{36} + \frac{1}{16} = \frac{205}{144}.$

3) Sometimes, Always, or Never:

a) If $\displaystyle\lim_{n\to\infty} a_n = 0$, $\displaystyle\sum_{n=1}^{\infty} a_n$ converges.

Sometimes.

b) If $\displaystyle\lim_{n\to\infty} a_n \neq 0$, $\displaystyle\sum a_n$ diverges.

Always.

4) Suppose $\displaystyle\sum_{n=1}^{\infty} a_n = 3$ and $\displaystyle\sum_{n=1}^{\infty} b_n = 4$. Evaluate each of the following:

a) $\displaystyle\sum_{n=1}^{\infty} (a_n + 2b_n)$

$3 + 2(4) = 11.$

b) $\displaystyle\sum_{n=1}^{\infty} \frac{a_n}{5}$

$\frac{3}{5}.$

c) $\displaystyle\sum_{n=1}^{\infty} \frac{1}{a_n}$

This diverges. If $\displaystyle\sum_{n=1}^{\infty} a_n = 3$, then $\displaystyle\lim_{n\to\infty} a_n = 0$. Thus $\displaystyle\lim_{n\to\infty} \frac{1}{a_n} \neq 0$, so $\displaystyle\sum_{n=1}^{\infty} \frac{1}{a_n}$ diverges.

5) Sometimes, Always, or Never:

a) $\displaystyle\sum_{n=1}^{\infty} a_n$ converges, but $\displaystyle\lim_{n\to\infty} a_n$ does not
exist.

Never. If $\displaystyle\sum_{n=1}^{\infty} a_n$ converges, then $\displaystyle\lim_{n\to\infty} a_n$
exists and is 0.

b) $\displaystyle\sum_{n=1}^{\infty} a_n$ converges, $\displaystyle\sum_{n=1}^{\infty} (a_n + b_n)$
converges, but $\displaystyle\sum_{n=1}^{\infty} b_n$ diverges.

Never. $\displaystyle\sum_{n=1}^{\infty} b_n = \sum_{n=1}^{\infty} (a_n + b_n) - \sum_{n=1}^{\infty} a_n$ is
the difference of two convergent series and
must converge.

6) Does $\displaystyle\sum_{n=1}^{\infty} \cos\left(\frac{1}{n}\right)$ converge?

No, the nth term does not approach 0.

**Page
706**

B. A *geometric series* has the form
$$\sum_{n=1}^{\infty} ar^{n-1} = a + ar + ar^2 + ar^3 + \dots$$
and converges (for $a \neq 0$) to $\frac{a}{1-r}$ if and only if $-1 < r < 1$.
The *harmonic series* $\displaystyle\sum_{n=1}^{\infty} \frac{1}{n} = 1 + \frac{1}{2} + \frac{1}{3} + \dots + \frac{1}{n} + \dots$ diverges. This is an
example of a series where $\displaystyle\lim_{n\to\infty} a_n = 0$ but $\displaystyle\sum_{n=1}^{\infty} a_n$ diverges.

7) a) $\displaystyle\sum_{n=1}^{\infty} \frac{2^n}{100}$ is a geometric series in which

$a = $ _____ and $r = $ _____.

$$\sum_{n=1}^{\infty} \frac{2^n}{100} = \frac{2}{100} + \frac{4}{100} + \frac{8}{100} + \dots$$

$a = \frac{2}{100}$ and $r = 2$.

b) Does the series converge?

No, since $r \geq 1$.

8) $\displaystyle\sum_{n=1}^{\infty} \left(\frac{-2}{9}\right)^n = $ _____.

This is a geometric series with $a = r = -\frac{2}{9}$
which converges to $\frac{a}{1-r} = -\frac{2}{11}$.

9) Does $\displaystyle\sum_{n=1}^{\infty} \frac{6}{n}$ converge?

It diverges, since it is a multiple of the harmonic series:

$$\sum_{n=1}^{\infty} \frac{6}{n} = 6\sum_{n=1}^{\infty} \frac{1}{n}.$$

10) $\displaystyle\sum_{n=1}^{\infty} \frac{2^n}{3^{n+1}} =$ _____.

$\displaystyle\sum_{n=1}^{\infty} \frac{2^n}{3^{n+1}} = \sum_{n=1}^{\infty} \frac{2}{9}\left(\frac{2}{3}\right)^{n-1}$, which is a

geometric series with $a = \frac{2}{9}$, $r = \frac{2}{3}$ that

converges to $\displaystyle\frac{\frac{2}{9}}{1 - \frac{2}{3}} = \frac{2}{3}$.

11) Achilles gives the tortoise a 100 m head start. If Achilles runs at 5 m/s and the tortoise at $\frac{1}{2}$ m/s how far has the tortoise traveled by the time Achilles catches him?

Achilles

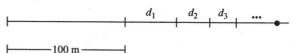

Tortoise

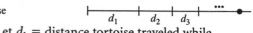

Let $d_1 =$ distance tortoise traveled while Achilles was running the 100 m to the tortoise's starting point.

$$d_1 = \left(\frac{100 \text{ m}}{5 \text{ m/s}}\right)\left(\frac{1}{2} \text{ m/s}\right) = 10 \text{ m}.$$

Let $d_2 =$ distance tortoise traveled while Achilles was running the distance d_1. Since $d_1 = 10$ and Achilles runs at 5 m/s,

$$d_2 = \left(\frac{10 \text{ m}}{5 \text{ m/s}}\right)\left(\frac{1}{2} \text{ m/s}\right) = 1 \text{ m}.$$

In general, for each n, $d_n = \left(\frac{d_{n-1}}{5}\right)\left(\frac{1}{2}\right)$

$= \frac{d_{n-1}}{10}.$

The total distance traveled by the tortoise is

$\displaystyle\sum_{n=1}^{\infty} d_n = 10 + 1 + \frac{1}{10} + \dots$ which is a

geometric series with $a = 10$, $r = \frac{1}{10}$.

This converges to $\displaystyle\frac{10}{1 - \frac{1}{10}} = \frac{100}{9}$ m.

Section 11.3   The Integral Test and Estimates of Sums

This section and the next three sections provide tests to determine whether a series converges without explicitly finding the sum. (It is a good first step to determine whether something exists before trying to calculate or estimate it.) The Integral Test in this section is a natural first choice, for it relates infinite series $\left(\sum\limits_{n=1}^{\infty} \ldots\right)$ to improper integrals $\left(\int_{1}^{\infty} \ldots dn\right)$(; however it only applies to certain series of positive terms. The test, when applicable, also permits us to estimate the sum of the convergent series.

Concepts to Master

A. Integral Test for convergence; *p*-series

B. Estimate of the sum of a convergent series using the Integral Test

Summary and Focus Questions

Page 715

A. *Integral Test:* Let $\{a_n\}$ be a sequence of positive, decreasing terms $(a_{n+1} \le a_n$ for all $n)$. Suppose $f(x)$ is a positive, continuous, and decreasing function on $[1, \infty)$ such that $a_n = f(n)$ for all n. Then $\sum\limits_{n=1}^{\infty} a_n$ converges if and only if $\int_{1}^{\infty} f(x)\,dx$ converges.

The Integral Test works well if a_n has the form of a function whose antiderivative is easily found. It is one of the few general tests that gives necessary and sufficient conditions for convergence.

A *p-series* has the form $\sum\limits_{n=1}^{\infty} \dfrac{1}{n^p}$, where p is a constant. A *p*-series converges (by the Integral Test) if and only if $p > 1$.

1) True or False:
The Integral Test will determine to what value a series converges.

False. The test only indicates whether a series converges.

2) Test $\displaystyle\sum_{n=1}^{\infty} \frac{n}{e^n}$ for convergence.

$\frac{n}{e^n}$ suggests the function $f(x) = xe^{-x}$.
On the interval $[1, \infty]$, f is continuous and decreasing.

$$\int_1^{\infty} xe^{-x} = \lim_{t \to \infty} \int_1^t xe^{-x}\, dx.$$

Using integration by parts,
$(u = x,\ dv = e^{-x}\, dx)$

$$\int_1^t xe^{-x}\, dx = -xe^{-x} - e^{-x}\Big]_1^t = \frac{2}{e} - \frac{t+1}{e^t}.$$

$$\lim_{t \to \infty}\left(\frac{2}{e} - \frac{t+1}{e^t}\right) = \frac{2}{e} - 0 = \frac{2}{e}.$$

(By L'Hospital's Rule, $\frac{t+1}{e^t} \to 0$.)

Thus $\int_1^{\infty} xe^{-x}\, dx = \frac{2}{e}$, so $\displaystyle\sum_{n=1}^{\infty} \frac{n}{e^n}$ converges.

Note: We can *not* conclude that $\displaystyle\sum_{n=1}^{\infty} \frac{n}{e^n} = \frac{2}{e}$.

3) Which of these converge?

a) $\displaystyle\sum_{n=1}^{\infty} \frac{1}{n^2}.$

Converges (*p*-series with $p = 2$).

b) $\displaystyle\sum_{n=1}^{\infty} \frac{1}{\sqrt{n}}.$

Diverges $\left(p\text{-series with } p = \frac{1}{2}\right).$

c) $\displaystyle\sum_{n=1}^{\infty} \frac{3}{2n^3}.$

Converges. $\displaystyle\sum_{n=1}^{\infty} \frac{3}{2n^3} = \frac{3}{2}\sum_{n=1}^{\infty} \frac{1}{n^3}$ is a multiple of a *p*-series with $p = 3$.

d) $\displaystyle\sum_{n=1}^{\infty} \left(\frac{1}{n^3} + \frac{1}{8^n}\right).$

Converges. $\displaystyle\sum_{n=1}^{\infty} \left(\frac{1}{n^3} + \frac{1}{8^n}\right) = \sum_{n=1}^{\infty} \frac{1}{n^3} + \sum_{n=1}^{\infty} \frac{1}{8^n}$, the sum of a convergent *p*-series
$(p = 3)$ and a convergent geometric series
$\left(r = \frac{1}{8}\right).$

Page 717

B. If $\displaystyle\sum_{n=1}^{\infty} a_n = s$ has been found to converge by the Integral Test using $f(x)$, then
the error $R_n = s - s_n$ between the series value and the nth partial sum
satisfies

$$\int_{n+1}^{\infty} f(x)\, dx \le s - s_n \le \int_n^{\infty} f(x)\, dx.$$

4) a) Find the sixth partial sum of $\displaystyle\sum_{n=1}^{\infty}\frac{1}{n^2}$.

$s_6 = 1 + \frac{1}{4} + \frac{1}{9} + \frac{1}{16} + \frac{1}{25} + \frac{1}{36}$

$\quad = \frac{5369}{3600} \approx 1.491.$

b) Estimate the difference between your answer to part a) and the exact value of

$s = \displaystyle\sum_{n=1}^{\infty}\frac{1}{n^2}$.

We know by the Integral Test using

$f(x) = \frac{1}{x^2}$ that $\displaystyle\sum_{n=1}^{\infty}\frac{1}{n^2}$ converges. Thus

$\displaystyle\int_{7}^{\infty}\frac{1}{x^2}\,dx \le s - s_6 \le \int_{6}^{\infty}\frac{1}{x^2}\,dx.$

$\displaystyle\int_{7}^{\infty}\frac{1}{x^2}\,dx = \lim_{t\to\infty}\int_{7}^{t}x^{-2}\,dx$

$\quad = \displaystyle\lim_{t\to\infty}\left(\frac{1}{7} - \frac{1}{t}\right) = \frac{1}{7}.$

Likewise $\displaystyle\int_{6}^{\infty}\frac{1}{x^2}\,dx = \frac{1}{6}.$

Therefore $\frac{1}{7} \le s - s_6 \le \frac{1}{6}.$

c) Estimate $s = \displaystyle\sum_{n=1}^{\infty}\frac{1}{n^2}$ using the results of parts a) and b).

$\frac{1}{7} \le s - s_6 \le \frac{1}{6}.$

$\frac{1}{7} + s_6 \le s \le \frac{1}{6} + s_6.$

$\frac{1}{7} + \frac{5369}{3600} \le s \le \frac{1}{6} + \frac{5369}{3600}$

$\frac{41183}{25200} \le s \le \frac{5969}{3600}$

$1.634 \le s \le 1.658$

It turns out that $s = \frac{\pi^2}{6} \approx 1.645$, so s_6 is a rather good estimate.

5) How many terms of $\displaystyle\sum_{n=1}^{\infty}\frac{1}{n^2+1}$ are necessary to find the sum to within 0.01?

Let $f(x) = \frac{1}{x^2+1}$.

$\displaystyle\sum_{n=1}^{\infty}\frac{1}{n^2+1}$ converges because $\displaystyle\int_{1}^{\infty}\frac{1}{x^2+1}\,dx$

$= \displaystyle\lim_{t\to\infty}\int_{1}^{t}\frac{1}{x^2+1}\,dx$

$= \displaystyle\lim_{t\to\infty}(\tan^{-1}(t) - \tan^{-1}(1)) = \frac{\pi}{2} - \frac{\pi}{4} = \frac{\pi}{4}.$

Since $s - s_n \le \displaystyle\int_{n}^{\infty}f(x)\,dx$, it is sufficient to

find n such that $\displaystyle\int_{n}^{\infty}f(x)\,dx \le 0.01$

From above, we see that

$\displaystyle\int_{n}^{\infty}\frac{1}{x^2+1}\,dx = \frac{\pi}{2} - \tan^{-1}(n).$

$\frac{\pi}{2} - \tan^{-1}(n) < 0.01$

$\tan^{-1}(n) > \frac{\pi}{2} - 0.01$

$n > \tan\left(\frac{\pi}{2} - 0.01\right) \approx 99.997.$

The first 100 terms are sufficient.

Section 11.4   The Comparison Tests

This section deals only with series of positive terms. If such a series converges, then any other series that is term for term smaller than that one will also converge. This section makes this notion of comparing series of positive terms precise and discusses estimating limits.

Concepts to Master

A. Comparison Test; Limit Comparison Test
B. Estimate the sum of a convergent series

Summary and Focus Questions

Page 722

A. Let $\sum\limits_{n=1}^{\infty} a_n$ and $\sum\limits_{n=1}^{\infty} b_n$ be series whose terms are all *positive*.

Comparison Test:

   1. If $a_n \leq b_n$ for all n and $\sum\limits_{n=1}^{\infty} b_n$ converges, then $\sum\limits_{n=1}^{\infty} a_n$ converges.

   2. If $a_n \geq b_n$ for all n and $\sum\limits_{n=1}^{\infty} b_n$ diverges, then $\sum\limits_{n=1}^{\infty} a_n$ diverges.

Limit Comparison Test:

   1. If $\lim\limits_{n\to\infty} \dfrac{a_n}{b_n} = c > 0$, then $\sum\limits_{n=1}^{\infty} a_n$ and $\sum\limits_{n=1}^{\infty} b_n$ either both converge or both diverge.

   2. If $\lim\limits_{n\to\infty} \dfrac{a_n}{b_n} = 0$ and $\sum\limits_{n=1}^{\infty} b_n$ converges, then $\sum\limits_{n=1}^{\infty} a_n$ converges.

   3. If $\lim\limits_{n\to\infty} \dfrac{a_n}{b_n} = 0$ and $\sum\limits_{n=1}^{\infty} a_n$ diverges, then $\sum\limits_{n=1}^{\infty} b_n$ diverges.

For a given series $\sum\limits_{n=1}^{\infty} a_n$, success using either comparison test to determine whether $\sum\limits_{n=1}^{\infty} a_n$ converges will depend on coming up with another series $\sum\limits_{n=1}^{\infty} b_n$ (geometric, *p*-series, ...) whose convergence is known and for which a_n and b_n may be compared.

Example: Determine whether $\displaystyle\sum_{n=1}^{\infty} \frac{1}{\sqrt{n}+4}$ converges.

You need a hunch beforehand whether $\displaystyle\sum_{n=1}^{\infty} \frac{1}{\sqrt{n}+4}$ converges. Because

$\frac{1}{\sqrt{n}+4}$ is "like" $\frac{1}{\sqrt{n}}$ and $\displaystyle\sum_{n=1}^{\infty} \frac{1}{\sqrt{n}}$ is a divergent p-series $\left(p = \frac{1}{2}\right)$, we have

reason to believe that $\displaystyle\sum_{n=1}^{\infty} \frac{1}{\sqrt{n}+4}$ diverges. We also have an idea of the type

of b_n to look for—one for which $a_n \geq b_n$:

Let $b_n = \frac{1}{3\sqrt{n}}$. Then for all n, $\frac{1}{\sqrt{n}+4} \geq \frac{1}{3\sqrt{n}}$. Therefore, $\displaystyle\sum_{n=1}^{\infty} \frac{1}{\sqrt{n}+4}$ diverges.

1) Determine whether or not each of the following converge. Find a series $\displaystyle\sum_{n=1}^{\infty} b_n$ to use with the Comparison Test or Limit Comparison Test.

a) $\displaystyle\sum_{n=1}^{\infty} \frac{1}{n^2+2n}$, $b_n = $ _____.

Because $\frac{1}{n^2+2n}$ is "like" $\frac{1}{n^2}$ for large n, and $\displaystyle\sum_{n=1}^{\infty} \frac{1}{n^2}$ is a convergent p-series, we suspect that the given series converges.

Use $b_n = \frac{1}{n^2}$. Since $\frac{1}{n^2 + 2n} < \frac{1}{n^2}$ and

$\displaystyle\sum_{n=1}^{\infty} \frac{1}{n^2}$ converges, $\displaystyle\sum_{n=1}^{\infty} \frac{1}{n^2 + 2n}$ converges by the Comparison Test.

b) $\displaystyle\sum_{n=1}^{\infty} \frac{\sqrt[3]{n}}{n+4}$, $b_n = $ _____.

Since $\frac{\sqrt[3]{n}}{n+4}$ is "like" $\frac{\sqrt[3]{n}}{n} = \frac{1}{\sqrt[3]{n^2}}$,

use $b_n = \frac{1}{\sqrt[3]{n^2}}$.

Then $\displaystyle\lim_{n\to\infty} \frac{a_n}{b_n} = \lim_{n\to\infty} \frac{\frac{\sqrt[3]{n}}{n+4}}{\frac{1}{\sqrt[3]{n^2}}} = \lim_{n\to\infty} \frac{n}{n+4} = 1.$

Since $\displaystyle\sum_{n=1}^{\infty} \frac{1}{\sqrt[3]{n^2}}$ diverges $\left(\text{a } p\text{-series with}\right.$

$p = \frac{2}{3}\Big)$, $\displaystyle\sum_{n=1}^{\infty} \frac{\sqrt[3]{n}}{n+4}$ diverges by the Limit

Comparison Test.

c) $\displaystyle\sum_{n=1}^{\infty} \frac{1}{n+2^n}, b_n = \underline{\quad\quad}.$

Use $b_n = \dfrac{1}{2^n}$. Then for $n \geq 1, \dfrac{1}{n+2^n} \leq \dfrac{1}{2^n}$.

Since $\displaystyle\sum_{n=1}^{\infty} \frac{1}{2^n}$ converges $\left(\text{a geometric series}\right.$

with $\left. r = \dfrac{1}{2}\right), \displaystyle\sum_{n=1}^{\infty} \frac{1}{n+2^n}$ converges.

2) Suppose $\{a_n\}$ and $\{b_n\}$ are positive sequences. Sometimes, Always, Never:

a) If $\displaystyle\lim_{n\to\infty} \frac{a_n}{b_n} = 0$ and $\displaystyle\sum_{n=1}^{\infty} a_n$ converges,

then $\displaystyle\sum_{n=1}^{\infty} b_n$ converges.

Sometimes. True for $a_n = \dfrac{1}{n^3}$ and $b_n = \dfrac{1}{n^2}$.

False for $a_n = \dfrac{1}{n^2}$ and $b_n = \dfrac{1}{n}$.

b) If $\displaystyle\lim_{n\to\infty} \frac{a_n}{b_n} = \infty$ and $\displaystyle\sum_{n=1}^{\infty} b_n$ diverges,

then $\displaystyle\sum_{n=1}^{\infty} a_n$ diverges.

Always.

Page 724

B. If $\displaystyle\sum_{n=1}^{\infty} a_n = s$ converges by the Comparison Test using $t = \displaystyle\sum_{n=1}^{\infty} b_n$, then

$$s - s_n \leq t - t_n.$$

This means that $t - t_n$ (which may be easier to calculate) is an upper estimate for $s - s_n$.

3) In question 1c), $\displaystyle\sum_{n=1}^{\infty} \frac{1}{n+2^n}$ converges.

a) Find s_4, the fourth partial sum.

$$s_4 = \frac{1}{1+2} + \frac{1}{2+4} + \frac{1}{3+8} + \frac{1}{4+16}$$
$$= \frac{1}{3} + \frac{1}{6} + \frac{1}{11} + \frac{1}{20} = \frac{141}{220}.$$

b) Estimate the difference between this series and its fourth partial sum.

From 1c), $b_n = \dfrac{1}{2^n}$ may be used to show

$\displaystyle\sum \frac{1}{n+2^n}$ converges. Let $s = \displaystyle\sum_{n=1}^{\infty} \frac{1}{n+2^n}$ and

$t = \displaystyle\sum_{n=1}^{\infty} \frac{1}{2^n}$. Then $s - s_4 \leq t - t_4$.

Since t is the result of a geometric series we can calculate it:

$$t = \frac{\frac{1}{2}}{1 - \frac{1}{2}} = 1$$

$$t_4 = \frac{1}{2} + \frac{1}{4} + \frac{1}{8} + \frac{1}{16} = \frac{15}{16}.$$

Thus, $t - t_4 = 1 - \dfrac{15}{16} = \dfrac{1}{16}$ and

$s - s_4 \leq \dfrac{1}{16}$. Therefore, we know $s_4\left(\dfrac{141}{220}\right)$ is

within $\dfrac{1}{16}$ of the value of s.

Section 11.5   Alternating Series

An alternating series is one for which consecutive terms have opposite signs. This section shows you a test for convergence of these kinds of series and an especially simple method to estimate the sum.

Concepts to Master

A. Alternating Series Test
B. Estimating the sum of a convergent alternating series

Summary and Focus Questions

Page 727

A. An *alternating series* has successive terms of opposite sign—that is, it has one of these forms:

$$\sum_{n=1}^{\infty} (-1)^{n-1} a_n \ \text{ or } \ \sum_{n=1}^{\infty} (-1)^n a_n, \text{ where } a_n > 0.$$

The Alternating Series Test:

If $\{a_n\}$ is a decreasing sequence with $\lim_{n\to\infty} a_n = 0$, then $\sum_{n=1}^{\infty} (-1)^{n-1} a_n$ converges.

1) Is $\sum_{n=1}^{\infty} \dfrac{\sin n}{n}$ an alternating series?

No, although some terms are positive and others negative.

2) Determine whether each converge:

a) $\sum_{n=1}^{\infty} \dfrac{(-1)^{n-1}}{e^n}$.

This is an alternating series with $a_n = \dfrac{1}{e^n}$.

$\dfrac{1}{e^{n+1}} \leq \dfrac{1}{e^n}$, so the terms decrease.

Since $\lim_{n\to\infty} \dfrac{1}{e^n} = 0$, $\sum_{n=1}^{\infty} \dfrac{(-1)^n}{e^n}$ converges by the Alternating Series Test.

b) $\displaystyle\sum_{n=1}^{\infty} \frac{(-1)^n n}{n+1}$.

This is an alternating series but the other conditions for the test do not hold.

Since $\displaystyle\lim_{n\to\infty} \frac{n}{n+1} \neq 0$, $\displaystyle\sum_{n=1}^{\infty} \frac{(-1)^n n}{n+1}$ diverges by the Divergence Test.

c) $\displaystyle\sum_{n=1}^{\infty} \frac{(-1)^{n-1} \ln n}{n}$

It may not be obvious that $a_n = \dfrac{\ln n}{n}$ decreases. Let $f(x) = \dfrac{\ln x}{x}$.

Then $f'(x) = \dfrac{x\left(\frac{1}{x}\right) - \ln x\,(1)}{x^2} = \dfrac{1 - \ln x}{x^2}$.

For $x > 1$, $f'(x) < 0$. Thus $f(x)$, and hence $\{a_n\}$ is decreasing. Finally,

$$\lim_{n\to\infty} \frac{\ln n}{n} = \lim_{n\to\infty} \frac{\frac{1}{n}}{1} = 0 \text{ by L'Hospital's Rule.}$$

Therefore, $\displaystyle\sum_{n=1}^{\infty} \frac{(-1)^{n-1} \ln n}{n}$ converges.

B. If $\displaystyle\sum_{n=1}^{\infty} (-1)^{n-1} a_n$ is an alternating series with $0 \leq a_{n+1} \leq a_n$ and $\displaystyle\lim_{n\to\infty} a_n = 0$ then

Page 729

$$\left|s_n - s\right| \leq a_{n+1}.$$

Since the difference between the limit of the series and the nth partial sum does not exceed a_{n+1}, s_n may be used to approximate s to within an accuracy of a_{n+1} by s_n.

3) The series $\displaystyle\sum_{n=1}^{\infty} \frac{(-1)^n}{\sqrt{n+1}}$ converges.

Estimate the error between the sum of the series and its fifteenth partial sum.

Since the series alternates and $\dfrac{1}{\sqrt{n+1}}$ decreases to 0, the error $\left|s_{15} - s\right|$ does not exceed $a_{16} = \dfrac{1}{\sqrt{16+1}} = 0.2$.

4) For what value of n is s_n, the nth partial sum, within 0.001 of $s = \sum\limits_{n=1}^{\infty} \dfrac{(-1)^{n+1}}{2n+5}$?

$$|s_n - s| \leq a_{n+1} = \frac{1}{2(n+1)+5} = \frac{1}{2n+7}$$
$$\leq 0.001.$$

$2n + 7 \geq 1000,\ 2n \geq 993,\ n \geq 496.5.$

Let $n = 497$. Then s_{497} is within 0.001 of s.

5) Approximate $\sum\limits_{n=1}^{\infty} \dfrac{(-1)^{n+1}}{n^3}$ to within 0.005.

$$|s_n - s| \leq a_{n+1} = \frac{1}{(n+1)^3} \leq 0.005 = \frac{1}{200}.$$

$$(n+1)^3 \geq 200$$

$$n + 1 \geq \sqrt[3]{200} \approx 5.84$$

$$n \geq 4.84.\ \text{Let } n = 5.$$

Therefore s_5 is within 0.005 of s.

$$s_5 = \frac{1}{1^3} - \frac{1}{2^3} + \frac{1}{3^3} - \frac{1}{4^3} + \frac{1}{5^3}$$
$$= 1 - \frac{1}{8} + \frac{1}{27} - \frac{1}{64} + \frac{1}{125} \approx 0.9044.$$

We do not know to what the series converges, but that value is within 0.005 of 0.9044.

Section 11.6 Absolute Convergence and the Ratio and Root Tests

The concept of convergence of an infinite series may be split into two separate subconcepts: absolute convergence and conditional convergence. This section gives a precise definition of each. It also describes the Root and Ratio Tests, which may be used to check for absolute convergence.

Concepts to Master

A. Absolute convergence; Conditional convergence

B. Ratio Test; Root Test

C. Rearrangement of terms of a series

Summary and Focus Questions

Page 731

A. For any series (with a_n not necessarily positive or alternating) two types of convergence may be defined:

$$\sum_{k=1}^{\infty} a_k \text{ converges } \textit{absolutely} \text{ means that } \sum_{k=1}^{\infty} |a_k| \text{ converges.}$$

$$\sum_{k=1}^{\infty} a_k \text{ converges } \textit{conditionally} \text{ means that } \sum_{k=1}^{\infty} a_k \text{ converges but } \sum_{k=1}^{\infty} |a_k|$$

diverges.

Either one of absolute convergence or conditional convergence implies (ordinary) convergence. Conversely, convergence implies either absolute or conditional convergence. Thus *every* series must behave in exactly one of these three ways: diverge, converge absolutely, or converge conditionally.

Examples: $\displaystyle\sum_{n=1}^{\infty} \frac{(-1)^{n-1}}{n^3}$ is absolutely convergent because the series

$$\sum_{n=1}^{\infty} \left| \frac{(-1)^{n-1}}{n^s} \right| = \sum_{n=1}^{\infty} \frac{1}{n^3} \text{ converges } (p\text{-series}, p = 3). \text{ On the other hand,}$$

$\displaystyle\sum_{n=1}^{\infty} \frac{(-1)^{n-1}}{\sqrt[3]{n}}$ is conditionally convergent because it converges (by the

Alternating Series Test) but $\displaystyle\sum_{n=1}^{\infty} \left| \frac{(-1)^{n-1}}{\sqrt[3]{n}} \right| = \sum_{n=1}^{\infty} \frac{1}{\sqrt[3]{n}}$ diverges

$\left(p\text{-series}, p = \frac{1}{3} \right).$

1) Sometimes, Always, or Never:

a) If $\sum\limits_{n=1}^{\infty} |a_n|$ diverges, then $\sum\limits_{n=1}^{\infty} a_n$ diverges.

Sometimes. True for $a_n = n$ but false for $a_n = \dfrac{(-1)^n}{n}$.

b) If $\sum\limits_{n=1}^{\infty} a_n$ diverges, then $\sum\limits_{n=1}^{\infty} |a_n|$ diverges.

Always.

c) If $a_n \geq 0$ for all n, then $\sum\limits_{n=1}^{\infty} a_n$ is not conditionally convergent.

Always, since $|a_n| = a_n$ for all n.

2) Determine whether each converges conditionally, converges absolutely, or diverges.

a) $\sum\limits_{n=1}^{\infty} \dfrac{(-1)^n}{\sqrt{n}}$.

The series is alternating with $\dfrac{1}{\sqrt{n}}$ decreasing to 0 so it converges. It remains to check for absolute convergence:

$$\sum_{n=1}^{\infty} \left| \frac{(-1)^n}{\sqrt{n}} \right| = \sum_{n=1}^{\infty} \frac{1}{\sqrt{n}} \text{ diverges } (p\text{-series}$$

with $p = \tfrac{1}{2}$). Thus $\sum\limits_{n=1}^{\infty} \dfrac{(-1)^n}{\sqrt{n}}$ converges conditionally.

b) $\sum\limits_{n=1}^{\infty} \dfrac{\sin n + \cos n}{n^3}$.

Check for absolute convergence first:
$$\sum_{n=1}^{\infty} \left| \frac{\sin n + \cos n}{n^3} \right| = \sum_{n=1}^{\infty} \frac{|\sin n + \cos n|}{n^3}.$$
Since $|\sin n + \cos n| \leq 2$ and $\sum\limits_{n=1}^{\infty} \dfrac{2}{n^3}$ converges (it is a p-series with $p = 3$),

$\sum\limits_{n=1}^{\infty} \dfrac{|\sin n + \cos n|}{n^3}$ converges by the

Comparison Test. Thus $\sum\limits_{n=1}^{\infty} \dfrac{\sin n + \cos n}{n^3}$ converges absolutely.

3) Does $\sum_{n=1}^{\infty} \dfrac{\sin e^n}{e^n}$ converge?

Yes. The series is not alternating but does contain both positive and negative terms. Check for absolute convergence:

$$\sum_{n=1}^{\infty} \left| \frac{\sin e^n}{e^n} \right| = \sum_{n=1}^{\infty} \frac{\left| \sin e^n \right|}{e^n}.$$

Since $\left| \sin e^n \right| \le 1$, $\dfrac{\left| \sin e^n \right|}{e^n} \le \dfrac{1}{e^n}$.

$\sum_{n=1}^{\infty} \dfrac{1}{e^n}$ converges $\left(\text{geometric series with } r = \dfrac{1}{e}\right)$ so by the Comparison Test

$\sum_{n=1}^{\infty} \left| \dfrac{\sin e^n}{e^n} \right|$ converges. Therefore $\sum_{n=1}^{\infty} \dfrac{\sin e^n}{e^n}$

converges absolutely and hence converges.

Page 733

B. The following tests are very useful for determining whether a series converges absolutely. Neither involves another series or function, which makes them relatively easy to use; however, each has cases where it fails.

The Ratio Test works well when the nth term, a_n, contains exponentials or factorials or when a_n is defined recursively. It will fail when a_n is a rational function of n.

Ratio Test: Let a_n be a sequence of nonzero terms.

1. If $\lim_{n \to \infty} \left| \dfrac{a_{n+1}}{a_n} \right| = L < 1$, then $\sum_{n=1}^{\infty} a_n$ converges absolutely.

2. If $\lim_{n \to \infty} \left| \dfrac{a_{n+1}}{a_n} \right| = L > 1$ or $\lim_{n \to \infty} \left| \dfrac{a_{n+1}}{a_n} \right| = \infty$, then $\sum_{n=1}^{\infty} a_n$ diverges.

3. If $\lim_{n \to \infty} \left| \dfrac{a_{n+1}}{a_n} \right| = 1$, the Ratio Test fails: $\sum_{n=1}^{\infty} a_n$ may converge or diverge.

The Root Test works well when a_n contains an expression to the nth power.

Root Test:

1. If $\lim_{n \to \infty} \sqrt[n]{|a_n|} = L < 1$, then $\sum_{n=1}^{\infty} a_n$ converges absolutely.

2. If $\lim_{n \to \infty} \sqrt[n]{|a_n|} = L > 1$ or $\lim_{n \to \infty} \sqrt[n]{|a_n|} = \infty$, $\sum_{n=1}^{\infty} a_n$ diverges.

3. If $\lim_{n \to \infty} \sqrt[n]{|a_n|} = 1$, the Root Test fails: $\sum_{n=1}^{\infty} a_n$ may converge or diverge.

4) Determine whether each converges by the Ratio Test.

a) $\sum\limits_{n=1}^{\infty} \dfrac{n}{4^n}$.

$$\lim_{n\to\infty} \left| \frac{a_{n+1}}{a_n} \right| = \lim_{n\to\infty} \frac{\frac{n+1}{4^{n+1}}}{\frac{n}{4^n}} = \lim_{n\to\infty} \frac{n+1}{4n} = \frac{1}{4}.$$

Thus $\sum\limits_{n=1}^{\infty} \dfrac{n}{4^n}$ converges absolutely and therefore converges.

b) $\sum\limits_{n=1}^{\infty} \dfrac{(-4)^n}{n!}$.

$$\lim_{n\to\infty} \left| \frac{a_{n+1}}{a_n} \right| = \lim_{n\to\infty} \left| \frac{\frac{(-4)^{n+1}}{(n+1)!}}{\frac{(-4)^n}{n!}} \right| = \lim_{n\to\infty} \frac{4}{n+1} = 0.$$

Thus $\sum\limits_{n=1}^{\infty} \dfrac{(-4)^n}{n!}$ converges absolutely and therefore converges.

c) $\sum\limits_{n=1}^{\infty} \dfrac{(-1)^n}{\sqrt[3]{n}}$.

$$\lim_{n\to\infty} \left| \frac{a_{n+1}}{a_n} \right| = \lim_{n\to\infty} \left| \frac{\frac{(-1)^{n+1}}{\sqrt[3]{n+1}}}{\frac{(-1)^n}{\sqrt[3]{n}}} \right|$$

$$= \lim_{n\to\infty} \sqrt[3]{\frac{n}{n+1}} = 1.$$

The Ratio Test fails. The Alternating Series Test can be used to show that this series converges.

d) $\sum\limits_{n=1}^{\infty} a_n$, where $a_1 = 4$ and $a_{n+1} = \dfrac{3a_n}{2n+1}$.

$$\lim_{n\to\infty} \left| \frac{a_{n+1}}{a_n} \right| = \lim_{n\to\infty} \frac{\frac{3a_n}{2n+1}}{a_n} = \lim_{n\to\infty} \frac{3}{2n+1} = 0.$$

Thus $\sum\limits_{n=1}^{\infty} a_n$ converges absolutely and therefore converges.

5) Determine whether each converges by the Root Test:

a) $\displaystyle\sum_{n=1}^{\infty} \frac{1}{(n+1)^n}.$

$$\lim_{n\to\infty} \sqrt[n]{|a_n|} = \lim_{n\to\infty} \sqrt[n]{\frac{1}{(n+1)^n}} = \lim_{n\to\infty} \frac{1}{n+1} = 0.$$

Thus $\displaystyle\sum_{n=1}^{\infty} \frac{1}{(n+1)^n}$ converges absolutely and therefore converges.

b) $\displaystyle\sum_{n=1}^{\infty} \frac{3^n}{n^3}.$

$$\lim_{n\to\infty} \sqrt[n]{|a_n|} = \lim_{n\to\infty} \sqrt[n]{\frac{3^n}{n^3}} = \lim_{n\to\infty} \frac{3}{n^{3/n}} = \frac{3}{1} = 3.$$

Therefore by the Root Test $\displaystyle\sum_{n=1}^{\infty} \frac{3^n}{n^3}$ diverges.

To show $\displaystyle\lim_{n\to\infty} n^{3/n} = 1$ let $y = \lim_{n\to\infty} n^{3/n}.$

Then $\ln y = \displaystyle\lim_{n\to\infty} \ln n^{3/n} = \lim_{n\to\infty} \frac{3\ln n}{n}$

$\left(\text{form } \dfrac{\infty}{\infty}\right) = \displaystyle\lim_{n\to\infty} \frac{\frac{3}{n}}{1} = 0.$

Thus $y = e^0 = 1.\Big)$

c) $\displaystyle\sum_{n=1}^{\infty} \frac{(-1)^n}{n}$

$$\lim_{n\to\infty} \sqrt[n]{\left|\frac{(-1)^n}{n}\right|} = \lim_{n\to\infty} \frac{1}{\sqrt[n]{n}} = 1.$$

The Root Test fails, but we know this series converges because it is the alternating harmonic series.

C. A *rearrangement* of an infinite series is another series obtained by changing the order of the terms.

Page 736

If $\displaystyle\sum_{n=1}^{\infty} a_n$ converges absolutely, then all rearrangements converge to the same sum.

If $\displaystyle\sum_{n=1}^{\infty} a_n$ converges conditionally, then rearrangements do not all converge to the same sum.

6) Do all rearrangements of $\displaystyle\sum_{n=1}^{\infty} \frac{(-1)^{n-1}}{n^4}$ converge to the same value?

Yes. The series is absolutely convergent since $\displaystyle\sum_{n=1}^{\infty} \frac{1}{n^4}$ is a p-series ($p = 4$).

Section 11.7 Strategy for Testing Series

There are no new techniques in this section – just a summary of the strategies to follow for determining whether a given series converges.

Concepts to Master

Apply the various tests of convergence

Summary and Focus Questions

Page 738

The main strategy for determining whether a given series converges is to observe the form of the nth term—whether it matches a certain type (p-series, alternating, etc.), resembles a known form (comparison tests or Integral Test), or is amenable to calculation using the Ratio or Root Tests. Here is a brief summary of each test together with a representative example of its use:

Test Name	Form/Conditions	Conclusion(s)	Example
Divergence	$\lim\limits_{n\to\infty} a_n \neq 0$	$\sum a_n$ diverges	
Integral	Find $f(x)$, decreasing with $f(n) = a_n \geq 0$.	$\sum a_n$ converges if and only if $\int_1^\infty f(x)\,dx$ converges.	
p-series	$a_n = \dfrac{1}{n^p}$	$\sum \dfrac{1}{n^p}$ converges if and only if $p > 1$.	
Geometric Series	$a_n = ar^{n-1}$	$\sum ar^{n-1}$ converges if and only if $|r| < 1$.	
Alternating Series	$a_n = (-1)^n b_n$ $(b_n \geq 0)$	$\sum a_n$ converges if $\lim\limits_{n\to\infty} b_n = 0$ and b_n is decreasing.	
Comparison	$0 \leq a_n \leq b_n$	If $\sum b_n$ converges then $\sum a_n$ converges. If $\sum a_n$ diverges then $\sum b_n$ diverges.	
Limit Comparison	$\lim\limits_{n\to\infty} \dfrac{a_n}{b_n} = c$	If $0 < c < \infty$, $\sum a_n$ converges if and only if $\sum b_n$ does.	

Test Name	Form/Conditions	Conclusion(s)	Example
Ratio	$\lim\limits_{n\to\infty} \left| \dfrac{a_{n+1}}{a_n} \right| = L$	If $L < 1$, $\sum a_n$ converges absolutely. If $L > 1$, $\sum a_n$ diverges.	
Root	$\lim\limits_{n\to\infty} \sqrt[n]{|a_n|} = L$	If $L < 1$, $\sum a_n$ converges absolutely. If $L > 1$, $\sum a_n$ diverges.	

For each series, what is an appropriate test to apply for each?

1) $\sum\limits_{n=1}^{\infty} \dfrac{(n+1)^n}{4^n}$.

Root Test, because of the $(n+1)^n$ and 4^n exponentials.

2) $\sum\limits_{n=1}^{\infty} \dfrac{5}{n^2 + 6n + 3}$.

Comparison Test, compare to $\sum \dfrac{1}{n^2}$.

3) $\sum\limits_{n=1}^{\infty} \dfrac{(-1)^n n}{n^6 + 1}$.

Alternating Series Test for convergence, then Comparison Test (compare to $\sum \dfrac{1}{n^5}$).

4) $\sum\limits_{n=1}^{\infty} \dfrac{2^n n^3}{n!}$.

Ratio Test, because of the $n!$ term.

5) $\sum\limits_{n=1}^{\infty} \dfrac{n}{e^{n^2}}$.

The exponential e^{n^2} suggests the Ratio Test; the function $f(x) = \dfrac{x}{e^{x^2}} = xe^{-x^2}$ may be used in the Integral Test.

Section 11.8   Power Series

This section introduces a special type of function called a power series whose functional values are infinite series. As such, there will be values for its variable x for which the power series is defined (for that x, the resulting series converges) and others for which it does not exist (resulting series diverges). We shall see that the tests of convergence will help in determining the domain of a power series.

Concepts to Master

Power series; Interval of Convergence; Radius of convergence

Summary and Focus Questions

Page 740

A *power series in (x-a)* is an expression of the form

$$\sum_{n=0}^{\infty} c_n (x - a)^n, \text{ where } c_n \text{ and } a \text{ are constants.}$$

For any particular value of x, the value of the power series is an infinite series. Thus, for some values of x the expression converges and for others it may diverge.

The domain of a power series in $(x-a)$ is those values of x for which it converges and is called the *interval of convergence*. It consists of all real numbers from $a - R$ to $a + R$ for some number R, where $0 \le R \le \infty$. The power series diverges for all x outside the interval of convergence. The number a is called the *center* and R is the *radius* of convergence. When $R = 0$, the interval is the single point $\{a\}$; when $R = \infty$, the interval is all real numbers.

The value of R is found by applying the Ratio Test to $\sum_{n=0}^{\infty} c_n (x - a)^n$ and solving the resulting inequality, $\lim_{n\to\infty} \left| \frac{c_{n+1}(x - a)^{n+1}}{c_n(x - a)^n} \right| = \lim_{n\to\infty} \left| \frac{c_{n+1}}{c_n} \right| |x - a| < 1$, for $|x - a|$.

When $0 < R < \infty$, the two endpoints $a - R$ and $a + R$ pose special problems: the Ratio Test fails and the endpoint may or may not be in the interval of convergence. You must use some other test to determine convergence at each endpoint.

1) True or False:
A power series is an infinite series.

False. A power series is an expression for a function, $f(x)$. For any x in the domain of f, $f(x)$ is the sum of a convergent series.

2) For what value of x does $\displaystyle\sum_{n=0}^{\infty} c_n(x-a)^n$ always converge?

At $x = a$, $\displaystyle\sum_{n=0}^{\infty} c_n(x-a)^n = c_0$.

The power series always converges when x is the center.

3) Compute the value of:

a) $\displaystyle\sum_{n=0}^{\infty} n^2(x-3)^n$ at $x = 4$.

When $x = 4$ the power series is

$$\sum_{n=0}^{\infty} n^2 1^n = \sum_{n=0}^{\infty} n^2. \text{ This infinite series}$$

diverges so there is no value for $x = 4$.

b) $\displaystyle\sum_{n=0}^{\infty} 2^n(x-1)^n$ at $x = \frac{5}{6}$.

When $x = \frac{5}{6}$ the power series is

$$\sum_{n=0}^{\infty} 2^n\left(-\frac{1}{6}\right)^n = \sum_{n=0}^{\infty} \left(-\frac{1}{3}\right)^n$$

$$= 1 - \frac{1}{3} + \frac{1}{9} - \frac{1}{27} + \cdots$$

This is a geometric series with $a = 1$, $r = -\frac{1}{3}$. It converges to $\dfrac{1}{1 - \left(-\frac{1}{3}\right)} = \dfrac{3}{4}$.

4) Find the center, radius, and interval of convergence for:

a) $\displaystyle\sum_{n=1}^{\infty} (2n)!(x-1)^n$.

Use the Ratio Test:

$$\lim_{n\to\infty} \left| \frac{(2(n+1))!(x-1)^{n+1}}{(2n)!(x-1)^n} \right|$$

$$= \left(\lim_{n\to\infty} (2n+1)(2n+2)\right)|x-1| = \infty,$$

for all x except when $x = 1$. Thus the center is 1, the radius is 0, and the interval of convergence is $\{1\}$.

b) $\displaystyle\sum_{n=1}^{\infty} \frac{(x-4)^n}{2^n n}$.

Use the Ratio Test:

$$\lim_{n\to\infty} \left| \frac{(x-4)^{n+1}}{2^{n+1}(n+1)} \cdot \frac{2^n n}{(x-4)^n} \right|$$

$$= \lim_{n\to\infty} \frac{n}{2(n+1)} |x-4| = \tfrac{1}{2}|x-4|.$$

From $\tfrac{1}{2}|x-4| < 1$ we conclude

$|x-4| < 2$. The center is 4 and radius is 2.

When $|x-4| = 2$, $x = 2$ or $x = 6$.

At $x = 2$, the power series becomes

$\displaystyle\sum_{n=1}^{\infty} \frac{(-1)^n}{n}$ which converges by the

Alternating Series Test.

At $x = 6$, the power series is the divergent

harmonic series $\displaystyle\sum_{n=1}^{\infty} \frac{1}{n}$.

The interval of convergence is $[2, 6)$.

c) $\displaystyle\sum_{n=1}^{\infty} \frac{x^n}{n!}$

Use the Ratio Test:

$$\lim_{n\to\infty} \left| \frac{\frac{x^n+1}{(n+1)!}}{\frac{x^n}{n!}} \right| = \lim_{n\to\infty} \frac{|x|}{n+1} = 0$$

for all values of x. Thus the interval of convergence is all real numbers, $(-\infty, \infty)$.

5) Suppose $\displaystyle\sum_{n=0}^{\infty} c_n(x-5)^n$ converges for $x = 8$. For what other values must it converge?

The interval of convergence is $5 - R$ to $5 + R$.

Because 8 is in this interval $8 \le 5 + R$.
Thus $3 \le R$ which means the interval at least contains all numbers between $5 - 3$ and $5 + 3$. Thus the power series converges for at least all $x \in (2, 8]$.

6) True, False: The two power series

$\displaystyle\sum_{n=0}^{\infty} c_n(x-a)^n$ and $\displaystyle\sum_{n=0}^{\infty} 2c_n(x-a)^n$ have

the same interval of convergence.

True.

Section 11.9 Representations of Functions as Power Series

In the last section we saw that a power series is a function. In this section we shall turn things around and see that for many functions, such as $f(x) = \dfrac{2x}{1 + x^2}$, it is possible to write them as power series.

Concepts to Master

A. Obtaining a power series expression for a function by manipulating known series

B. Differentiation and integration of power series

Summary and Focus Questions

Page 745

A. To write a given function as a power series we start with a known power series for a similar function. Then by substitutions (such as x^2 for x, etc.) and multiplication by constants and powers of x, we turn the power series into one for the given function. An example will help:

Example: Find a power series for $f(x) = \dfrac{2x}{1 + x^2}$.

We start with a known series expression for the similar function $\dfrac{1}{1 - x}$:

$$\frac{1}{1 - x} = \sum_{n = 0}^{\infty} x^n$$

Then substitute $-x^2$ for x:

$$\frac{1}{1 + x^2} = \sum_{n = 0}^{\infty} (-x^2)^n = \sum_{n = 0}^{\infty} (-1)^n x^{2n}$$

Then multiply by $2x$:

$$\frac{2x}{1 + x^2} = \sum_{n = 0}^{\infty} (-1)^n (2x) x^{2n} = \sum_{n = 0}^{\infty} (-1)^n 2 x^{2n+1}.$$

The interval of convergence for $\dfrac{1}{1 - x} = \sum_{n = 0}^{\infty} x^n$ is $(-1, 1)$. From $|x| < 1$, the substitution $-x^2$ for x yields $\left|-x^2\right| < 1$, which is still $|x| < 1$. Therefore, the interval of convergence for $\dfrac{2x}{1 + x^2} = \sum_{n = 0}^{\infty} (-1)^n 2 x^{2n+1}$ is $(-1, 1)$.

1) Given $\dfrac{1}{1-x} = \displaystyle\sum_{n=0}^{\infty} x^n, |x| < 1$, find a power series expression for:

a) $\dfrac{1}{2-x}$.

$$\frac{1}{2-x} = \frac{1}{2\left(1 - \frac{x}{2}\right)} = \frac{1}{2}\left(\frac{1}{1 - \frac{x}{2}}\right).$$

Thus $\dfrac{1}{2-x} = \dfrac{1}{2} \displaystyle\sum_{n=0}^{\infty} \left(\frac{x}{2}\right)^n = \sum_{n=0}^{\infty} \frac{1}{2} \cdot \frac{x^n}{2^n}$

$$= \sum_{n=0}^{\infty} \frac{x^n}{2^{n+1}}.$$

From $\left|\dfrac{x}{2}\right| < 1, |x| < 2$, so $(-2, 2)$ is the interval of convergence.

b) $\dfrac{x}{x^3 - 1}$.

We note that $\dfrac{x}{x^3 - 1} = -\dfrac{x}{1 - x^3}$.

First substitute x^3 for x:

$$\frac{1}{1 - x^3} = \sum_{n=0}^{\infty} (x^3)^n = \sum_{n=0}^{\infty} x^{3n}.$$

Now multiply by $-x$:

$$\frac{-x}{1 - x^3} = -x \sum_{n=0}^{\infty} x^{3n} = -\sum_{n=0}^{\infty} x^{3n+1}.$$

Originally $|x| < 1$, hence $|x^3| < 1$. Thus the interval of convergence is still $(-1, 1)$.

Page 746

B. If $f(x) = \displaystyle\sum_{n=0}^{\infty} a_n (x - c)^n$ has interval of convergence with radius R then:

1. f is continuous on $(c - R, c + R)$.

2. For all $x \in (c - R, c + R)$, f may be differentiated term by term:

$$f'(x) = \sum_{n=1}^{\infty} na_n (x - c)^{n-1}.$$

3. For all $x \in (c - R, c + R)$, f may be integrated term by term:

$$\int f(x)\,dx = C + \sum_{n=0}^{\infty} \frac{a_n(x - c)^{n+1}}{n + 1}.$$

Both $f'(x)$ and $\int f(x)\,dx$ have radius of convergence R but the endpoints $c - R$ and $c + R$ must be checked individually.

Term by term integration and differentiation may be used to find power series for functions.

Example: To find a power series for $\ln (1 + x^2)$ recall from part A that the series for $\frac{1}{1-x}$ was used to find that $\frac{2x}{1+x^2} = \sum_{n=0}^{\infty} (-1)^n 2x^{2n+1}$.

Thus $\int \frac{2x}{1+x^2}\,dx = \int \sum_{n=0}^{\infty} (-1)^n 2x^{2n+1}\,dx = \sum_{n=0}^{\infty} (-1)^n 2 \int x^{2n+1}\,dx$

$= \sum_{n=0}^{\infty} (-1)^n 2 \frac{x^{2n+2}}{2n+2} = \sum_{n=0}^{\infty} (-1)^n \frac{x^{2n+2}}{n+1}.$

Since $\ln (1 + x^2) = \int \frac{2x}{1+x^2}\,dx$, $\ln (1 + x^2) = \sum_{n=0}^{\infty} (-1)^n \frac{x^{2n+2}}{n+1}.$

This series has the same interval of convergence $(-1, 1)$ as the series for $\frac{1}{1-x}$ but we need to check the endpoints. At both $x = 1$ and $x = -1$, $x^{2n+2} = 1$ and the resulting series is $\sum_{n=0}^{\infty} \frac{(-1)^n}{n+1}$ converges. Therefore, the interval of convergence for

$\ln (1 + x^2) = \sum_{n=0}^{\infty} (-1)^n \frac{x^{2n+2}}{n+1}$ is $[-1, 1]$.

2) Find $f'(x)$ for

$f(x) = \sum_{n=0}^{\infty} 2^n(x-1)^n.$

$f'(x) = \sum_{n=0}^{\infty} 2^n n(x-1)^{n-1}.$

3) Find $\int f(x)\,dx$ where

$f(x) = \sum_{n=1}^{\infty} \frac{(x-3)^n}{n!}.$

$\int f(x)\,dx = C + \sum_{n=0}^{\infty} \frac{1}{n!} \frac{(x-3)^{n+1}}{n+1}$

$= C + \sum_{n=0}^{\infty} \frac{(x-3)^{n+1}}{(n+1)!}.$

4) Is $f(x) = \sum_{n=0}^{\infty} \frac{x^n}{4^n}$ continuous at $x = 2$?

Yes. 2 is in $[-4, 4]$, the interval of convergence for $f(x)$.

5) Give $\dfrac{1}{1-x} = \displaystyle\sum_{n=0}^{\infty} x^n, |x| < 1$, find a power

series for $\dfrac{1}{(1+x)^2}$.

Substitute $-x$ for x:

$$\frac{1}{1+x} = \sum_{n=0}^{\infty} (-x)^n = \sum_{n=0}^{\infty} (-1)^n x^n.$$

Differentiate term by term:

$$\frac{-1}{(1+x)^2} = \sum_{n=0}^{\infty} (-1)^n n x^{n-1}.$$

Thus $\dfrac{1}{(1+x)^2} = (-1)\displaystyle\sum_{n=0}^{\infty} (-1)^n n x^{n-1}.$

$$= \sum_{n=0}^{\infty} (-1)^{n+1} n x^{n-1}.$$

Since $|-x| < 1$ is the same as $|x| < 1$, the radius of convergence is $R = 1$. At $x = 1$ we

have $\displaystyle\sum_{n=0}^{\infty} (-1)^{n+1} n$ and at $x = -1$ we have

$$\sum_{n=0}^{\infty} (-1)^{n+1} n (-1)^{n-1} = \sum_{n=0}^{\infty} n.$$

Both series diverge so the interval of convergence is $(-1, 1)$.

6) Find $\displaystyle\int \frac{1}{1+x^5}\, dx$ as a power series.

From $\dfrac{1}{1-x} = \displaystyle\sum_{n=0}^{\infty} x^n,$

$$\frac{1}{1+x} = \sum_{n=0}^{\infty} (-x)^n = \sum_{n=0}^{\infty} (-1)^n x^n.$$

Thus $\dfrac{1}{1+x^5} = \displaystyle\sum_{n=0}^{\infty} (-1)^n x^{5n}.$

$$\int \frac{1}{1+x^5}\, dx = \sum_{n=0}^{\infty} \int (-1)^n x^{5n}\, dx$$

$$= \sum_{n=0}^{\infty} \frac{(-1)^n}{5n+1} x^{5n+1} + C.$$

7) From $\dfrac{1}{1-x} = \displaystyle\sum_{n=0}^{\infty} x^n$, find the sum of the

series $\displaystyle\sum_{n=1}^{\infty} n\left(\tfrac{1}{3}\right)^{n-1}.$

By differentiation, $\left(\dfrac{1}{1-x}\right)' = \displaystyle\sum_{n=1}^{\infty} n x^{n-1}.$

Thus $\displaystyle\sum_{n=1}^{\infty} n x^{n-1} = \frac{1}{(1-x)^2}.$

At $x = \dfrac{1}{3}, \dfrac{1}{\left(1 - \frac{1}{3}\right)^2} = \dfrac{9}{4}.$

Section 11. 10   Taylor and Maclaurin Series

This section describes which functions may be written as power series and how to find such representations. The second partial sum of this representation will turn out to be our old friend—the equation of the tangent line. Partial sums of the series for a function will be useful for approximating values of the function.

Concepts to Master

A. Taylor series; Maclaurin series; Analytic functions

B. Taylor polynomials of degree n

C. Remainder of a Taylor series; Taylor's Inequality

D. Multiplication and division of power series

Summary and Focus Questions

Page 752

A. If a function $y = f(x)$ can be expressed as a power series in $(x - a)$, that series must have the form:

$$f(x) = \sum_{n=0}^{\infty} \frac{f^{(n)}(a)}{n!}(x-a)^n$$

$$= f(a) + \frac{f'(a)}{1!}(x-a) + \frac{f''(a)}{2!}(x-a)^2 + \frac{f^3(a)}{3!}(x-a)^3 + \ldots$$

This is called the *Taylor series for f at a.* In the special case of $a = 0$, it is called the *Maclaurin series for f.*

Not all functions can be represented by their power series; if f can be expressed as a power series in $(x - a)$, we say that f is *analytic about a.* In such a case f is equal to its Taylor series at a.

Here are some basic Taylor series about zero (Maclaurin series):

$e^x = \sum_{n=0}^{\infty} \frac{x^n}{n!}$ for all x.

$\ln(1 + x) = \sum_{n=0}^{\infty} \frac{(-1)^{n+1}}{n} x^n$ for $x \in (-1, 1]$.

$\sin x = \sum_{n=0}^{\infty} \frac{(-1)^n}{(2n + 1)!} x^{2n+1}$ for all x.

$\cos x = \sum_{n=0}^{\infty} \frac{(-1)^n}{(2n)!} x^{2n}$ for all x.

$\frac{1}{1 - x} = \sum_{n=0}^{\infty} x^n$ for $|x| < 1$.

$\tan^{-1} x = \sum_{n=0}^{\infty} (-1)^n \frac{x^{2n+1}}{2n + 1}$ for $|x| \le 1$.

To find a Taylor series for a given $y = f(x)$, you first find a formula for $f^{(n)}(a)$, usually in terms of n. Computing the first few derivatives $f'(a)$, $f''(a), f'''(a), f^{(4)}(a), \ldots$ often helps.

For some functions f it is easier to find the Taylor series for $f'(x)$ or $\int f(x)\,dx$ then integrate or differentiate term by term to obtain the series for f. In some other cases, substitutions in the basic Taylor series can be used to find the Taylor series for f.

1) Find the Taylor series for $\frac{1}{x}$ at $a = 1$.

   a) directly from the definition.

First find the general form of $f^{(n)}(1)$:

$$f(x) = x^{-1} \qquad\qquad f(1) = 1$$
$$f'(x) = -x^{-2} \qquad\quad f'(1) = -1$$
$$f''(x) = 2x^{-3} \qquad\quad f''(1) = 2$$
$$f'''(x) = -6x^{-4} \qquad f'''(1) = -6$$
$$f^{(4)}(x) = 24x^{-5} \qquad f^{(4)}(1) = 24$$

In general, $f^{(n)}(1) = (-1)^n n!$.
Thus the Taylor series is

$$\sum_{n=0}^{\infty} \frac{(-1)^n n!}{n!} (x - 1)^n = \sum_{n=0}^{\infty} (-1)^n (x - 1)^n$$

$$= \sum_{n=0}^{\infty} (1 - x)^n.$$

   b) using substitution in a geometric series
$$\left(\text{Hint: } \frac{1}{x} = \frac{1}{1 - (1 - x)}\right).$$

In the form $\frac{1}{1 - (1 - x)}$, this is the value of a geometric series with $a = 1$, $r = 1 - x$.

Thus $\frac{1}{x} = \sum_{n=1}^{\infty} 1(1 - x)^{n-1} = \sum_{n=0}^{\infty} (1 - x)^n.$

   c) What is the interval of convergence for your answer to part a)?

$$\lim_{n\to\infty} \left| \frac{(1 - x)^{n+1}}{(1 - x)^n} \right| = \lim_{n\to\infty} |1 - x| = |1 - x|.$$

$|1 - x| < 1$ is equivalent to $0 < x < 2$.
At $x = 0$ and $x = 2$ the terms of

$$\sum_{n=0}^{\infty} (1 - x)^n \text{ do no approach zero so those}$$

series diverge. The interval of convergence is $(0, 2)$.

2) Find directly the Maclaurin series for
$f(x) = e^{4x}$.

$$f(x) = e^{4x} \qquad\qquad f(0) = 1$$
$$f'(x) = 4e^{4x} \qquad\qquad f'(0) = 4$$
$$f''(x) = 16e^{4x} \qquad\qquad f''(0) = 16$$
In general $f^{(n)}(0) = 4^n$.

Thus $e^{4x} = \displaystyle\sum_{n=0}^{\infty} \frac{4^n}{n!} x^n$.

3) True, False:
If f is analytic at a, then

$$f(x) = \sum_{n=0}^{\infty} \frac{f^{(n)}(a)}{n!} (x - a)^n.$$

True.

4) Obtain the Maclaurin series for $\dfrac{x}{1 + x^2}$ from
the series for $\dfrac{1}{1 - x} = \displaystyle\sum_{n=0}^{\infty} x^n$.

$$\frac{1}{1 - x} = 1 + x + x^2 + \ldots x^n + \ldots$$
Substitute $-x^2$ for x:
$$\frac{1}{1 + x^2} = 1 - x^2 + x^4 - x^6 + \ldots$$
$$+ (-1)^n x^{2n} + \ldots$$
Now multiply by x:
$$\frac{x}{1 + x^2} = x - x^3 + x^5 - x^7 + \ldots$$
$$+ (-1)^n x^{2n+1} + \ldots$$
$$= \sum_{n=0}^{\infty} (-1)^n x^{2n+1}.$$

5) Using $e^x = \displaystyle\sum_{n=0}^{\infty} \frac{x^n}{n!}$ find the Maclaurin series
for $\sinh x$.

$\left(\text{Remember, } \sinh x = \dfrac{e^x - e^{-x}}{2}.\right)$

$$e^x = 1 + x + \frac{x^2}{2!} + \ldots \frac{x^n}{n!} + \ldots$$
Thus $e^{-x} = 1 - x + \dfrac{x^2}{2!} + \ldots \dfrac{(-1)^n x^n}{n!} + \ldots$

Subtracting term by term (the even terms
cancel)

$$e^x - e^{-x} = 2x + \frac{2x^3}{3!} + \ldots + \frac{2x^{2n+1}}{(2n+1)!} + \ldots$$

Thus $\sinh x = x + \dfrac{x^3}{3!} + \ldots + \dfrac{x^{2n+1}}{(2n+1)!} + \ldots$

$$= \sum_{n=0}^{\infty} \frac{x^{2n+1}}{(2n+1)!}.$$

6) Find the infinite series expression for

$$\int_0^{1/2} \frac{1}{1-x^3}\, dx.$$

$$\frac{1}{1-x} = \sum_{n=0}^{\infty} x^n. \text{ Thus } \frac{1}{1-x^3} = \sum_{n=0}^{\infty} x^{3n}.$$

$$\int \frac{1}{1-x^3}\, dx = \int \sum_{n=0}^{\infty} x^{3n}\, dx$$

$$= \sum_{n=0}^{\infty} \frac{x^{3n+1}}{3n+1}.$$

$$\int_0^{1/2} \frac{1}{1-x^3}\, dx = \sum_{n=0}^{\infty} \frac{x^{3n+1}}{3n+1}\bigg]_0^{1/2}$$

$$= \sum_{n=0}^{\infty} \frac{\left(\frac{1}{2}\right)^{3n+1}}{3n+1} - 0 = \sum_{n=0}^{\infty} \frac{1}{(6n+2)8^n}.$$

B. If $f^{(n)}(a)$ exists, we may define the *Taylor polynomial of degree n for f about a* to be the *n*th partial sum of the Taylor series:

Page 753

$$T_n(x) = f(a) + \frac{f'(a)}{1!}(x-a) + \frac{f''(a)}{2!}(x-a)^2 + \frac{f^{(3)}(a)}{3!}(x-a)^3 + \dots \frac{f^{(n)}(a)}{n!}(x-a)^n.$$

The Taylor polynomial of degree one, $T_1(x) = f(a) + f'(a)(x-a)$, is the familiar equation for the tangent line to $y = f(x)$ at a point a.

$T_2(x) = f(a) + f'(a)(x-a) + \frac{1}{2}f''(a)(x-a)^2$ is the "tangent parabola" to $y = f(x)$.

Since the first n derivatives of $T_n(x)$ at a are equal to the corresponding first n derivatives of $f(x)$ at a, $T_n(x)$ is an approximation for $f(x)$ if x is near a.

7) Let $f(x) = \sqrt{x}$.

  a) Construct the Taylor polynomial of degree 3 for $f(x)$ about $x = 1$.

$$\begin{aligned} f(x) &= x^{1/2} & f(1) &= 1 \\ f'(x) &= \tfrac{1}{2}x^{-1/2} & f'(1) &= \tfrac{1}{2} \\ f''(x) &= -\tfrac{1}{4}x^{-3/2} & f''(1) &= -\tfrac{1}{4} \\ f^{(3)}(x) &= \tfrac{3}{8}x^{-5/2} & f^{(3)}(1) &= \tfrac{3}{8} \end{aligned}$$

$T_3(x)$

$$= 1 + \tfrac{1}{2}(x-1) - \tfrac{1/4}{2!}(x-1)^2 + \tfrac{3/8}{3!}(x-1)^3$$

$$= 1 + \tfrac{1}{2}(x-1) - \tfrac{1}{8}(x-1)^2 + \tfrac{1}{16}(x-1)^3$$

  b) Approximate $\sqrt{\tfrac{3}{2}}$ using the Taylor polynomial of degree 3 from part a).

$$T_3\!\left(\tfrac{3}{2}\right) = 1 + \tfrac{1}{2}\!\left(\tfrac{1}{2}\right) - \tfrac{1}{8}\!\left(\tfrac{1}{2}\right)^2 + \tfrac{1}{16}\!\left(\tfrac{1}{2}\right)^3 = \tfrac{157}{128}$$

$$\approx 1.2265.$$

(To 4 decimals, $\sqrt{\tfrac{3}{2}} = 1.2247$.)

8) a) Find the Taylor polynomial of degree 6 for $f(x) = \cos x$ about $x = 0$.

$$f(0) = \cos 0 = 1$$
$$f'(x) = -\sin x, f'(0) = 0$$
$$f''(x) = -\cos x, f''(0) = -1$$
$$f^{(3)}(x) = \sin x, f^{(3)}(0) = 0$$
$$f^{(4)}(x) = \cos x, f^{(4)}(0) = 1$$
$$f^{(5)}(0) = 0 \text{ and } f^{(6)}(0) = -1.$$

Thus $T_6(x) = 1 + \frac{-1}{2!}x^2 + \frac{1}{4!}x^4 - \frac{1}{6!}x^6$

$$= 1 - \frac{x^2}{2} + \frac{x^4}{24} - \frac{x^6}{720}.$$

Note, of course, that this is the first few terms of the Maclaurin series for cosine.

b) Use your answer to part a) to approximate $\cos 1$.

$$\cos 1 \approx T_6(1) = 1 - \frac{(1)^2}{2} + \frac{(1)^4}{24} - \frac{(1)^6}{720}$$
$$= 1 - 0.5 + 0.041667 - 0.001389$$
$$= 0.540278$$

(Note: $\cos 1 = 0.540302$ to 6 decimal places.)

Page 754

C. If $f^{(n)}(x)$ exists for $|x - a| < d$, the nth *remainder of the Taylor series of degree n for f* is:

$$R_n(x) = f(x) - T_n(x)$$

(The R for $R_n(x)$ should not be confused with the R used earlier in the chapter for the radius of convergence of an infinite series.)

Taylor's Inequality: If $|f^{(n+1)}(x)| \leq M$ for $|x - a| \leq d$, then the nth remainder $R_n(x)$ of the Taylor series satisfies the inequality

$$|R_n(x)| \leq \frac{M}{(n+1)!}|x - a|^{n+1} \text{ for } |x - a| \leq d.$$

If f has derivatives of all orders and $\lim\limits_{n\to\infty} R_n(x) = 0$ for $|x - a| < d$, then $f(x)$ is equal to its Taylor series for $|x - a| \leq d$. This is one way to show that a function is equal to its Taylor series—show $\lim\limits_{n\to\infty} R_n(x) = 0$. You will often need to use this limit:

$$\lim_{n\to\infty} \frac{x^n}{n!} = 0 \text{ for all real numbers } x.$$

9) True or False:
In general, the larger the number n, the closer the Taylor polynomial approximation is to the actual functional value.

True.

10) a) Find an expression for the nth Taylor polynomial. $T_n(x)$ for $f(x) = 2^x$ at $x = 0$.

$$f(x) = 2^x \qquad\qquad f(0) = 1$$
$$f'(x) = (\ln 2)\, 2^x \qquad f'(0) = \ln 2$$
$$f''(x) = (\ln 2)^2\, 2^x \qquad f''(0) = (\ln 2)^2$$
and, in general,
$$f^n(x) = (\ln 2)^n\, 2^x \qquad f^n(0) = (\ln 2)^n.$$
$$T_n(x) = 1 + (\ln 2)x + \frac{(\ln 2)^2}{2!}x^2 + \dots$$
$$+ \frac{(\ln 2)^n}{n!}\, x^n.$$

b) Find an upper bound for $\left|f^{(n+1)}(x)\right|$ for $|x| \le 1$.

$$\left|f^{(n+1)}(x)\right| = \left|(\ln 2)^{n+1} \cdot 2^x\right|$$
$$= (\ln 2)^{n+1} \cdot 2^x \le (\ln 2)^{n+1} \cdot 2 \text{ since}$$
$$|x| \le 1.$$

c) Show that $f(x) = 2^x$ is equal to its Taylor series for $|x| < 1$.

We need to show that $\lim\limits_{n\to\infty} R_n(x) = 0$ for $|x| \le 1$. From part b)
$$\left|f^{(n+1)}(x)\right| \le 2\,(\ln 2)^{n+1}.$$
By Taylor's Inequality
$$0 \le |R_n(x)| \le \frac{2(\ln 2)^{n+1}}{(n+1)!}|x|^{n+1} \le \frac{2(\ln 2)^{n+1}}{(n+1)!}$$
since $|x| \le 1$.
$$\lim_{n\to\infty} 2\frac{(\ln 2)^{n+1}}{(n+1)!} = 2 \lim_{n\to\infty} \frac{(\ln 2)^{n+1}}{(n+1)!} = 2(0) = 0.$$
Therefore, by the Squeeze Theorem for limits, $\lim\limits_{n\to\infty} R_n(x) = 0$.

11) Find $\int_0^1 xe^{-x}\,dx$ using series to within 0.01.

$\left(\text{We could use integration by parts:}\right.$

$\left.\int_0^1 xe^{-x}\,dx = (-xe^{-x} - e^{-x})\Big]_0^1 = 1 - \frac{2}{e}.\right)$

$e^x = \sum_{n=0}^{\infty} \frac{x^n}{n!},\ e^{-x} = \sum_{n=0}^{\infty} \frac{(-1)^n x^n}{n!},$

so $xe^{-x} = \sum_{n=0}^{\infty} \frac{(-1)^n x^{n+1}}{n!}$. Thus

$\int_0^1 xe^{-x}\,dx = \int_0^1 \sum_{n=0}^{\infty} \frac{(-1)^n x^{n+1}}{n!}\,dx$

$= \sum_{n=0}^{\infty} \frac{(-1)^n x^{n+2}}{n!(n+2)}\Big]_0^1 = \sum_{n=0}^{\infty} \frac{(-1)^n}{n!(n+2)} - 0$

$= \sum_{n=0}^{\infty} \frac{(-1)^n}{n!(n+2)}$

$= \frac{1}{0!2} - \frac{1}{1!3} + \frac{1}{2!4} - \frac{1}{3!5} + \frac{1}{4!6} - \frac{1}{5!7} + \dots$

$= \frac{1}{2} - \frac{1}{3} + \frac{1}{8} - \frac{1}{30} + \frac{1}{144} + \dots$

This is an alternating series and the fifth term $\frac{1}{144}$ is less than 0.01. Thus we may use $\int_0^1 xe^{-x}\,dx \approx \frac{1}{2} - \frac{1}{3} + \frac{1}{8} - \frac{1}{30} \approx 0.2583.$ This is within 0.01 of the actual value $1 - \frac{2}{e} \approx 0.2642.$

D. Convergent series may be multiplied and divided like polynomials. Often finding the first few terms of the result is sufficient.

Page 759

12) Find the first three terms of the Maclaurin series for $e^x \sin x$.

$e^x = 1 + x + \frac{x^2}{2} + \frac{x^3}{6} + \dots$

$\sin x = x - \frac{x^3}{3!} + \frac{x^5}{5!} - \dots$

Thus

$e^x \sin x = x\left(1 + x + \frac{x^2}{2} + \frac{x^3}{6} + \dots\right)$
$\qquad - \frac{x^3}{6}\left(1 + x + \frac{x^2}{2} + \frac{x^3}{6} + \dots\right)$
$\qquad + \frac{x^5}{120}\left(1 + x + \frac{x^2}{2} + \frac{x^3}{6} + \dots\right) + \dots$

$= \left(x + x^2 + \frac{x^3}{2} + \frac{x^4}{6} + \dots\right)$
$\qquad - \left(\frac{x^3}{6} + \frac{x^4}{6} + \frac{x^5}{12} + \frac{x^6}{36} + \dots\right)$
$\qquad + \left(\frac{x^5}{120} + \frac{x^6}{120} + \frac{x^7}{240} + \dots\right)$

$= x + x^2 + \frac{x^3}{6} + \dots$

Section 11. 11   The Binomial Series

You learned in your algebra class that the Binomial Theorem may be used to expand $(1 + x)^k$:

$$(1 + x)^k = 1^k + \binom{k}{1}1^{k-1}x + \binom{k}{2}1^{k-2}x^2 + \binom{k}{3}1^{k-3}x^3 + \dots + \binom{k}{k-1}1x^{k-1} + x^k$$

$$= 1 + \binom{k}{1}x + \binom{k}{2}x^2 + \binom{k}{3}x^3 + \dots + \binom{k}{k-1}x^{k-1} + x^k$$

where $\binom{k}{r} = \frac{k!}{r!(k-r)!}$ is the binomial coefficient. This section extends the

concept of a binomial coefficient and the expansion of $(1 + x)^k$ to the case where k is no longer a positive integer. In the general case, the expansion of $(1 + x)^k$ is an infinite series. The Binomial Theorem becomes a special case in which only the first few terms of the infinite series are non-zero.

Concepts to Master

Binomial coefficient; Binomial series for $(1 + x)^k$

Page
762

Summary and Focus Questions

For any real number k and a non-negative integer n, the *binomial coefficient* is

$$\binom{k}{n} = \begin{cases} 1 & \text{if } n = 0 \\ \dfrac{k(k-1)(k-2)\dots(k-n+1)}{n!} & \text{if } n \geq 1 \end{cases}$$

When k is a positive integer, this is the binomial coefficient that you have seen in an algebra class.

For any real number k the *binomial series for* $(1 + x)^k$ is the Maclaurin series for $(1 + x)^k$:

$$(1 + x)^k = \sum_{n=0}^{\infty} \binom{k}{n}x^n = 1 + kx + \frac{k(k-1)}{2!}x^2 + \dots + \binom{k}{n}x^n + \dots$$

The binomial series converges to $(1 + x)^k$ for these cases:

Condition on k	Interval of Convergence
$k \leq -1$	$(-1, 1)$
$-1 < k < 0$	$(-1, 1]$
k nonnegative integer	$(-\infty, \infty)$
$k > 0$, but not an integer	$[-1, 1]$

1) Find $\binom{\frac{4}{3}}{5}$.

$$\binom{\frac{4}{3}}{5} = \frac{\left(\frac{4}{3}\right)\left(\frac{1}{3}\right)\left(-\frac{2}{3}\right)\left(-\frac{5}{3}\right)\left(-\frac{8}{3}\right)}{5 \cdot 4 \cdot 3 \cdot 2 \cdot 1} = \frac{-8}{3^6} = -\frac{8}{729}.$$

2) True or False:
If k is a nonnegative integer, then the binomial series for $(1 + x)^k$ is a finite sum.

True, only the first $k + 1$ terms of the infinite series are non-zero.

3) Find the binomial series for $\sqrt[3]{1 + x}$.

$$\sum_{n=0}^{\infty} \binom{\frac{1}{3}}{n} x^n.$$

4) Find a power series for $f(x) = \sqrt{4 + x}$.

$$\sqrt{4 + x} = 2\left(1 + \frac{x}{4}\right)^{1/2} = 2\sum_{n=0}^{\infty} \binom{\frac{1}{2}}{n}\left(\frac{x}{4}\right)^n$$
$$= 2\sum_{n=0}^{\infty} \binom{\frac{1}{2}}{n}\frac{x^n}{4^n}.$$

5) Find an infinite series for $\sin^{-1} x$. (Hint: Start with $(1 + x)^{-1/2}$.)

$$\frac{1}{\sqrt{1 + x}} = (1 + x)^{-1/2} = \sum_{n=0}^{\infty} \binom{-\frac{1}{2}}{n} x^n.$$

Substitute $-x$ for x:

$$\frac{1}{\sqrt{1 - x}} = \sum_{n=0}^{\infty} \binom{-\frac{1}{2}}{n}(-x)^n$$
$$= \sum_{n=0}^{\infty} (-1)^n\binom{-\frac{1}{2}}{n}x^n.$$

Substitute x^2 for x:

$$\frac{1}{\sqrt{1 - x^2}} = \sum_{n=0}^{\infty} (-1)^n\binom{-\frac{1}{2}}{n}x^{2n}.$$

Now integrate term by term:

$$\sin^{-1} x = \sum_{n=0}^{\infty} \frac{(-1)^n\binom{-\frac{1}{2}}{n}x^{2n+1}}{2n + 1}.$$

Section 11.12 Applications of Taylor Polynomials

This last section uses Taylor polynomials to estimate values of functions and uses Taylor's Inequality to measure the accuracy of such approximations.

Concepts to Master

A. Approximate functional values using a Taylor polynomial

B. Estimate the error in Taylor polynomial approximations

Summary and Focus Questions

Page 766

A. The Taylor polynomial of degree n for f about a.

$$T_n(x) = f(a) + \frac{f'(a)}{1!}(x - a) + \frac{f''(a)}{2!}(x - a)^2 + \frac{f^{(3)}(a)}{3!}(x - a)^3 + \ldots \frac{f^{(n)}(a)}{n!}(x - a)^n$$

may be used to approximate $f(x)$ for any given x in the interval of convergence for the Taylor series for f about a.

1) a) For $f(x) = \frac{1}{3 + x}$, find $T_3(x)$ where $a = 2$.

$$f(x) = (3 + x)^{-1} \qquad f(2) = \frac{1}{5}$$
$$f'(x) = -(3 + x)^{-2} \qquad f'(2) = \frac{-1}{25}$$
$$f''(x) = 2(3 + x)^{-3} \qquad f''(2) = \frac{2}{125}$$
$$f'''(x) = -6(3 + x)^{-4} \qquad f'''(2) = \frac{-6}{625}$$

$$T_3(x) = \frac{1}{5} - \frac{1}{25}(x - 2) + \frac{2/125}{2!}(x - 2)^2$$
$$\qquad - \frac{6/625}{3!}(x - 2)^3$$
$$= \frac{1}{5} - \frac{1}{25}(x - 2) + \frac{1}{125}(x - 2)^2$$
$$\qquad - \frac{1}{625}(x - 2)^3.$$

b) Estimate $f(3)$ with $T_3(3)$.

$$T_3(3) = \frac{1}{5} - \frac{(3 - 2)}{25} + \frac{(3 - 2)^2}{125} - \frac{(3 - 2)^3}{625}$$
$$= \frac{1}{5} - \frac{1}{25} + \frac{1}{125} - \frac{1}{625}$$
$$= \frac{104}{625} \approx 0.1664$$

We note that this is a lot of work to approximate $f(3) = \frac{1}{6}$.

2) **a)** For $f(x) = x^{5\backslash 2}$, find $T_3(x)$ where $a = 4$.

$$f(x) = x^{5/2} \qquad\qquad f(4) = 32$$

$$f'(x) = \tfrac{5}{2}x^{3/2} \qquad\quad f'(4) = 20$$

$$f''(x) = \tfrac{15}{4}x^{1/2} \qquad\quad f''(4) = \tfrac{15}{2}$$

$$f'''(x) = \tfrac{15}{8}x^{-1/2} \qquad f'''(4) = \tfrac{15}{16}$$

$$T_3(x) = 32 + 20(x - 4) + \tfrac{15/2}{2!}(x - 4)^2$$
$$+ \tfrac{15/16}{3!}(x - 4)^3$$
$$= 32 + 20(x - 4) + \tfrac{15}{4}(x - 4)^2$$
$$+ \tfrac{5}{32}(x - 4)^3.$$

b) Estimate $(5)^{5/2}$ with $T_3(x)$.

$$T_3(5) = 32 + 20(1)^1 + \tfrac{15}{4}(1)^2 + \tfrac{5}{32}(1)^3$$
$$= \tfrac{1789}{32} \approx 55.9063.$$

To four decimals $(5)^{5/2}$ is approximately 55.9017.

B. The accuracy of the approximation $f(x) \approx T_n(x)$ will, in general, depend on:

n — the larger the value of n, the better the estimate

x and a — the closer that x is to a, the better the estimate.

Page 767

From section 12.10, we may use Taylor's Inequality:

If $\left|f^{(n+1)}(x)\right| \le M$ for $|x - a| \le d$, then $\left|R_n(x)\right| \le \dfrac{M}{(n + 1)!}|x - a|^{n+1}$ for $|x - a| \le d$.

to estimate the error. The approximation $T_n(x)$ will be accurate to within any given number ϵ by selecting n such that

$$\frac{M}{(n + 1)!}|x - a|^{n+1} < \epsilon.$$

In the special case where the Taylor series is an alternating series, $\left|R_n(x)\right|$ may be estimated with $\left|\dfrac{f^{(n+1)}(x)}{(n + 1)!}(x - a)^{n+1}\right|$, the $(n + 1)^{\text{st}}$ term of the Taylor series for f.

3) a) Use Taylor's Inequality to estimate the accuracy of $f(x) \approx T_3(x)$ for $1 < x < 4$ for question 1).

From 1 a), $f^{(4)}(x) = 24(3 + x)^{-5}$.
If $1 \leq x \leq 4$, then $4 \leq x + 3 \leq 7$.
Therefore $4^5 \leq (x + 3)^5 \leq 7^5$, and

$\frac{1}{7^5} \leq (x + 3)^{-5} \leq \frac{1}{4^5}$ and

$\frac{24}{7^5} \leq 24(x + 3)^{-5} \leq \frac{24}{4^5}$

Thus $\left| f^{(4)}(x) \right| \leq \frac{24}{4^5}$.

From $1 \leq x \leq 4$, $-1 \leq x - 2 \leq 2$.
Thus $(x - 2)^4 \leq 2^4 = 16$.
Finally,

$\left| R_3(x) \right| \leq \frac{\frac{24}{4^5}}{4!} (x - 2)^4$

$\leq \frac{24}{4^5 \cdot 4!} \cdot 16 = 0.0156.$

b) Check the accuracy of the estimate when $x = 3$.

$f(3) = \frac{1}{3 + 3} = \frac{1}{6} \approx 0.1667.$
$T_3(3) \approx 0.1664$ by 1 b).
Thus $R_3 (3) \approx 0.1667 - 0.1664 = 0.0003$,
which is within 0.0156.

4) a) For $f(x) = x^{5/2}$, estimate the error between $f(5)$ and $T_3(5)$. (See question 2.)

From 2 a), $f^{(4)}(x) = -\frac{15}{16}x^{-3/2}$.

For $|x - 4| \leq 2, 2 \leq x \leq 6.$

$6^{-3/2} \leq x^{-3/2} \leq 2^{-3/2}.$

Thus $\left| f^{(4)}(x) \right| \leq \frac{15}{16} \cdot 2^{-3/2} = \frac{15}{32\sqrt{2}}.$

Therefore $\left| R_3(x) \right| \leq \frac{15}{32\sqrt{2}} |x - 4|^4$

$\leq \frac{\frac{15}{32\sqrt{2}}}{4!}(2)^4 \approx 0.221.$

b) Check the accuracy of the estimate $f(5) \approx T_3(5)$.

$f(5) = 5^{5/2} \approx 55.9017.$
$T_3(5) \approx 55.9063$ (question 2b)).
$R_3(5) \approx 55.9017 - 55.9063 = -0.0046$,
which is within 0.221.

5) a) What degree Maclaurin polynomial is needed to approximate cos 0.5 accurate to within 0.00001?

Rephrased, the question asks: For what n is $|R_n(0.5)| < 0.00001$ when $f(x) = \cos x$ and $a = 0$?

Because $f^{(n)}(x) = \pm\sin x$ or $\pm\cos x$ for all n, $|f^{(n+1)}(x)| \leq 1$.

Thus $|R_n(x)| \leq \frac{1}{(n+1)!}|x|^{n+1}$

$= \frac{|x|^{n+1}}{(n+1)!} \leq \frac{1}{(n+1)!}$ if $|x| \leq 1$.

$\frac{1}{(n+1)!} \leq 0.00001$ means

$(n+1)! \geq 100{,}000$.

Therefore $n = 8$, since $8! = 40{,}320$ and $9! = 362{,}800$.

b) Use your answer to part a) to estimate cos 0.5 to within 0.00001.

$T_8(x) = 1 - \frac{x^2}{2} + \frac{x^4}{24} - \frac{x^6}{720} + \frac{x^8}{40320}$

$T_8(0.5) = 1 - \frac{(0.5)^2}{2} + \frac{(0.5)^4}{24} - \frac{(0.5)^6}{720} + \frac{(0.5)^8}{40320}$

$\approx 0.877583.$

To 6 decimals, $\cos 0.5 = 0.877583$ so the approximation is within 0.00001.

Technology Plus for Chapter 11

1) Using a spreadsheet or calculator, find the first 20 partial sums of $\sum\limits_{n=1}^{\infty} \dfrac{1}{n^4 + 1}$. Estimate the sum of the series.

$$\sum_{n=1}^{\infty} \frac{1}{n^4 + 1} \approx 0.57844.$$

2) Use the 15th partial sum to estimate $\displaystyle\sum_{n=1}^{\infty} \frac{(-1)^{n-1}}{n^3}$. How accurate is the estimate?

We see that $\displaystyle\sum_{n=1}^{15} \frac{(-1)^{n-1}}{n^3} \approx 0.90168$.

This is within $\left| \dfrac{(-1)^{n-1}}{16^3} \right| = 0.00024$

of the sum.

3) On the same screen sketch a graph of $f(x)\ \dfrac{1}{\sqrt{x}}$ and its first 3 Taylor polynomials about $x = 1$.

$f(x) = x^{-1/2}$, $f(1) = 1$

$f'(x) = -\frac{1}{2}x^{-3/2}$, $f'(1) = -\frac{1}{2}$

$f''(x) = \frac{3}{4}x^{-5/2}$, $f''(1) = \frac{3}{4}$

$f'''(x) = -\frac{15}{8}x^{-7/2}$, $f'''(1) = -\frac{15}{8}$

$TP_1(x) = 1 + (-\frac{1}{2})(x-1) = 1 - \frac{1}{2}(x-1)$

$TP_2(x) = 1 - \frac{1}{2}(x-1) + \dfrac{\frac{3}{4}}{2!}(x-1)^2$

$= 1 - \frac{1}{2}(x-1) + \frac{3}{8}(x-1)^2$

$TP_3(x) = 1 - \frac{1}{2}(x-1) + \frac{3}{8}(x-1)^2$

$\qquad + \dfrac{-\frac{15}{8}}{3!}(x-1)^3$

$= 1 - \frac{1}{2}(x-1) + \frac{3}{8}(x-1)^2 - \frac{15}{48}(x-1)^3$

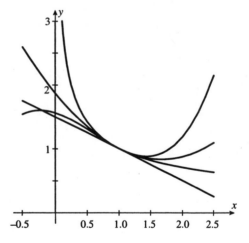

Section 1.1

_____ **1.** True, False:

$x^2 + 6x + 2y = 1$ defines y as a function of x.

_____ **2.** True, False:

In $V = \frac{4}{3}\pi r^3$, r is the dependent variable.

_____ **3.** The implied domain of $f(x) = \frac{1}{\sqrt{1-x}}$ is:

a) $(1, \infty)$ c) $x \neq 1$

b) $(-\infty, 1)$ d) $(-1, 1)$

_____ **4.** True, False:

The graph below is the graph of an even function.

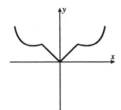

_____ **5.** Which graph best represents $y = x + \frac{1}{x}$

a)   b)   c)

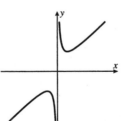 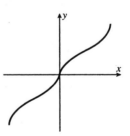

_____ **6.** True, False:

$f(x) = x^2$ is decreasing for $-10 \leq x \leq -1$.

Section 1.2

_____ **1.** True, False:

$f(x) = \frac{x^2 + \sqrt{x}}{2x+1}$ is a rational function.

_____ **2.** True, False:

The degree of $f(x) = 4x^3 + 7x^6 + 1$ is 7.

_____ **3.** Which graph is best described by a linear model?

a)

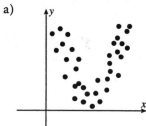

b)

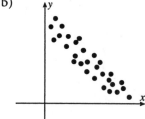

c)

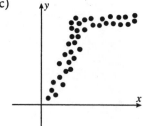

____ **1.** For $f(x) = 3x + 1$ and $g(x) = x$, $\dfrac{f}{g}(x) = $ ____?

a) $\dfrac{3 + x}{x}$

b) $\dfrac{x}{3x + 1}$

c) $\dfrac{3x + 1}{x}$

d) $\dfrac{3x + 1}{x}$

____ **2.** For $f(x) = \sqrt{x^2 + x}$, we may write $f(x) = h \circ g(x)$, where:

a) $h(x) = \sqrt{x}$ and $g(x) = x^2 + x$

b) $h(x) = x^2 + x$ and $g(x) = \sqrt{x}$

c) $h(x) = x^2$ and $g(x) = \sqrt{x}$

d) $h(x) = x^2 + x$ and $g(x) = x^2$

____ **3.** Let $f(x) = 2 + \sqrt{x}$ and $g(x) = x + 3$. Then $g \circ f(x) = $ ____?

a) $2 + \sqrt{x} + 3$

b) $2 + \sqrt{x + 3}$

c) $3 + \sqrt{x + 2}$

d) $(x + 3)(2 + \sqrt{x})$

____ **4.** Sometimes, Always, or Never:
$f \circ g(x) = g \circ f(x)$.

____ **5.** For $f(x) = \cos(x^2 + 1)$ we may write $f(x) = h \circ g(x)$, where:

a) $h(x) = \cos x^2$ and $g(x) = x + 1$

b) $h(x) = \cos x$ and $g(x) = x^2 + 1$

c) $h(x) = x^2 + 1$ and $g(x) = \cos x$

d) $h(x) = x^2$ and $g(x) = \cos(x + 1)$

____ **6.** The graph of $y = f(3x)$ is obtained from the graph of $y = f(x)$ by:

a) stretching vertically by a factor of 3.

b) compressing vertically by a factor of 3.

c) stretching horizontally by a factor of 3.

d) compressing horizontally by a factor of 3.

____ **7.** True, False:
Given the graph of $y = f(x)$ on the left below, the graph on the right below
is $y = -f(x)$.

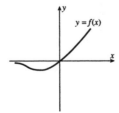
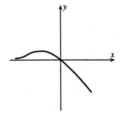

Section 1.4

_____ **1.** Which is the best viewing rectangle for the graph at the right?

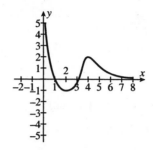

   a) $[-6, 6]$ by $[-4, 4]$
   b) $[-1, 6]$ by $[-2, 4]$
   c) $[0, 8]$ by $[0, 4]$
   d) $[-4, 4]$ by $[0, 8]$

_____ **2.** Which is the best viewing rectangle for $y = \dfrac{1}{x^2 - 9}$?

   a) $[0, 10]$ by $[0, 10]$
   b) $[-3, 3]$ by $[-10, 10]$
   c) $[-5, 5]$ by $[-10, 10]$
   d) $[-5, 5]$ by $[0, 10]$

Section 1.5

____ **1.** Which is the best graph of $y = 1.5^x$? (The graph of $y = 2^x$ is shown as a dashed curve for comparison.)

a)

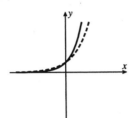

b)

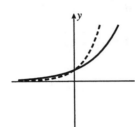

c)

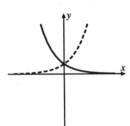

d)

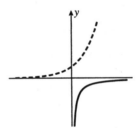

____ **2.** A sketch of $y = e^{-x^2}$ looks like:

a)   b)   c)   d)

____ **3.** What is the number a in the following graph?

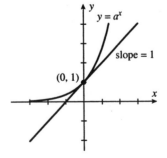

$y = a^x$

slope = 1

(0, 1)

a) e b) 1   c) 2   d) 3

____ **4.** $e^x(e^x)^2 =$

a) e^{x^3} b) e^{3x} c) $3e^x$ d) e^{4x}

Section 1.6

____ **1.** A function f is one-to-one means:

 a) if $x_1 = x_2$, then $f(x_1) = f(x_2)$
 b) if $x_1 \neq x_2$, then $f(x_1) = f(x_2)$
 c) if $x_1 \neq x_2$, then $f(x_1) \neq f(x_2)$
 d) if $f(x_1) \neq f(x_2)$, then $x_1 \neq x_2$

____ **2.** If $f(x) = \sqrt[3]{x + 3}$, then $f^{-1}(x) =$

 a) $\dfrac{1}{\sqrt[3]{x + 3}}$ b) $x^3 + 3$ c) $(x + 3)^3$ d) $x^3 - 3$

____ **3.** Sometime, Always, or Never:

 If f is one-to-one and (a, b) is on the graph of $y = f(x)$ then (b, a) is on the graph of $y = f^{-1}(x)$.

____ **4.** If f is one-to-one, differentiable, $f(4) = 7$ and $f'(4) = \frac{1}{2}$ then $(f^{-1})'(7) =$

 a) $\frac{1}{7}$ b) $\frac{1}{4}$ c) $\frac{1}{2}$ d) 2

____ **5.** $\log_{27} 9 =$

 a) $\frac{1}{3}$ b) $\frac{2}{3}$ c) $\frac{3}{2}$ d) $\frac{1}{2}$

____ **6.** True, False:
 $\ln(a + b) = \ln a + \ln b$.

____ **7.** Simplified, $\log_3 9x^3$ is:

 a) $2 + 3 \log_3 x$ b) $6 \log_3 x$ c) $9 \log_3 x$ d) $9 + \log_3 x$

____ **8.** Solve for x: $e^{2x-1} = 10$.

 a) $\frac{1}{2}(1 + 10^e)$ b) $3 + \ln 10$ c) $2 + \ln 10$ d) $\ln \sqrt{10e}$

____ **9.** Solve for x: $\ln(e + x) = 1$.
 a) 0 b) 1 c) $-e$ d) $e^e - e$

Section 2.1

_____ 1. For $f(x) = 5x^2 + 1$ the slope of the secant line between the points corresponding to $x = 3$ and $x = 4$ is:

  a) 7   b) 35   c) 81   d) 1

_____ 2. A guess for the slope of the tangent line $f(x) = 5x^2 + 1$ at $P(3, 46)$ based on the chart below is:

x	4	3.1	3.01	3.001
$f(x)$	81	49.05	46.3005	46.030005

  a) 26   b) 30   c) 35   d) 46

_____ 3. True, False:

The slope of a tangent line may be interpreted as average velocity.

Section 2.2

1. In the graph at the right;
   $$\lim_{x \to 2} f(x) = \_\_\_\_.$$

   a) 2
   b) 4
   c) 0
   d) does not exist

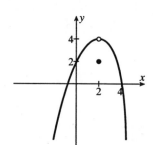

2. $\lim_{x \to 6} (x^2 - 10x) = \_\_\_\_.$
   a) -24 b) 6   c) 26   d) does not exist

3. In the graph at the right,
   $$\lim_{x \to 1^-} f(x) = \_\_\_\_.$$
   a) -2
   b) 1
   c) 3
   d) does not exist

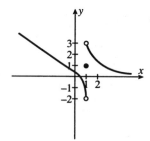

4. For $f(x) = \begin{cases} 2 - x^2 & \text{if } x > 1 \\ 3 & \text{if } x \leq 1 \end{cases}$, which limit does not exist?

   a) $\lim_{x \to 3} f(x)$ b) $\lim_{x \to 1^+} f(x)$ c) $\lim_{x \to 1^-} f(x)$ d) $\lim_{x \to 1} f(x)$

5. $\lim_{x \to -1} \dfrac{-1}{(1 + x)^2} = \_\_\_\_.$
   a) ∞ b) $-\infty$ c) 1   d) -1

6. True, False:

   The graph in question 3 has a vertical asymptote at $x = 1$.

Section 2.3

_____ **1.** True, False:

If $h(x) = g(x)$ for all $x \neq a$ and $\lim\limits_{x \to a} h(x) = L$, then $\lim\limits_{x \to a} g(x) = L$.

_____ **2.** $\lim\limits_{x \to 3} \dfrac{7x + 6}{5x - 12} =$ _____.

     a) $-\dfrac{1}{2}$ b) $\dfrac{7}{5}$ c) 9   d) 4

_____ **3.** Sometimes, Always, or Never:

$$\lim\limits_{x \to a} \frac{f(x)}{g(x)} = \frac{\lim\limits_{x \to a} f(x)}{\lim\limits_{x \to a} g(x)}.$$

_____ **4.** $\lim\limits_{x \to 2} \dfrac{x^2(x - 1)}{1 - x} =$ _____.

     a) -4 b) -1 c) 0   d) does not exist

_____ **5.** $\lim\limits_{x \to 1} \dfrac{x^3 - x^2}{1 - x} =$ _____.

     a) 1   b) -1 c) 0   d) does not exist

_____ **6.** If $2x + 2 \leq f(x) \leq x^2 + 3$, then $\lim\limits_{x \to 1} f(x) =$ _____.

     a) 1

     b) 4

     c) does not exist

     d) exists but cannot be determined from the information given

Section 2.4

_____ **1.** True, False:

$|f(x) - L| < \epsilon$, means that the distance between $f(x)$ and L is less than ϵ.

_____ **2.** For a given $\epsilon > 0$, the corresponding δ in the definition of $\lim_{x \to 6} (3x + 5) = 23$ is:

   a) $\frac{\epsilon}{6}$ b) $\frac{\epsilon}{3}$ c) $\frac{\epsilon}{5}$ d) $\frac{\epsilon}{23}$

_____ **3.** For a given $M > 0$, the corresponding δ in the definition of $\lim_{x \to 0} \frac{1}{2x^4} = \infty$ is:

   a) $\sqrt[4]{2M}$ b) $2\sqrt[4]{M}$ c) $\frac{1}{2\sqrt[4]{M}}$ d) $\frac{1}{\sqrt[4]{2M}}$

On Your Own

Section 2.5

_____ **1.** Sometimes, Always, or Never:

If $\lim_{x \to a} f(x)$ and $f(a)$ both exist, then f is continuous at a.

_____ **2.** $f(x) = \frac{x + 2}{x(x - 5)}$ is continuous for all x except:

 a) $x = -2$
 b) $x = -2, 0, 5$
 c) $x = 0, 5$
 d) f is continuous for all x.

_____ **3.** True, False:

$f(x) = \frac{x^2(x + 1)}{x}$ is continuous for all real numbers.

_____ **4.** For the continuous function at the right, how many points c does the Intermediate Value Theorem guarantee must be between a and b and have $f(c) = N$?

 a) 1
 b) 3
 c) 5
 d) none

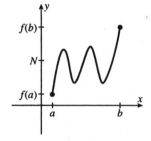

_____ **5.** True, False:

Let $f(x) = x^2$. The Intermediate Value Theorem guarantees there is $w \in (-1, 5)$ for which $f(w) = 0$.

Section 2.6

_____ **1.** Sometime, Always, Never:

$$\lim_{x \to \infty} f(x) = \lim_{x \to -\infty} f(x).$$

_____ **2.** $\lim\limits_{x \to \infty} \dfrac{\sqrt{x^2 + 1}}{3x + 2} =$

    a) 0   b) 1   c) $\dfrac{1}{3}$ d) does not exist

_____ **3.** True, False:

If the limits exist, then $\lim\limits_{x \to \infty} (f(x) + g(x)) = \lim\limits_{x \to \infty} f(x) + \lim\limits_{x \to \infty} g(x)$.

_____ **4.** The horizontal asymptote(s) for $f(x) = \dfrac{|x|}{x + 1}$ is(are):

    a) $y = 0$ b) $y = 1$

    c) $y = -1$ d) $y = 1$ and $y = -1$

_____ **5.** True, False:

$$\lim_{x \to \infty} \frac{6x^3 + 8x^2 - 9x + 7}{5x^3 - 4x + 30} = \lim_{x \to \infty} \frac{6x^3}{5x^3}.$$

_____ **6.** $\lim\limits_{x \to \infty} \dfrac{6 - x^3}{3 + x} =$

    a) ∞ b) $-\infty$ c) 0   d) 2

_____ **7.** Sometimes, Always, or Never:

$$\lim_{x \to -\infty} a^x = 0.$$

_____ **8.** $\lim\limits_{x \to -\infty} \left(\frac{1}{2}\right)^{x^2} =$

    a) 0   b) 1   c) ∞ d) does not exist

_____ **9.** $\lim\limits_{x \to \infty} 2^{x/(x + 1)} =$

    a) 0   b) 1   c) 2   d) ∞

Section 2.7

____ 1. True, False:
$$\lim_{x \to a} \frac{f(x) - f(a)}{x - a} = \lim_{k \to 0} \frac{f(a + k) - f(a)}{k}.$$

____ 2. True, False:
$\dfrac{f(x) - f(a)}{x - a}$ may be interpreted as the instantaneous velocity of a particle at time a.

____ 3. The slope of the tangent line to $y = x^3$ at $x = 2$ is:

    a) 18   b) 12   c) 6   d) 0

____ 4. True, False:
Slope of a tangent line, instantaneous velocity, and instantaneous rate of change of a function are each an interpretation of the same limit concept.

____ 5. Given the table of function values below, which of the following is the best estimate for the instantaneous rate of change of $y = f(x)$ at $x = 4$?

x	6	5	4.5	4.1	4.05	4.01	4
$f(x)$	17	13.3	11.7	10.32	10.16	10.031	10

    a) 17   b) 10   c) 3   d) 0

_____ **1.** True, False:

$$f'(a) = \lim_{a \to 0} \frac{f(a + h) - f(a)}{h}.$$

_____ **2.** True, False:

$$f'(x) = \lim_{x \to a} \frac{f(x) - f(a)}{x - a}.$$

_____ **3.** For $f(x) = 10x^2$, $f'(3) =$ _____.

a) 10   b) 20   c) 30   d) 60

_____ **4.** The instantaneous rate of change of $y = x^3 + 3x$ at the point corresponding to $x = 2$ is:

a) 10   b) 15   c) 2   d) 60

_____ **5.** $\lim_{h \to 0} \dfrac{\sqrt{4 + h} - 2}{h}$ is the derivative of:

a) $f(x) = \sqrt{4 + x}$ at $x = 2$

b) $f(x) = \sqrt{x}$ at $x = 4$

c) $f(x) = \sqrt{x}$ at $x = 2$

d) $f(x) = \dfrac{\sqrt{4 + h} - 2}{h}$ at $x = 0$

_____ **6.** True, False:

$\frac{dm}{dv}$ stands for the derivative of the function m with respect to the variable v.

Section 2.9

_____ **1.** Is $f(x) = \tan x$ differentiable at $x = \frac{\pi}{2}$?

_____ **2.** True, False:
If $y = f(x)$ is not differentiable at a point $x = a$, then $y = f(x)$ is not continuous at $x = a$.

_____ **3.** Which is the best graph of $y = f'(x)$ for the given graph of $y = f(x)$?

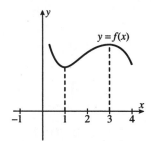

a)
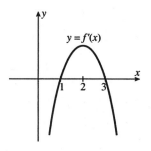

b)
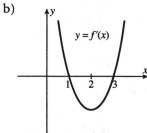

c)
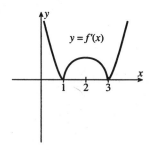

d) None of these

Section 3.1

____ **1.** For $f(x) = 6x^4$, $f'(x) =$ ____.

     a) $24x$ b) $6x^3$ c) $10x^3$ d) $24x^3$

____ **2.** For $y = \frac{1}{2}gt^2$ (where g is a constant), $\frac{dy}{dt} =$ ____.

     a) $\frac{1}{4}g^2$ b) $\frac{1}{2}g$ c) gt d) gt^2

____ **3.** For $f(x) = 5 + g(x)$, $f'(x) =$ ____.

     a) $5g'(x)$ b) $5 + g'(x)$ c) $0 \cdot g'(x)$ d) $g'(x)$

____ **4.** $\lim\limits_{h \to 0} \frac{a^h - 1}{h} = 1$ for $a =$

     a) 0 b) 1 c) e d) The limit is never 1

____ **5.** For $f(x) = e^x - x$, $f'(x) =$

     a) $e^x - x$ b) $e^x - 1$ c) $e^1 - x$ d) e^1

Section 3.2

—— **1.** True, False:

$(f(x)g(x))' = f'(x)g'(x).$

—— **2.** For $f(x) = \frac{x}{x+1}, f'(x) = $ ——.

    a) 1   b) $\frac{x^2}{(x+1)^2}$ c) $\frac{-1}{(x+1)^2}$ d) $\frac{1}{(x+1)^2}$

—— **3.** For $f(x) = \frac{e^x}{x}, f'(x) = $

    a) e^x b) $\frac{xe^x - 1}{x^2}$ c) $\frac{e^x}{x} - \frac{e^x}{x^2}$ d) $\frac{xe^x - x}{x^2}$

Section 3.3

_____ **1.** True, False:

$f'(t)$ is used to measure the average rate of change of f with respect to t.

_____ **2.** Air is being pumped into a chamber so that after t seconds the pressure in the chamber is $3 + 3x + x^2$ pounds/in^2. The rate of change of the pressure at $t = 2$ second is:

a) 7 pounds/in^2/sec

b) $3\frac{1}{6}$ pounds/in^2/sec

c) $3\frac{1}{2}$ pounds/in^2/sec

d) $2\frac{5}{6}$ pounds/in^2/sec

_____ **3.** A bacteria population grows in such a way that after t hours its population is $1000 + 10t + t^3$. The growth rate after 3 hours is:

a) 13 bacteria/hour

b) 37 bacteria/hour

c) 10 bacteria/hour

d) 1057 bacteria/hour

_____ **4.** A circle is increasing in area. The rate of change of area with respect to its radius r is:

a) $2\pi r$ b) πr^2 c) πr d) $2\pi r^2$

Section 3.4

_____ **1.** $\lim\limits_{x \to 0} \dfrac{2 \sin x}{3x} =$ _____.

    a)   1   b) $\dfrac{2}{3}$ c) $\dfrac{3}{2}$ d)   0

_____ **2.** If $\lim\limits_{x \to 0} \dfrac{\sin x}{x} = 1$, the angle x must be measured in:

    a)   radians   b)   degrees   c)   it does not matter

_____ **3.** For $y = x \sin x$, $y' =$ _____.

    a) $x \cos x$
    b) $x \cos x + 1$
    c) $\cos x$
    d) $x \cos x + \sin x$

_____ **4.** For $y = 3x + \tan x$, $y' =$ _____.

    a) $3 + \sec x$
    b) $3x \sec^2 x + 3 \tan x$
    c) $3 + \sec^2 x$
    d) $x \cos x + \sin x$

_____ **5.** For $y = \sin^2 x + \cos^2 x$, $y' =$ _____.

    a) $2 \sin x - 2 \cos x$
    b) $2 \sin x \cos x$
    c) $4 \sin x \cos x$
    d)   0

Section 3.5

____ **1.** True, False:

If $y = f(g(x))$ then $y' = f'(g'(x))$.

____ **2.** For $y = (x^2 + 1)^2$, $y' =$ ____.

     a) $2(x^2 + 1)$ b) $4(x^2 + 1)$ c) $4x(x^2 + 1)$ d) $2x(x^2 + 1)$

____ **3.** For $y = \sin 6x$, $\dfrac{dy}{dx} =$ ____.

     a) $\cos 6x$ b) $6\cos x$ c) $\cos 6$ d) $6\cos 6x$

____ **4.** For $y = \dfrac{4}{\sqrt{3 + 2x}}$, $y' =$ ____.

     a) $-4(3 + 2x)^{-3/2}$

     b) $4(3 + 2x)^{-3/2}$

     c) $-2(3 + 2x)^{-1/2}$

     d) $2(3 + 2x)^{-1/2}$

____ **5.** True, False:

For $y = f(u)$ and $u = g(x)$, $\dfrac{dy}{dx} = \dfrac{dy}{du} \cdot \dfrac{du}{dx}$.

____ **6.** Sometimes, Always, or Never:

$(a^x)' = a^x$.

____ **7.** If $f(x) = 2e^{3x}$, $f'(x) =$

     a) $2e^{3x}$ b) $2xe^{3x}$ c) $6e^{3x}$ d) $6xe^{3x}$

On Your Own

Section 3.6

_____ 1. True, False:

$y = \sqrt{4 - \sqrt{x^2 + 1}}$ defines y implicitly as a function of x.

_____ 2. If y is defined as a function of x by $y^3 = x^2$, $y' =$ _____.

a) $\frac{2x}{y^3}$ b) $\frac{2x}{3y^2}$ c) $\frac{x^2}{3y^2}$ d) $\frac{x^2}{3y}$

_____ 3. The slope of the line tangent to the circle $x^2 + y^2 = 100$ at the point $(-6, 8)$ is:

a) $\frac{3}{4}$ b) $-\frac{3}{4}$ c) $\frac{4}{3}$ d) $-\frac{4}{3}$

_____ 4. The range of $f(x) = \cos^{-1} x$ is:

a) $\left[-\frac{\pi}{2}, \frac{\pi}{2}\right]$

b) $[0, \pi]$

c) $\left[0, \frac{\pi}{2}\right] \cup \left[\pi, \frac{3\pi}{2}\right]$

d) all reals

_____ 5. $\sec^{-1}\left(\frac{2}{\sqrt{3}}\right) =$

a) $-\frac{\pi}{3}$ b) $\frac{\pi}{3}$ c) $-\frac{\pi}{6}$ d) $\frac{\pi}{6}$.

_____ 6. For $f(x) = \cos^{-1}(2x)$, $f'(x) =$

a) $\frac{-1}{\sqrt{1 - 4x^2}}$

b) $\frac{-2}{\sqrt{1-4x^2}}$

c) $\frac{-4}{\sqrt{1-4x^2}}$

d) $\frac{-8}{\sqrt{1-4x^2}}$

Section 3.7

____ **1.** True, False:

$y^{(6)}$ is the derivative of $y^{(5)}$.

____ **2.** For $y = \dfrac{5}{x^3}$, $y'' =$ ____.

a) $\dfrac{5}{6x}$　　　　b) $\dfrac{5}{12x}$　　　　c) $-\dfrac{60}{x^5}$　　　　d) $60x^{-5}$

____ **3.** For $y = x^n$ (n a positive integer), $y^{(n)} =$ ____.

a) 0　　　　b) $n!$　　　　c) nx^{n-1}　　　　d) $n!x$

____ **4.** For a particle whose position at time t (seconds) is
$t^3 - 3t^2 + 10t + 1$ meters, its acceleration at time $t = 3$ is:

a) 18 m/s^2　　　　b) 12 m/s^2　　　　c) 0 m/s^2　　　　d) -5 m/s^2

____ **5.** For $y = \cos x$, $y^{(14)} =$ ____.

a) $\cos x$　　　　b) $\sin x$　　　　c) $-\cos x$　　　　d) $-\sin x$

Section 3.8

____ **1.** For $y = \ln(3x^2 + 1)$, $y' =$

a) $\dfrac{1}{3x^2 + 1}$ b) $\dfrac{6}{x}$ c) $\dfrac{6x}{3x^2 + 1}$ d) $\dfrac{1}{6x}$

____ **2.** For $y = 3^x$, $y' =$

a) $3^x \log_3 e$ b) $3^x \ln 3$ c) $\dfrac{3x}{\log_3 e}$ d) $\dfrac{3x}{\ln 3}$

____ **3.** By logarithmic differentiation, if $y = \dfrac{\sqrt{x}}{x + 1}$, then $y' =$

a) $\dfrac{\sqrt{x}}{x + 1}\left(\dfrac{1}{\sqrt{x}} - \dfrac{1}{x + 1}\right)$ b) $\dfrac{\sqrt{x}}{x + 1}\left(\dfrac{1}{x} - \dfrac{1}{x + 1}\right)$

c) $\dfrac{\sqrt{x}}{x + 1}\left(\dfrac{2}{x} - \dfrac{1}{x + 1}\right)$ d) $\dfrac{\sqrt{x}}{x + 1}\left(\dfrac{1}{2x} - \dfrac{1}{x + 1}\right)$

____ **4.** $\lim\limits_{x \to 0^+} (1 + x)^{-1/x} =$

a) e b) $-e$ c) $\dfrac{1}{e}$ d) $-\dfrac{1}{e}$

Section 3.9

_____ **1.** True, False:

$\sinh^2 x + \cosh^2 x = 1$.

_____ **2.** $\cosh 0 =$

a) 0   b) 1   c) $\frac{e^2}{2}$ d) is not defined

_____ **3.** For $f(x) = \sinh x$, $f''(x) =$

a) $\cosh^2 x$ b) $\sinh^2 x$ c) $\tanh x$ d) $\sinh x$

_____ **4.** For $f(x) = \tanh^{-1} 2x$, $f'(x) =$

a) $2(\text{sech}^{-1} 2x)^2$

b) $2(\text{sech}^2 2x)^{-1}$

c) $\dfrac{2}{1 - 4x^2}$

d) $\dfrac{1}{1 - 4x^2}$

Section 3.10

_____ 1. A right triangle has one leg with constant length 8 cm. The length of the other leg is decreasing at 3 cm/sec. The rate of change of the hypotenuse when the variable leg is 6 cm is:

a) $-\frac{9}{5}$ cm/s

b) 3 cm/s

c) $-\frac{5}{3}$ cm/s

d) $\frac{5}{3}$ cm/s

_____ 2. How fast is the angle between the hands of a clock increasing at 3:30 pm?

a) 2π radians/hr

b) $\frac{\pi}{6}$ radians/hr

c) $\frac{11\pi}{6}$ radians/hr

d) $\frac{5\pi}{6}$ radians/hr

Section 3.11

_____ **1.** For $f(x) = x^3 + 7x$, $df = $ _____.

    a) $(x^3 + 7x)\, dx$

    b) $(3x^2 + 7)\, dx$

    c) $3x^2 + 7$

    d) $x^3 + 7x + dx$

_____ **2.** The best approximation of Δy using differentials for $y = x^2 - 4x$ at $x = 6$ when $\Delta x = dx = 0.03$ is:

    a) 0.36   b) 0.24   c) 0.20   d) 0.30

_____ **3.** The linearization of $f(x) = \dfrac{1}{\sqrt{x}}$ at $x = 4$ is:

    a) $L(x) = \frac{1}{2} + \frac{1}{\sqrt{x}}(x - 2)$

    b) $L(x) = \frac{1}{4} + \frac{1}{2}(x - 2)$

    c) $L(x) = \frac{1}{2} + \frac{1}{2}(x - 4)$

    d) $L(x) = \frac{1}{2} - \frac{1}{16}(x - 4)$

_____ **4.** Using differentials, an approximation to $(3.04)^3$ is:

    a) 28.094464   b) 28.08   c) 28.04   d) 28

On Your Own

Section 4.1

_____ **1.** True, False:
If c is a critical number, then $f'(c) = 0$.

_____ **2.** True, False:
If $f'(c) = 0$, then c is a critical number.

_____ **3.** The critical numbers of $f(x) = 3x^4 + 20x^3 - 36x^2$ are:

    a) $0, 1, 6$ b) $0, -1, 6$ c) $0, 1, -6$ d) $0, -1, -6$

_____ **4.** True, False:
The absolute extrema of a continuous function on a closed interval always exist.

_____ **5.** The graph at the right has a local maximum at $x =$ _____.

    a) 1, 3, and 5
    b) 2
    c) 4
    d) 2 and 4

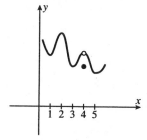

_____ **6.** The absolute maximum value of $f(x) = 6x - x^2$ on $[1, 7]$ is:

    a) 3   b) 9   c) 76   d) -7

Section 4.2

_____ **1.** Which of the following is not a hypothesis of Rolle's Theorem?

    a) f is continuous on $[a, b]$.

    b) $f(a) = f(b)$.

    c) $f(x) \geq 0$ for all $x \in (a, b)$.

    d) f is differentiable on (a, b).

_____ **2.** Yes, No:

Do all the hypotheses for the Mean Value Theorem hold for
$f(x) = 1 - |x|$ on $[-1, 2]$?

_____ **3.** A number c that satisfies the Mean Value Theorem for $f(x) = x^3$ on $[1, 4]$ is:

    a) $\sqrt{7}$ b) $\sqrt{21}$ c) $\sqrt{63}$ d) 63

_____ **4.** True, False:

If $f'(x) = g'(x)$ for all x then $f(x) = g(x)$.

Section 4.3

____ **1.** If 4 is a critical number for f and $f'(x) < 0$ for $x < 4$ and $f'(x) > 0$ for $x > 4$, then:

    a) f has a local maximum at 4.

    b) f has a local minimum at 4.

    c) 4 is either a local maximum or minimum, but not enough information is given to decide which.

    d) 4 is neither a local maximum nor local minimum.

____ **2.** True, False:

    $f(x) = 10x - x^2$ is monotonic on $(4, 8)$.

____ **3.** True, False:

    $f(x) = 4x^2 - 8x + 1$ is increasing on $(0, 1)$.

____ **4.** $f(x) = 12x - x^3$ has a local maximum at:

    a) $x = 0$

    b) $x = 2$

    c) $x = -2$

    d) f does not have a local maximum

____ **5.** True, False:

    If a function is not increasing on an interval, then it is decreasing on the interval.

____ **6.** True, False:

    If $f''(c) = 0$, then c is a point of inflection for f.

____ **7.** $f(x) = (x^2 - 3)^2$ has points of inflection at $x =$ ____.

    a) $0, \sqrt{3}, -\sqrt{3}$ b) $-1, 1$

    c) $0, -1, 1$ d) $\sqrt{3}, -\sqrt{3}$

____ **8.** $f(x) = x^3 - 2x^2 + x - 1$ has a local maximum at:

    a) 0 b) 1 c) $\frac{1}{3}$ d) f has no local maxima

_____ **9.** What graph has $f'(2) > 0$ and $f''(2) < 0$?

a)

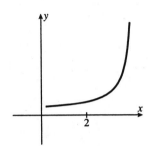

b)

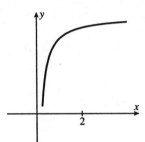

c)

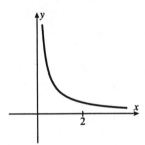

d)

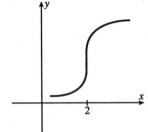

Section 4.4

_____ **1.** $\lim\limits_{x\to 5} \dfrac{x^2 - 25}{x^2 - 9x + 20} =$

    a) 1   b) 10   c) ∞ d) does not exist

_____ **2.** $\lim\limits_{x\to 0^+} \sqrt[3]{x}\, \ln x =$

    a) 0   b) 1   c) $-\infty$ d) does not exist

_____ **3.** $\lim\limits_{x\to\infty} (\ln x)^{1/x} =$

    a) 0   b) 1   c) e d) ∞

Section 4.5

_____ **1.** True, False:
   The domain is all x for which $f(x)$ is defined.

_____ **2.** True, False:
   The x-intercepts are the values of x for which $f(x) = 0$.

_____ **3.** True, False:
   $f(x)$ is symmetric about the origin if $f(x) = -f(x)$.

_____ **4.** True, False:
   If $f'(x) > 0$ for all x in an interval I then f is increasing on I.

_____ **5.** True, False:
   If $f'(c) = 0$ and $f''(c) < 0$, then c is a local minimum.

_____ **6.** True, False:
   If $f''(x) > 0$ for $x < c$ and $f''(x) < 0$ for $x > c$, then c is a point of inflection.

_____ **7.** Which is the graph of $f(x) = x + \frac{1}{x}$?

a)
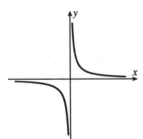

b)
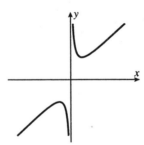

c)
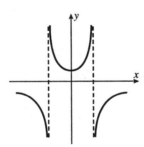

d)
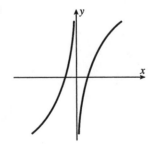

Section 4.6

____ **1.** The best graph of $f(x) = 3x^5 - 5x^3$ is:

a)

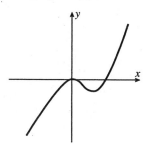

b)

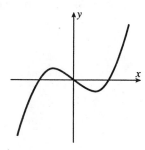

c)

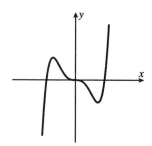

d)

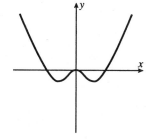

____ **2.** For the family of functions $f(x) = x^2 - ax$, each function is:

    a)   a parabola through $(0, 0)$ and $(a, 0)$.

    b)   is concave up.

    c)   both a) and b) are correct.

____ 1. For two nonnegative numbers, twice the first plus the second is 12. What is the maximum product of two such numbers?

    a) 12

    b) 18

    c) 36

    d) There is no such maximum.

____ 2. A carpenter has a 10 foot long board to mark off a triangular area on the floor in the corner of a room. See the figure. What is the function for the triangular area in terms of x that may be used to determine the maximum such area?

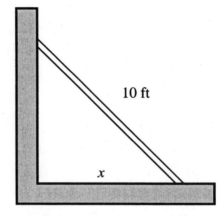

    a) $A = \frac{1}{2} x^2$

    b) $A = \frac{1}{2} (x^2 + 10)$

    c) $A = \frac{1}{2} x \sqrt{100 - x^2}$

    d) $A = \frac{1}{2} x \sqrt{100 - 2x^2}$

Section 4.8

_____ **1.** True, False:

Average cost is the derivative of total cost.

_____ **2.** True, False:

Marginal revenue is the derivative of total revenue.

_____ **3.** Average cost is minimum when:

   a) average cost equals total cost
   b) average cost equals marginal cost
   c) marginal cost equals total cost
   d) marginal cost equals zero

_____ **4.** Profit is maximum when:

   a) total revenue equals total cost
   b) marginal revenue equals marginal cost
   c) marginal revenue or marginal cost is zero
   d) marginal revenue is maximum

_____ **5.** Suppose the price (demand function) for an item is $p = 16 - 0.01x$, where x is the number of items. If the cost function is $C(x) = 1000 + 10x + 0.02x^2$, how many items should be made and sold to maximize profits?

   a) 50   b) 100   c) 200   d) 250

Section 4.9

_____ **1.** Sometimes, Always, or Never:
The value of x_2 in Newton's Method will be a closer approximation to the root than the initial value x_1.

_____ **2.** With an initial estimate of -1, use Newton's Method once to estimate a zero of $f(x) = 5x^2 + 15x + 9$.

a) $-\frac{6}{5}$ 　　　 b) $-\frac{4}{5}$ 　　　 c) $-\frac{7}{5}$ 　　　 d) $-\frac{3}{5}$

_____ **3.** Suppose in an estimation of a root of $f(x) = 0$ we make an initial guess x_1, so the $f'(x_1) = 0$. Which of the following will be true?

a) x_1 is a root of $f(x) = 0$.
b) Newton's Method will produce a sequence $x_1, x_2, \ldots$ that does approximate a zero.
c) $x_2 = x_1$
d) Newton's Method cannot be used.

Section 4.10

____ **1.** True, False:
If $h'(x) = k(x)$, then $k(x)$ is an antiderivative of $h(x)$.

____ **2.** Find $f(x)$ if $f'(x) = 10x^2 + \cos x$.

     a) $20x - \sin x + C$

     b) $20x - \cos x + C$

     c) $\frac{10}{3}x^3 - \cos x + C$

     d) $\frac{10}{3}x^3 + \sin x + C$

____ **3.** Find $f(x)$ if $f'(x) = \frac{1}{x^2}$ and $f(2) = 0$.

     a) $-\frac{1}{x} + \frac{1}{2}$

     b) $-\frac{1}{x} - \frac{1}{2}$

     c) $-\frac{3}{x^3} + \frac{3}{8}$

     d) $-\frac{3}{x^3} - \frac{3}{8}$

____ **4.** True, False:
If $f'(x) = g'(x)$, then $f(x) = g(x)$.

____ **5.** If velocity is $v(t) = 4t + 4$ and $s(1) = 2$, then $s(t) =$
     a) $2t^2 + 4t - 4$
     b) $2t^2 + 4t + 6$
     c) $t^2 + 4$
     d) $t^2 - 4$

Section 5.1

_____ 1. Find the sum of the areas of approximating rectangles for the area under $f(x) = 48 - x^2$, between $x = 1$ and $x = 5$. Use $\Delta x = 1$ and the right endpoints of each subinterval for x_i^*.

    a) 138   b) 99   c) 15   d) 192

_____ 2. True, False:
In general, better approximations to the area under the curve $y = f(x)$ between $x = a$ and $x = b$ are obtained by selection of partitions with smaller values of Δx.

_____ 3. Selecting midpoints of subintervals for x_i^* to approximate the area under $f(x) = 10x + 5$ between $x = 1$ and $x = 4$ will generate a Riemann sum which is:

    a) greater than the actual area
    b) less than the actual area
    c) equal to the actual area

_____ 4. The shaded area at the right is

    a) $\lim\limits_{n\to\infty} \sum\limits_{i=1}^{n} (x_i^2 + 1)\, \Delta x$

    b) $\lim\limits_{n\to\infty} \sum\limits_{i=1}^{n} \left((3 - 1)^2 + 1\right) \Delta x$

    c) $\lim\limits_{n\to\infty} \sum\limits_{i=1}^{n} (3x_i^2 + x_i)\, \Delta x$

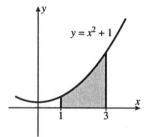

Section 5.2

___ 1. Sometimes, Always, or Never:

$\int_a^b f(x)\, dx$ equals the area between $y = f(x)$, the x-axis, $x = a$, and $x = b$.

___ 2. Sometimes, Always, or Never:

If $f(x)$ is continuous on $[a, b]$, then $\int_a^b f(x)\, dx$ exists.

___ 3. Using $n = 4$ and midpoints for x_i^*, the Riemann sum for $\int_{-1}^7 x^2\, dx$ is

    a) $\dfrac{344}{3}$ b) 72 c) 168 d) 112

___ 4. If $\int_1^3 f(x)\, dx = 10$ and $\int_1^3 g(x)\, dx = 6$, the $\int_1^3 (2f(x) - 3g(x))dx =$

    a) 2 b) 4 c) 18 d) 38

___ 5. If $f(x) \geq 5$ for all $x \in [2, 6]$ then $\int_2^6 f(x)\, dx \geq$ _____.

(Choose the best answer.)

    a) 4 b) 5 c) 20 d) 30

___ 6. $\int_3^5 6\, dx =$

    a) 2 b) 8 c) 6 d) 12

___ 7. $\sum_{i=1}^n (4i^2 + i) =$

    a) $4\sum_{i=1}^n (i^2 + i)$ c) $\sum_{i=1}^n 5i^2$

    b) $4\sum_{i=1}^n i^2 + \sum_{i=1}^n i$ d) $4\sum_{i=1}^n i^2 + 4\sum_{i=1}^n i$

___ 8. True, or False:

$\int_0^7 2\sqrt[3]{x}\, dx = 2\int_0^7 \sqrt[3]{x}\, dx.$

Section 5.3

_____ **1.** If $h(x) = \int_3^x (8t^3 + 2t)dt$, then $h'(x) =$

a) $8t^3 + 2t$ b) $24x^2 + 2$ c) $8x^3 + 2x - 3$ d) $8x^3 + 2x$

_____ **2.** Evaluate $\int_0^3 4x\,dx$.

a) 6   b) 12   c) 18   d) 24

_____ **3.** True, False:

If $f'(x) = F(x)$, then $\int_b^a f(x)\,dx = F(b) - F(a)$.

___ **1.** $\int f(x)\,dx = F(x)$ means

    a) $f'(x) = F(x)$
    b) $f(x) = F'(x)$
    c) $f(x) = F(b) - F(a)$
    d) $f(x) = F(x) + C$

___ **2.** $\int x^{-4}dx =$

    a) $-4x^{-5} + C$
    b) $\frac{-4}{5}x^{-5} + C$
    c) $-3x^{-3} + C$
    d) $\frac{x^{-3}}{-3} + C$

___ **3.** $\int (x - \cos x)dx =$

    a) $1 - \sin x + C$

    b) $1 + \sin x + C$

    c) $\frac{x^2}{2} - \sin x + C$

    d) $\frac{x^2}{2} + \sin x + C$

___ **4.** Evaluate $\int_{-1}^{1} (x^2 - x^3)dx$

    a) $\frac{3}{2}$
    b) $\frac{5}{6}$
    c) $\frac{1}{2}$
    d) $\frac{2}{3}$

___ **5.** If A_1 and A_2 are the areas at
the right then $\int_a^b f(x)\,dx =$

    a) $A_1 - A_2$
    b) $-A_1 + A_2$
    c) $A_1 + A_2$
    d) $|A_1 - A_2|$

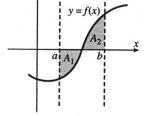

___ **6.** $\int \frac{1}{2x+1}\,dx =$

    a) $\ln|2x+1| + C$
    b) $\frac{1}{2}\ln|2x+1| + C$

    c) $2\ln|2x+1| + C$
    d) $\ln\left|\frac{2x+1}{2}\right| + C$

Section 5.5

_____ 1. Suppose $\int f(x) = F(x) + C.$ Then $\int f(g(x))g'(x)\,dx =$
   a) $F(x)g(x) + C$
   b) $f'(g(x)) + C$
   c) $F(g'(x)) + C$
   d) $F(g(x)) + C$

_____ 2. What substitution should be made to evaluate $\int \sqrt{x^2 + 2x}\,(x + 1)\,dx$?

   a) $u = \sqrt{x^2 + 2x}$
   b) $u = x^2 + 2x$
   c) $u = x + 1$
   d) $u = \sqrt{x^2 + 2x}\,(x + 1)$

_____ 3. What substitution should be made to evaluate $\int \sin^2 x \cos x\,dx$?
   a) $u = \sin x$
   b) $u = \cos x$
   c) $u = \sin x \cos x$
   d) $u = \sin^2 x \cos x$

_____ 4. Evaluate $\int_0^1 (x^3 + 1)^2 x^2\,dx.$
   a) $\frac{1}{9}$ b) $\frac{1}{3}$ c) $\frac{7}{9}$ d) $\frac{8}{9}$

_____ 5. True False:
   If $f(x) = f(-x)$ for all x then $\int_{-2}^2 f(x) = 2\int_0^2 f(x).$

_____ 6. $\int_0^1 e^{2x}\,dx =$ a) $2e^2 - 1$ b) $e^2 - 1$ c) $\frac{1}{2}e^2$ d) $\frac{1}{2}(e^2 - 1)$

Section 5.6

_____ 1. $\int_1^7 \frac{1}{t}\, dt =$

    a) $\frac{1}{7}$ b) $\ln 7$ c) $\ln \frac{1}{7}$ d) $\ln 6$

_____ 2. For what number k is the shaded area equal to 2?

    a) $\ln 2$

    b) $\ln \frac{1}{2}$

    c) $2e$

    d) e^2

_____ 3. True, False:
$\ln(a + b) = \ln a + \ln b.$

_____ 4. For $y = \ln(3x^2 + 1), y' =$

    a) $\frac{1}{3x^2 + 1}$ b) $\frac{6}{x}$ c) $\frac{6x}{3x^2 + 1}$ d) $\frac{1}{6x}$

_____ 5. By definition, $\sqrt{2}^{\sqrt{3}} =$

    a) $\sqrt{6}$ b) $(\ln \sqrt{3})e^{\sqrt{2}}$ c) $e^{\sqrt{3}\ln\sqrt{2}}$ d) $e^{\sqrt{2}\ln\sqrt{3}}$

_____ 6. Solve for x: $e^{2x-1} = 10.$

    a) $\frac{1}{2}(a + 10^e)$ b) $3 + \ln 10$ c) $2 + \ln 10$ d) $\ln\sqrt{10e}$

_____ 7. Solve for x: $\ln(e + x) = 1.$

    a) 0 b) 1 c) $-e$ d) $e^e - e$

_____ 8. $\lim\limits_{x \to 0^+} (1 + x)^{-1/x} =$

    a) e b) $-e$ c) $\frac{1}{e}$ d) $-\frac{1}{e}$

Section 6.1

1. A definite integral for the area shaded at the right is:

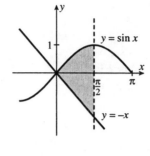

a) $\displaystyle\int_0^{\pi/2} (\sin x - x)\,dx$

b) $\displaystyle\int_0^{\pi/2} (\sin x + x)\,dx$

c) $\displaystyle\int_0^1 (\sin x - x)\,dx$

d) $\displaystyle\int_0^1 (\sin x + x)\,dx$

2. For the area of the shaded region, which integral form, $\int \ldots dx$ or $\int \ldots dy$, should be used?

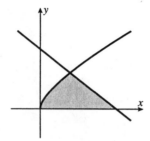

a) $\displaystyle\int \ldots dx$

b) $\displaystyle\int \ldots dy$

c) Either may be used with equal ease.

3. A definite integral for the area of the region bounded by $y = 2 - x^2$ and $y = x^2$ is:

a) $\displaystyle\int_{-1}^1 (2 - 2x^2)\,dx$

b) $\displaystyle\int_{-1}^1 (2x^2 - 2)\,dx$

c) $\displaystyle\int_0^2 (2 - 2x^2)\,dx$

d) $\displaystyle\int_0^2 (2x^2 - 2)\,dx$

On Your Own

Section 6.2

_____ 1. Find a definite integral for the volume of a solid with a circular base of radius 4 inches such that the cross section of any slice perpendicular to a certain diameter in the base is a square.

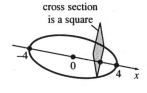

cross section is a square

a) $\int_{-4}^{4} (64 - 4x^2)\,dx$

b) $\int_{-4}^{4} 4x^2\,dx$

c) $\int_{-4}^{4} 4x^4\,dx$

d) $\int_{-4}^{4} (4 - x^2)^2\,dx$

_____ 2. An integral for the solid obtained by rotating the region at the right about the x-axis is:

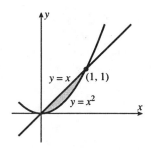

a) $\int_{0}^{1} \pi(x^4 - x^2)\,dx$

b) $\int_{0}^{1} \pi(x^2 - x^4)\,dx$

c) $\int_{0}^{1} \pi(x - x^2)\,dx$

d) $\int_{0}^{1} \pi(x^2 - x)\,dx$

_____ 3. An integral for the solid obtained by rotating the region at the right about the y-axis is:

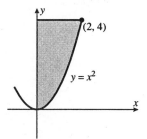

a) $\int_{0}^{4} \pi y\,dy$

b) $\int_{0}^{2} \pi y^2\,dy$

c) $\int_{0}^{4} \pi\sqrt{y}\,dy$

d) $\int_{0}^{2} \pi\sqrt{y}\,dy$

Section 6.3

1. Which definite integral is the volume of the solid obtained by revolving the region at the right about the y-axis using cylindrical shells?

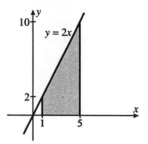

   a) $\displaystyle\int_1^5 2\pi x^2 \, dx$

   b) $\displaystyle\int_1^5 4\pi x^2 \, dx$

   c) $\displaystyle\int_1^5 2\pi x \, dx$

   d) $\displaystyle\int_2^{10} 4\pi x \, dx$

2. Which definite integral is the volume of the solid obtained by revolving the region at the right about the x-axis using cylindrical shells?

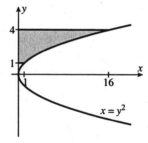

   a) $\displaystyle\int_1^{16} 2\pi y^3 \, dy$

   b) $\displaystyle\int_1^{16} 2\pi(4 - \sqrt{x}) \, dy$

   c) $\displaystyle\int_1^4 2\pi y^2 \, dy$

   d) $\displaystyle\int_1^4 2\pi y^3 \, dy$

Section 6.4

_____ **1.** How much work is done in lifting a 60-pound child 3 feet in the air?

    a)   20 ft-lbs

    b) $20g$ ft-lbs

    c)   180 ft-lbs

    d) $180g$ ft-lbs

_____ **2.** A particle moves along an x-axis from 2 m to 3 m pushed by a force of x^2 newtons. The integral that determines the amount of work done is:

    a) $\displaystyle\int_2^3 gx^3 \, dx$

    b) $\displaystyle\int_2^3 x \, dx$

    c) $\displaystyle\int_2^3 2\pi x^2 \, dx$

    d) $\displaystyle\int_2^3 x^2 \, dx$

Section 6.5

_____ **1.** The average value of $f(x) = 3x^2 + 1$ on the interval $[2, 4]$ is:

  a) 29   b) 66   c) 58   d) 36

_____ **2.** Which point x_1, x_2, x_3, or x_4, on the graph is the best choice to serve as the point guaranteed by the Mean Value Theorem for Integrals?

  a) x_1 b) x_2 c) x_3 d) x_4

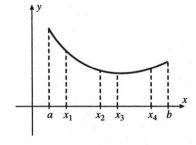

_____ **3.** Find a point c for the Mean Value Theorem for Integrals for $f(x) = x^2 - 2x$ on $[2, 5]$.

  a) 18   b) 6   c) $1 + \sqrt{7}$ d) no such c exists

Section 7.1

_____ **1.** Find the correct u and dv for integration by parts of $\int x^2 \cos 2x \, dx$.

    a) $u = x^2, dv = \cos 2x \, dx$
    b) $u = \cos 2x, dv = x^2 \, dx$
    c) $u = x^2, dv = 2x \, dx$
    d) $u = x \cos 2x, dv = x \, dx$

_____ **2.** Evaluate $\int 16t^3 \ln t \, dt$.

    a) $4t^4 + C$
    b) $4t^3 \ln t - 16t^3 + C$
    c) $4t^4 - t^3 \ln t + C$
    d) $4t^4 \ln t - t^4 + C$

_____ **3.** The integration by parts rule comes from which differentiation formula?

    a) $(u + v)' = u' + v'$
    b) $(uv)' = u'v'$
    c) $(uv)' = uv' + u'v$
    d) $u \, dv = v \, du$

Section 7.2

_____ **1.** By the methods of trigonometric integrals, $\int \sin^3 x \cos^2 x \, dx$ should be rewritten as:

    a) $\int \sin^3 x \, (1 - \sin^2 x) dx$
    b) $\int (1 - \cos^2 x)^{3/2} \cos^2 x \, dx$
    c) $\int (\sin^3 x)(\cos x) \cos x \, dx$
    d) $\int (1 - \cos^2 x) \cos^2 x \sin x \, dx$

_____ **2.** Using $\tan^2 x = \sec^2 x - 1$, $\int \tan^3 x \, dx =$

    a) $\frac{3}{2} \ln |\sec^2 x - 1| + C$
    b) $\frac{1}{2} \tan^2 x - \ln |\sec x| + C$
    c) $\frac{1}{2} \tan^2 x - \frac{1}{4} \sec^2 x + C$
    d) $\frac{1}{4} \tan^2 x + \ln |\sec x| + C$

Section 7.3

____ **1.** What trigonometric substitution should be made for $\int \frac{x^2}{\sqrt{x^2 - 25}} \, dx$?

    a) $x = 5 \sin \theta$
    b) $x = 5 \tan \theta$
    c) $x = 5 \sec \theta$
    d)   No such substitution will evaluate the integral

____ **2.** Rewrite $\int \frac{\sqrt{10 - x^2}}{x} \, dx$ using a trigonometric substitution.

    a) $\int \frac{\cos \theta}{\sqrt{10} \sin \theta} \, d\theta$
    b) $\int \frac{1}{\sqrt{10}} \sin \theta \cos^2 \theta \, d\theta$
    c) $\int \frac{\sqrt{10} \cos^2 \theta}{\sin \theta} \, d\theta$
    d) $\int \sqrt{10} \sin \theta \cos \theta \, d\theta$

____ **3.** Is a trigonometric substitution necessary for $\int \frac{x}{\sqrt{x^2 + 10}} \, dx$?

    a)   Yes   b)   No

Section 7.4

_____ **1.** What is the partial fraction form of $\dfrac{2x+3}{(x-2)(x+2)}$?

    a) $\dfrac{Ax}{x-2} + \dfrac{B}{x+2}$

    b) $\dfrac{A}{x-2} + \dfrac{B}{x+2}$

    c) $\dfrac{Ax}{x-2} + \dfrac{Bx}{x+2}$

    d) $\dfrac{Ax+B}{x-2} + \dfrac{Cx+D}{x+2}$

_____ **2.** What is the partial fraction form of $\dfrac{2x+3}{(x^2+1)^2}$

    a) $\dfrac{Ax+B}{(x^2+1)^2}$

    b) $\dfrac{A}{x^2+1} + \dfrac{B}{(x^2+1)^2}$

    c) $\dfrac{A}{x^2+1} + \dfrac{Bx+C}{(x^2+1)^2}$

    d) $\dfrac{Ax+B}{x^2+1} + \dfrac{Cx+D}{(x^2+1)^2}$

_____ **3.** Evaluate $\displaystyle\int \dfrac{2}{(x+1)(x+2)}\,dx$, using partial fractions.

    a) $2\ln|x+1| - 2\ln|x+2| + C$
    b) $\ln|x+1| + 3\ln|x+2| + C$
    c) $3\ln|x+1| - \ln|x+2| + C$
    d) $\ln|x+1| + \ln|x+2| + C$

_____ **4.** What rationalizing substitution should be made for $\displaystyle\int \dfrac{\sqrt[3]{x}+2}{\sqrt[3]{x}+1}\,dx$?

    a) $u = x$

    b) $u = \sqrt[3]{x}$

    c) $u = \sqrt[3]{x} + 2$

    d) $u = \sqrt[3]{x} + 1$

_____ **5.** What rationalizing substitution should be made for $\displaystyle\int \dfrac{1}{\sqrt{x}+\sqrt[5]{x}}\,dx$?

    a) $u = x$

    b) $u = \sqrt{x}$

    c) $u = \sqrt[5]{x}$

    d) $u = \sqrt[10]{x}$

Section 7.5

_____ **1.** True, False:

$\int x^2 \sqrt{x^2 - 4} \, dx$ should be evaluated using integration by parts.

_____ **2.** True, False:

A straight (simple) substitution may be used to evaluate $\int \frac{2 + \ln x}{x} \, dx$.

_____ **3.** True, False:

Partial fractions may be used for $\int \frac{1}{\sqrt{x + 2}\sqrt{x + 3}} \, dx$.

_____ **4.** True, False:

$\int \frac{\sqrt{x^2 + 1}}{x^3} \, dx$ may be solved using a trigonometric substitution.

_____ **5.** Sometimes, Always, or Never:

The antiderivative of an elementary function is elementary.

Section 7.6

_____ **1.** What is the number of the integral formula in your text's Table of Integrals that may be used to evaluate $\int \frac{dx}{x\sqrt{2-x^2}}$?

   a) #35   b) #36   c) #18   d) #24

_____ **2.** What is the number of the integral formula in your text's Table of Integrals that may be used to evaluate $\int \frac{x^2+2x+1}{\sqrt{x^2+2x+5}}\,dx$?

   a) #37   b) #38   c) #26   d) #46

Section 7.7

_____ **1.** If $f(x)$ is integrable and concave upward on $[a, b]$, then an estimate of $\int_a^b f(x)\, dx$ using the Trapezoidal Rule will always be:

   a) too large
   b) too small
   c) exact
   d) within $\frac{b-a}{12n}$ of the exact value

_____ **2.** An estimate of $\int_{-1}^3 x^4\, dx$ using Simpson's Rule with $n = 4$ gives:

   a) $\frac{242}{5}$ b) $\frac{148}{3}$ c) $\frac{152}{3}$ d) $\frac{244}{5}$

_____ **3.** Using $\dfrac{K(b-a)^3}{24n^2}$ find the maximum error in estimating $\int_{-1}^3 x^4\, dx$ with the Midpoint Rule and 8 subintervals.

   a) $\frac{1}{2}$ b) 9   c) $\frac{9}{2}$ d) $\frac{9}{8}$

____ **1.** True, False:

$\int_{-2}^{3} |x|\, dx$ is an improper integral.

____ **2.** By definition the improper integral $\int_{1}^{e} \frac{1}{x \ln x}\, dx =$

a) $\lim_{t \to 1^+} \int_{t}^{e} \frac{1}{x \ln x}\, dx$

b) $\lim_{t \to e^-} \int_{1}^{t} \frac{1}{x \ln x}\, dx$

c) $\int_{1}^{2} \frac{1}{x \ln x}\, dx + \int_{2}^{e} \frac{1}{x \ln x}\, dx$

d) It is not an improper integral.

____ **3.** $\int_{0}^{2} \frac{x - 2}{\sqrt{4x - x^2}}\, dx =$

a) 0   b) 2   c) -2 d) does not exist

____ **4.** Suppose we know $\int_{1}^{\infty} f(x)\, dx$ diverges and that $f(x) \geq g(x) \geq 0$ for all $x \geq 1$. what conclusion can be made about $\int_{1}^{\infty} g(x)dx$?

a) it converges to the value of $\int_{1}^{\infty} f(x)dx$

b) it converges, but we don't know to what value

c) it diverges

d) no conclusion (it could diverge or converge)

Section 8.1

_____ **1.** A definite integral for the length of $y = x^3$, $1 \le x \le 2$ is:

a) $\int_1^2 \sqrt{1 + x^3}\, dx$

b) $\int_1^2 \sqrt{1 + x^6}\, dx$

c) $\int_1^2 \sqrt{1 + 9x^4}\, dx$

d) $\int_1^2 \sqrt{1 + 3x^2}\, dx$

_____ **2.** The arc length function for $y = x^2$, $1 \le x \le 3$ is $s(x) =$

a) $\int_1^x \sqrt{1 + 4x^2}\, dx$

b) $\int_1^x \sqrt{1 + 4t^2}\, dt$

c) $\int_1^x \sqrt{1 + t^4}\, dt$

d) $\int_1^x \sqrt{1 + 4t^4}\, dt$

Section 8.2

____ **1.** An integral for the surface area obtained by rotating $y = \sin x + \cos x$, $0 \le x \le \frac{\pi}{2}$, about the x-axis is:

a) $\displaystyle\int_0^{\pi/2} 2\pi(\sin x + \cos x)\sqrt{2 - 2\sin x \cos x}\, dx$

b) $\displaystyle\int_0^{\pi/2} 2\pi(\sin x + \cos x)\sqrt{1 + \sin^2 x - \cos^2 x}\, dx$

c) $\displaystyle\int_0^{\pi/2} 2\pi(\cos x - \sin x)\sqrt{1 + \sin^2 x - \cos^2 x}\, dx$

d) $\displaystyle\int_0^{\pi/2} 2\pi(\cos x - \sin x)\sqrt{2 - 2\sin x \cos x}\, dx$

____ **2.** A hollow cylinder of radius 3 in. and height 4 in. is a surface of revolution. Its round surface area may be expressed as:

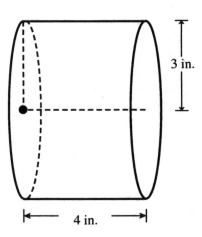

3 in.

4 in.

a) $\displaystyle\int_0^3 2\pi(4)\sqrt{1 + 0}\, dx$

b) $\displaystyle\int_0^3 2\pi(3)\sqrt{1 + 0}\, dx$

c) $\displaystyle\int_0^4 2\pi(3)\sqrt{1 + 0}\, dx$

d) $\displaystyle\int_0^4 2\pi(4)\sqrt{1 + 0}\, dx$

Section 8.3

____ **1.** Find a definite integral for the hydrostatic force exerted by a liquid with density 800 kg/m³ on the semicircular plate in the center of the figure.

a) $\displaystyle\int_{-2}^{2} 800(9.8)[8 - y]\left[2\sqrt{4 - y^2}\right] dy$

b) $\displaystyle\int_{-2}^{2} 800(9.8)[8 - y]\sqrt{4 - y^2}\, dy$

c) $\displaystyle\int_{-2}^{2} 800(9.8)[10 - y]\left[2\sqrt{4 - y^2}\right] dy$

d) $\displaystyle\int_{-2}^{2} 800(9.8)[10 - y]\sqrt{4 - y^2}\, dy$

____ **2.** The x-coordinate of the center of mass of the region bounded by $y = \frac{1}{x}$, $x = 1, x = 2, y = 0$ is:

   a) $\ln 2$ b) $\frac{1}{\ln 2}$ c) 1 d) $2\ln 2$

____ **3.** The y-coordinate of the center of mass of the region bounded by $y = \sqrt{x - 1}, x = 1, x = 5, y = 0$ is:

   a) $\frac{3}{14}$ b) $\frac{3}{4}$ c) $\frac{14}{3}$ d) 4

____ **4.** True, False:
$$\bar{x} = \frac{M_y}{\rho A} \text{ and } \bar{y} = \frac{M_x}{\rho A}.$$

Section 8.4

_____ **1.** True, False:

In many applications of definite integrals, the integral is used to compute the total amount of a varying quantity.

_____ **2.** Find a definite integral for the consumer surplus if 180 units are available and the demand function is $p(x) = 3000 - 10x$.

a) $\displaystyle\int_0^{180} (3000 - 10x)\,dx$

b) $\displaystyle\int_0^{180} (1800 - 10x)\,dx$

c) $\displaystyle\int_0^{180} (1200 - 10x)\,dx$

d) $\displaystyle\int_0^{180} (180 - 10x)\,dx$

_____ **3.** The dye dilution method is used to measure the cardiac output of a pig's heart with 10 mg of dye. After t seconds the dye concentration is modeled by $c(t) = -\frac{1}{2}t^2 + 4t$ for $0 \le t \le 6$ seconds. Find the cardiac output.

a) $\frac{5}{18}$ L/s

b) $\frac{3}{4}$ L/s

c) $\frac{10}{39}$ L/s

d) 36 L/s

Section 8.5

____ **1.** True, False:
The function whose graph is given
at the right is a probability density
function.

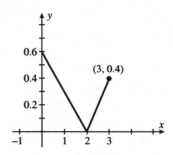

____ **2.** For the probability density function

$$f(x) = \begin{cases} 0 & \text{for } x \le 0 \\ 3\,e^{-3x} & \text{for } x > 0 \end{cases}$$

the mean is

a) $\displaystyle\int_0^\infty 3e^{-3x}dx$ b) $\displaystyle\int_0^\infty 3e^{-3x+1}dx$ c) $\displaystyle\int_0^\infty 3e^{-3x-1}$ d) $\displaystyle\int_0^\infty 3xe^{-3x}dx$

____ **3.** For a normal probability density function $f(x) = \dfrac{1}{\sigma\sqrt{2\pi}}\,e^{-(x-\mu)^2/(2\sigma^2)}$,
the probability that $x \ge \mu$ is:

a) 1   b) $\dfrac{1}{2}$ c) $\dfrac{1}{\sigma}$ d) $\dfrac{1}{2\sigma^2}$

Section 9.1

_____ **1.** $(y''')^2 + 6xy - 2y' + 7x = 0$ has order:

    a) 1   b) 2   c) 3   d) 4

_____ **2.** True, False:

$y = xe^x$ is a solution to $y' = y + \dfrac{y}{x}$.

_____ **3.** The general solution to $y' = \dfrac{2x}{y}$ is $y = \sqrt{2x^2 + C}$. Find a solution to the initial value problem $y' = \dfrac{2x}{y}$, $y(0) = 4$.

    a) $y = 4\sqrt{2x^2 + 1}$

    b) $y = \sqrt{2x^2 + 4}$

    c) $y = \sqrt{2x^2 + 4}$

    d) $y = \sqrt{2x^2 + 16}$

Section 9.2

_____ **1.** Is the figure at the right the
graph of the direction field
for $y' = (y - 1)x^2$?

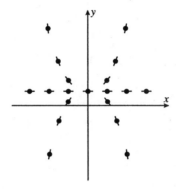

_____ **2.** The graph at the right
is the direction field for:

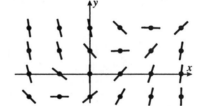

a) $y' = x - y$
b) $y' = xy$
c) $y' = x + y$
d) $y' = \frac{x}{y}$

_____ **3.** The first point determined by Euler's method for $\frac{dy}{dx} = x - 4y$, $y(11) = 2$
with step size 0.5 is:

a) $(11.0, 3.5)$ b) $(11.5, 3.5)$ c) $(11.0, 3.75)$ d) $(11.5, 3.75)$

Section 9.3

_____ **1.** A differential equation is separable if it can be written in the form:

  a) $\frac{dy}{dx} = f(x) + g(y)$

  b) $\frac{dy}{dx} = f(g(x, y))$

  c) $\frac{dy}{dx} = \frac{f(x)}{g(y)}$

  d) $\frac{dy}{dx} = f(x)^{g(y)}$

_____ **2.** True, False:
$y' + xe^y = e^{x + y}$ is separable.

_____ **3.** The general solution to $\frac{dy}{dx} = \frac{1}{xy}$ (for $x, y > 0$) is:

  a) $y = \ln x + C$

  b) $y = \sqrt{\ln x + C}$

  c) $y = \ln(\ln x + C)$

  d) $y = \sqrt{\ln x^2 + C}$

_____ **4.** The solution to $y' = -y^2$ with $y(1) = \frac{1}{3}$ is:

  a) $y = \frac{1}{x + 2}$

  b) $y = \ln x + \frac{1}{3}$

  c) $y = \frac{1}{2x^2} - \frac{1}{6}$

  d) $y = \frac{1}{2x} - \frac{1}{6}$

Section 9.4

_____ **1.** The general solution to $\frac{dy}{dt} = ky$ is:

  a) $y(t) = y(0)e^{kt}$

  b) $y(t) = y(k)e^t$

  c) $y(t) = y(t)e^k$

  d) $y(t) = e^{y(0)kt}$

_____ **2.** A bacteria culture starts with 50 organisms and after 2 hours there are 100. How many will there be after 5 hours?

  a) $50 \ln 5$

  b) $50e^5$

  c) $200\sqrt{2}$

  d) 300

_____ **3.** If $5,000 is deposited in an account that accumulates interest compounded continuously at 4% then after 6 years the amount accumulated is

  a) $5000\, e^{(-0.04)6}$

  b) $5000\, e^{(0.04)6}$

  c) $5000\, e^{(-4)6}$

  d) $5000\, e^{(4)6}$

Section 9.5

_____ **1.** True, False:
A population that follows a logistic model will never reach its carrying capacity.

_____ **2.** A population of ants on an island is modeled by $P(t) = \dfrac{80000}{1 + 2000e^{-0.5t}}$, where t is the number of years since 1970. How long did it take the population to reach half that of the island's carrying capacity?

    a) 7.6 years   b) 15.2 years   c) 22.8 years   d) 30.4 years

Section 9.6

____ **1.** True, False:
$xy' + 6x^2y = 10 - x^3$ is linear.

____ **2.** What is the integrating factor for $xy' + 6x^2y = 10 - x^3$?

a) e^{3x^2} b) e^{6x} c) e^{3x^3} d) e^{6x^2}

____ **3.** The solution to $x\dfrac{dy}{dx} - 2y = x^3$ is:

a) $y = e^{x^3} + Cx^2$ b) $y = e^{x^3} + x^2 + C$

c) $y = x^3 + x^2 + C$ d) $y = x^3 + Cx^2$

Section 9.7

_____ **1.** True, False:
Predator-prey population models are the solutions to two differential equations.

_____ **2.** The populations of sheep and wolves on an island is governed by a predator-prey model with constants $k = 0.03$, $r = 0.2$, $a = 0.0005$, and $b = 0.0001$. What is the constant solution for these populations?

a) 60 wolves and 2000 sheep
b) 30 wolves and 3000 sheep
c) 30 wolves and 2000 sheep
d) 60 wolves and 3000 sheep

Section 10.1

_____ **1.** The graph of $x = 2 + 3t,\ y = 4 - t$ is a:

a) circle
b) ellipse
c) line
d) parabola

_____ **2.** The best graph of $x = \cos t,\ y = \sin^2 t$ is:

a)

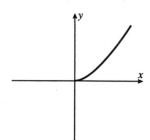

b)

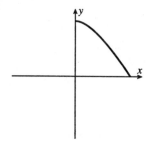

c)

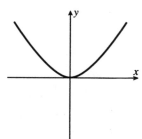

d)

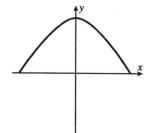

_____ **3.** Elimination of the parameter in $x = 2t^{3/2},\ y = t^{2/3}$ gives:

a) $x^4 = 16y^9$
b) $16x^4 = y^4$
c) $x^3 = 8y^4$
d) $8x^3 = y^4$

Section 10.2

_____ **1.** Find $\frac{dy}{dx}$ if $x = \sqrt{t}$, $y = \sin 2t$.

   a) $\frac{4\cos t}{\sqrt{t}}$ b) $\frac{\cos 2t}{\sqrt{t}}$ c) $\frac{\cos 2t}{2\sqrt{t}}$ d) $4\sqrt{t}\cos 2t$

_____ **2.** Find $\frac{d^2y}{dx^2}$ for $x = 3t^2 + 1$, $y = t^6 + 6t^5$.

   a) $t^4 + 5t^3$ b) $4t^3 + 15t^2$

   c) $\frac{2}{3}t^2 + \frac{5}{2}t$ d) $t^3 + \frac{1}{2}t^2$

_____ **3.** The slope of the tangent line at the point where $t = \frac{\pi}{6}$ to the curve $y = \sin 2t$, $x = \cos 3t$ is:

   a) $\frac{1}{3}$ b) $-\frac{1}{3}$ c) 3 d) -3

_____ **4.** A definite integral for the area under the curve described by $x = t^2 + 1$, $y = 2t$, $0 \le t \le 1$ is:

   a) $\int_0^1 (2t^3 + 2t)\,dt$ b) $\int_0^1 4t^2\,dt$

   c) $\int_0^1 (2t^2 + 2)\,dt$ d) $\int_0^1 4t\,dt$

On Your Own

Section 10.3

____ **1.** The length of the curve given by $x = 3t^2 + 2$, $y = 2t^3$, $t \in [0, 1]$ is:

a) $4\sqrt{2} - 2$

b) $8\sqrt{2} - 1$

c) $\frac{2}{3}\left(2\sqrt{2} - 1\right)$

d) $\sqrt{2} - 1$

____ **2.** Find a definite integral for the area of the surface of revolution about the x-axis obtained by rotating the curve $y = t^2$, $x = 1 + 3t$, $0 \le t \le 2$.

a) $\displaystyle\int_0^2 2\pi t^2 \sqrt{t^4 + 9t^2 + 6t + 1}\ dt$

b) $\displaystyle\int_0^2 2\pi t^2 \sqrt{4t^2 + 9}\ dt$

c) $\displaystyle\int_0^2 2\pi(2t)\sqrt{t^4 + 9t^2 + 6t + 1}\ dt$

d) $\displaystyle\int_0^2 2\pi(2t)\sqrt{4t^4 + 9}\ dt$

Section 10.4

1. The polar coordinates of the
point P plotted at the right are:

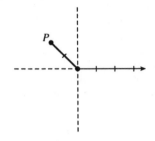

a) $\left(-2, \frac{\pi}{4}\right)$

b) $\left(-2, \frac{3\pi}{4}\right)$

c) $\left(2, -\frac{\pi}{4}\right)$

d) $\left(2, \frac{3\pi}{4}\right)$

2. Polar coordinates of the point with rectangular coordinates $(5, 5)$ are:

a) $(25, 0)$ b) $(5, \frac{\pi}{4})$

c) $(5\sqrt{2}, \frac{\pi}{4})$ d) $(50, -\frac{\pi}{4})$

3. Rectangular coordinates of the point with polar coordinates $\left(-1, \frac{3\pi}{2}\right)$ are:

a) $(-1, 0)$ b) $(0, 1)$

c) $(0, -1)$ d) $(1, 0)$

4. The graph of $\theta = 2$ in polar coordinates is a:

a) circle   b) line   c) spiral   d) 3-leaved rose

5. The slope of the tangent line to $r = \cos\theta$ at $\theta = \frac{\pi}{3}$ is:

a) $\sqrt{3}$ b) $\frac{1}{\sqrt{3}}$ c) $-\sqrt{3}$ d) $-\frac{1}{\sqrt{3}}$

Section 10.5

_____ **1.** The area of the region bounded by $\theta = \frac{\pi}{3}$, $\theta = \frac{\pi}{4}$, and $r = \sec \theta$ is:

a) $\frac{1}{2}\left(\sqrt{3} - 1\right)$

b) $\sqrt{3}$

c) $2(\sqrt{3} - 1)$

d) $2\sqrt{3}$

_____ **2.** The area of the shaded region is given by:

a) $\int_0^{\pi/2} \sin 3\theta \, d\theta$

b) $\int_0^{\pi/2} 2 \sin^2 3\theta \, d\theta$

c) $\int_0^{\pi/3} \sin 3\theta \, d\theta$

d) $\int_0^{\pi/3} 2 \sin^2 3\theta \, d\theta$

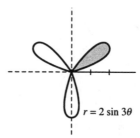

$r = 2 \sin 3\theta$

_____ **3.** The length of the arc $r = e^\theta$ for $0 \leq \theta \leq \pi$ is:

a) $e^\pi - 1$

b) $2(e^\pi - 1)$

c) $\sqrt{2}(e^\pi - 1)$

d) $2\sqrt{2}(e^\pi - 1)$

____ **1.** The best graph of $\frac{y^2}{16} = 1 + \frac{x^2}{25}$ is:

a)

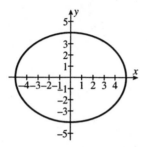

b)

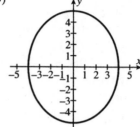

c)

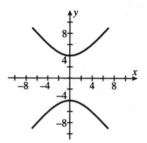

d)

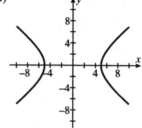

____ **2.** The conic section whose equation is
$y(3 - y) + 4x^2 = 2x(1 + 2x) - y$ is a

a) parabola   b) ellipse   c) hyperbola   d) None of these

____ **1.** The figure at the right shows
one point P on a conic and the
distances of P to the focus and
the directrix of a conic.
What type of conic is it?

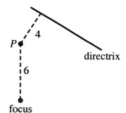

a) parabola
b) ellipse
c) hyperbola
d) not enough information is
provided to answer the
question

____ **2.** The polar equation of the conic with eccentricity 3 and directrix $x = -7$ is:

a) $r = \dfrac{21}{1 + 3 \cos \theta}$ b) $r = \dfrac{21}{1 - 3 \cos \theta}$

c) $r = \dfrac{21}{1 + 3 \sin \theta}$ d) $r = \dfrac{21}{1 - 3 \sin \theta}$

____ **3.** The directrix of the conic given by $r = \dfrac{6}{2 + 10 \sin \theta}$ is:

a) $x = \dfrac{5}{3}$ b) $x = \dfrac{3}{5}$ c) $y = \dfrac{5}{3}$ d) $y = \dfrac{3}{5}$

____ **4.** The polar form for the graph
at the right is:

a) $r = \dfrac{ed}{1 + e \cos \theta}$

b) $r = \dfrac{ed}{1 - 3 \cos \theta}$

c) $r = \dfrac{ed}{1 + e \sin \theta}$

d) $r = \dfrac{ed}{1 - e \sin \theta}$

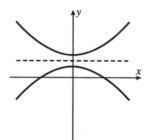

Section 11.1

_____ 1. $\lim\limits_{n\to\infty} \dfrac{n^2 + 3n}{2n^2 + n + 1} =$

     a)   0   b) $\dfrac{1}{2}$ c)   1   d) ∞

_____ 2. Sometimes, Always, or Never:
If $a_n \geq b_n \geq 0$ and $\{b_n\}$ diverges, then $\{a_n\}$ diverges.

_____ 3. Sometimes, Always, or Never:
If $a_n \geq b_n \geq c_n$ and both $\{a_n\}$ and $\{c_n\}$ converge, then $\{b_n\}$ converges.

_____ 4. Sometimes, Always, or Never:
If $\{a_n\}$ is increasing and bounded above, then $\{a_n\}$ converges.

_____ 5. True, False:
$a_n = \dfrac{(-1)^n}{n^2}$ is monotonic.

_____ 6. $\lim\limits_{n\to\infty} \dfrac{\arctan n}{2} =$

     a) $\dfrac{\pi}{4}$ b) $\dfrac{\pi}{2}$ c) π d)   does not exist

_____ 7. The fourth term of $\{a_n\}$ defined by $a_1 = 3$, and $a_{n+1} = \frac{2}{3}a_n$, $n = 1, 2, 3, \ldots$, is:

     a) $\dfrac{16}{81}$ b $\dfrac{16}{27}$ c) $\dfrac{8}{27}$ d) $\dfrac{8}{9}$

Section 11.2

_____ **1.** True, False:

$$\sum_{n=1}^{\infty} a_n \text{ converges if } \lim_{n \to \infty} a_n = 0.$$

_____ **2.** True, False:

If $\sum_{n=1}^{\infty} a_n$ converges and $\sum_{n=1}^{\infty} b_n$ converges, then $\sum_{n=1}^{\infty} (a_n - b_n)$ converges.

_____ **3.** The harmonic series is:

    a) $1 + 2 + 3 + 4 + \ldots$

    b) $1 + \frac{1}{2} + \frac{1}{3} + \frac{1}{4} + \ldots$

    c) $1 + \frac{1}{2} + \frac{1}{4} + \frac{1}{8} + \ldots$

    d) $1 - \frac{1}{2} + \frac{1}{4} - \frac{1}{8} + \ldots$

_____ **4.** True, False:

$$\sum_{n=1}^{\infty} \frac{n}{n+1} \text{ converges.}$$

_____ **5.** $\sum_{n=1}^{\infty} 2(\frac{1}{4})^n$ converges to:

    a) $\frac{9}{4}$ b) 2 c) $\frac{2}{3}$ d) the series diverges

_____ **6.** True, False:

$-3 + 1 - \frac{1}{3} + \frac{1}{9} - \frac{1}{27} + \ldots$ is a geometric series.

Section 11.3

_____ **1.** For what values of p does the series $\sum\limits_{n=1}^{\infty} \dfrac{1}{(n^2)^p}$ converge?

       a) $p > -\dfrac{1}{2}$ b) $p < -\dfrac{1}{2}$ c) $p > \dfrac{1}{2}$ d) $p < \dfrac{1}{2}$

_____ **2.** True, False:

       If $f(x)$ is continuous and decreasing, $f(n) = a_n$ for all $n = 1, 2, 3, \ldots$,
and $\displaystyle\int_{1}^{\infty} f(x)\,dx = M$, then $\sum\limits_{n=1}^{\infty} a_n = M$.

_____ **3.** Does $\sum\limits_{n=1}^{\infty} \dfrac{1}{n^{2/3}}$ converge?

_____ **4.** Does $\sum\limits_{n=1}^{\infty} \dfrac{n + 2}{(n^2 + 4n + 1)^2}$ converge?

_____ **5.** For $s = \sum\limits_{n=1}^{\infty} \dfrac{1}{n^3}$, an upper bound estimate for $s - s_6$ (where s_6 is the sixth partial sum) is:

       a) $\displaystyle\int_{1}^{\infty} x^{-3}\,dx$ b) $\displaystyle\int_{6}^{\infty} x^{-3}\,dx$

       c) $\displaystyle\int_{7}^{\infty} x^{-3}\,dx$ d)   the series does not converge

_____ **1.** Sometimes, Always, or Never:

If $0 \leq a_n \leq b_n$ for all n and $\displaystyle\sum_{n=1}^{\infty} a_n$ diverges, then $\displaystyle\sum_{n=1}^{\infty} b_n$ converges.

_____ **2.** Does $\displaystyle\sum_{n=1}^{\infty} \frac{n+1}{n^3}$ converge?

_____ **3.** Does $\displaystyle\sum_{n=1}^{\infty} \frac{\sqrt{n} + \sqrt[3]{n}}{n^{2/3} + n^{3/2} + 1}$ converge?

_____ **4.** Does $\displaystyle\sum_{n=1}^{\infty} \frac{\cos^2(2^n)}{2^n}$ converge?

_____ **5.** $s = \displaystyle\sum_{n=1}^{\infty} \frac{1}{n \cdot 2^n}$ converges by the Comparison Test, comparing it to $\displaystyle\sum_{n=1}^{\infty} \frac{1}{2^n}$.

Using this information, make an estimate of the difference between s and its third partial sum.

a) $\frac{1}{16}$ b) $\frac{1}{8}$ c) $\frac{1}{32}$ d) 1

Section 11.5

_____ **1.** Does $\displaystyle\sum_{n=1}^{\infty} \frac{(-1)^{n+1} \ln n}{n^2}$ converge?

_____ **2.** Does $\displaystyle\sum_{n=1}^{\infty} \frac{(-1)^n}{\sqrt[4]{n+1}}$ converge?

_____ **3.** For what value of n is the nth partial sum within 0.01 of the value of $\displaystyle\sum_{n=1}^{\infty} \frac{(-1)^n}{2^n}$? (Choose the smallest such n.)

    a) $n = 4$ b) $n = 6$ c) $n = 8$ d) $n = 10$

_____ **4.** True, False:

The Alternating Series Test may be applied to determine the convergence of $\displaystyle\sum_{n=1}^{\infty} \frac{2 + (-1)^n}{2n^2}$.

Section 11.6

_____ **1.** True, False:

If $\sum_{n=1}^{\infty} a_n$ converges absolutely, then it converges conditionally.

_____ **2.** True, False:

If $\lim_{n \to \infty} \left| \frac{a_n}{a_{n+1}} \right| = 3$, then $\sum_{n=1}^{\infty} a_n$ converge absolutely.

_____ **3.** True, False:

Every series must do one of these: converge absolutely, converge conditionally, or diverge.

_____ **4.** The series $\sum_{n=1}^{\infty} 2^{-n} n!$ _____.

    a) diverges

    b) converges absolutely

    c) converges conditionally

    d) converges, but not absolutely and not conditionally

_____ **5.** The series $\sum_{n=1}^{\infty} \frac{(-5)^{n+1}}{n^n}$ _____.

    a) diverges

    b) converges absolutely

    c) converges conditionally

    d) converges, but not absolutely and not conditionally

_____ **6.** True, False:

If a series converges absolutely, then all rearrangements of the terms of the series will also converge.

Section 11.7

_____ **1.** True, False:

$$\sum_{n=2}^{\infty} \frac{1}{(\ln n)^n} \text{ converges.}$$

_____ **2.** True, False:

$$\sum_{n=1}^{\infty} \frac{6}{7n + 8} \text{ converges.}$$

_____ **3.** True, False:

$$\sum_{n=1}^{\infty} \frac{(-1)^n \sqrt{n}}{n + 3} \text{ converges.}$$

_____ **4.** True, False:

$$\sum_{n=1}^{\infty} \frac{e^n}{n!} \text{ converges.}$$

Section 11.8

____ **1.** Sometimes, Always, Never:

The interval of convergence of a power series $\sum_{n=0}^{\infty} a_n(x-c)^n$ is an open interval
$(c - R, \ c + R)$. (When $R = 0$ we mean $\{0\}$ and when $R = \infty$ we mean $(-\infty, \infty)$.)

____ **2.** True, False:

If a number p is in the interval of convergence of $\sum_{n=0}^{\infty} a_n x^n$ then so is the number $\frac{p}{2}$.

____ **3.** For $f(x) = \sum_{n=0}^{\infty} \frac{(x-1)^n}{3^n}$, $f(3) =$

    a) 0   b) 2   c) 3   d) $f(3)$ does not exist

____ **4.** The interval of convergence of $\sum_{n=1}^{\infty} \frac{x^n}{\sqrt{n}}$ is:

    a) $[-1, 1]$ b) $[-1, 1)$ c) $(-1, 1]$ d) $(-1, 1)$

____ **5.** The radius of convergence of $\sum_{n=0}^{\infty} \frac{n(x-5)^n}{3^n}$ is:

    a) $\frac{1}{3}$ b) 1   c) 3   d) ∞

Section 11.9

_____ **1.** Given that $e^x = \sum_{n=0}^{\infty} \frac{x^n}{n!}$, a power series for xe^{x^2} is:

a) $\sum_{n=0}^{\infty} \frac{x^2}{n!}$

b) $\sum_{n=0}^{\infty} \frac{x^{2n}}{n!}$

c) $\sum_{n=0}^{\infty} \frac{x^{2n+1}}{n!}$

d) $\sum_{n=0}^{\infty} \frac{x^{2n}}{(n+1)!}$

_____ **2.** For $f(x) = \sum_{n=0}^{\infty} \frac{x^{2n}}{n!}$, $f'(x) =$

a) $\sum_{n=1}^{\infty} \frac{x^{2n-1}}{n!}$

b) $\sum_{n=1}^{\infty} \frac{2x^{2n-1}}{(n-1)!}$

c) $\sum_{n=1}^{\infty} \frac{2^n x^{2n-1}}{n!}$

d) $\sum_{n=1}^{\infty} \frac{(2n-1)x^{2n-1}}{(n-1)!}$

_____ **3.** Using $\frac{1}{1-x} = \sum_{n=0}^{\infty} x^n$, $\int \frac{x}{1-x^2}\, dx =$

a) $\sum_{n=0}^{\infty} \frac{x^{2n}}{2n}$

b) $\sum_{n=0}^{\infty} \frac{x^{2n+1}}{2n+1}$

c) $\sum_{n=0}^{\infty} \frac{x^{2n+2}}{2n+2}$

d) $\sum_{n=0}^{\infty} x^{2n+1}$

On Your Own

Section 11.10

_____ **1.** Given the Taylor Series $e^x = \sum_{n=0}^{\infty} \frac{x^n}{n!}$ a Taylor series for $e^{x/2}$ is:

a) $\sum_{n=0}^{\infty} \frac{2^n x^n}{n!}$

b) $\sum_{n=0}^{\infty} \frac{2x^n}{n!}$

c) $\sum_{n=0}^{\infty} \frac{x^n}{2^n n!}$

d) $\sum_{n=0}^{\infty} \frac{x^n}{2n!}$

_____ **2.** The nth term in the Taylor series about $x = 1$ for $f(x) = x^{-2}$ is:

a) $(-1)^{n+1}(n+1)!(x-1)^n$

b) $(-1)^n n!(x-1)^n$

c) $(-1)^{n+1}(n+1)(x-1)^n$

d) $(-1)^n n(x-1)^n$

_____ **3.** The Taylor polynomial of degree 3 for $f(x) = x(\ln x - 1)$ about $a = 1$ is $T_3(x) =$

a) $-1 + \frac{(x-1)^2}{2} - \frac{(x-1)^3}{6}$

b) $-2 + x + \frac{(x-1)^2}{2} - \frac{(x-1)^3}{6}$

c) $-1 + x - x^2 + x^3$

d) $-1 - x + x^2 - x^3$

_____ **4.** True, False:

If $T_n(x)$ is the nth Taylor polynomial for $f(x)$ at $x = c$, then $T_n^{(k)}(c) = f^{(k)}(c)$ for $k = 0, 1, ..., n$.

_____ **5.** If we know that $\left| f^{(n+1)}(x) \right| \le M$ for all $|x - a| \le d$, then the absolute value of the nth remainder of the Taylor series for $y = f(x)$ about a is less than or equal to

a) $\frac{M}{n!} |x - a|^n$

b) $\frac{M}{(n+1)!} |x - a|^{n+1}$

c) $\frac{M}{n!} |x - a|^{n+1}$

d) $\frac{M}{(n+1)!} |x - a|^n$

—— **1.** $\dbinom{\frac{1}{2}}{3} =$

a) $\dfrac{5}{16}$ b) $\dfrac{1}{16}$ c) $\dfrac{5}{8}$ d) $-\dfrac{5}{8}$

—— **2.** $\displaystyle\sum_{n=0}^{\infty} \dbinom{\frac{2}{3}}{n} x^n$ is the binomial series for:

a) $(1 + x)^{2/3}$ b) $(1 + x)^{-2/3}$

c) $(1 + x)^{3/2}$ d) $(1 + x)^{-3/2}$

—— **3.** Using a binomial series, the Maclaurin series for $\dfrac{1}{1 + x^2}$ is:

a) $\displaystyle\sum_{n=0}^{\infty} \dbinom{1}{n} x^n$ b) $\displaystyle\sum_{n=0}^{\infty} \dbinom{-1}{n} x^n$

c) $\displaystyle\sum_{n=0}^{\infty} \dbinom{1}{n} x^{2n}$ d) $\displaystyle\sum_{n=0}^{\infty} \dbinom{-1}{n} x^{2n}$

Section 11.12

____ **1.** The quadratic approximation for $f(x) = x^6$ at $a = 1$ is:

    a) $P(x) = 6(x - 1) + 30(x - 1)^2$

    b) $P(x) = 1 + 6(x - 1) + 30(x - 1)^2$

    c) $P(x) = 6(x - 1) + 15(x - 1)^2$

    d) $P(x) = 1 + 6(x - 1) + 15(x - 1)^2$

____ **2.** If the Maclaurin polynomial of degree 2 for $f(x) = e^x$ is used to approximate $e^{0.2}$ then the best estimate for the error with $0 < x < 1$ is:

    a) $\frac{1}{24}e$ b) $\frac{1}{6}e$ c) $\frac{1}{2}e$ d) e

____ **3.** What degree Taylor polynomial about $a = 1$ is needed to approximate $e^{1.05}$ accurate to within 0.0001?

    a) $n = 2$ b) $n = 3$ c) $n = 4$ d) $n = 5$

Answers

Chapter 1

Section 1.1

1. True
2. False
3. B
4. True
5. B
6. True

Section 1.2

1. False
2. False
3. B

Section 1.3

1. C
2. A
3. A
4. Sometimes
5. B
6. D
7. True

Section 1.4

1. B
2. C

Section 1.5

1. B
2. D
3. A
4. B

Section 1.6

1. C
2. D
3. Always
4. D
5. B
6. False
7. A
8. D
9. A

Chapter 2

Section 2.1

1. B
2. B
3. False

Section 2.2

1. B
2. A
3. A
4. D
5. B
6. False

Section 2.3

1. True
2. C
3. Sometimes
4. A
5. B
6. B

Section 2.4

1. True
2. B
3. D

Section 2.5

1. Sometimes
2. C
3. False
4. A
5. False

Section 2.6

1. Sometimes
2. C
3. True
4. D
5. True
6. B
7. Sometimes
8. A
9. C

Section 2.7

1. True
2. False
3. B
4. True
5. C

Section 2.8

1. False
2. True
3. D
4. B
5. B
6. True

Section 2.9

1. No
2. False
3. A

Chapter 3

Section 3.1

1. D
2. C
3. D
4. C
5. B

Section 3.2

1. False
2. D
3. C

Section 3.3

1. False
2. A
3. B
4. A

Section 3.4

1. B
2. A
3. D
4. C
5. D

Section 3.5

1. False
2. C
3. D
4. A
5. True
6. Sometimes
7. C

Section 3.6

1. False
2. B
3. A
4. B
5. D
6. B

Section 3.7

1. True
2. D
3. B
4. B
5. C

On Your Own

Answers

Section 3.8
1. C
2. B
3. D
4. C

Section 3.9
1. False
2. B
3. D
4. C

Section 3.10
1. A
2. C

Section 3.11
1. B
2. B
3. D
4. B

Chapter 4

Section 4.1
1. False
2. True
3. C
4. True
5. B
6. B

Section 4.2
1. C
2. No
3. A
4. False

Section 4.3
1. B
2. False
3. False
4. B
5. False
6. False
7. B

Section 4.4
1. B
2. A
3. B

Section 4.5
1. True
2. True
3. False
4. True
5. False
6. True
7. B

Section 4.6
1. C
2. C

Section 4.7
1. B
2. C

Section 4.8
1. False
2. True
3. B
4. B
5. B

Section 4.9
1. Sometimes
2. A
3. D

Section 4.10
1. False
2. D
3. A
4. False
5. A

Chapter 5

Section 5.1
1. A
2. True
3. C
4. A

Section 5.2
1. Sometimes
2. Always
3. D
4. A
5. C
6. D
7. B
8. True

Section 5.3
1. D
2. C
3. False

Section 5.4
1. B
2. D
3. C
4. D
5. B
6. B

Section 5.5
1. D
2. B
3. A
4. C
5. True
6. D

Section 5.6
1. B
2. D
3. False
4. C
5. C
6. D
7. A
8. C

Chapter 6

Section 6.1
1. B
2. B
3. A

Section 6.2
1. A
2. B
3. A

Section 6.3
1. B
2. D

Section 6.4
1. C
2. D

Section 6.5
1. A
2. D
3. C

Chapter 7

Section 7.1
1. A
2. D
3. C

Section 7.2
1. D
2. B

Section 7.3
1. C
2. C
3. B

On Your Own

Answers

Section 7.4
1. B
2. D
3. A
4. B
5. D

Section 7.5
1. False
2. True
3. False
4. True
5. Sometimes

Section 7.6
1. A
2. C

Section 7.7
1. A
2. B
3. C

Section 7.8
1. False
2. A
3. C
4. D

Chapter 8

Section 8.1
1. C
2. B

Section 8.2
1. A
2. C

Section 8.3
1. B
2. B
3. B
4. True

Section 8.4
1. True
2. B
3. A

Section 8.5
1. False
2. D
3. B

Chapter 9

Section 9.1
1. C
2. True
3. D

Section 9.2
1. No
2. A
3. B

Section 9.3
1. C
2. True
3. D
4. A

Section 9.4
1. A
2. C
3. B

Section 9.5
1. True
2. B

Section 9.6
1. True
2. A
3. D

Section 9.7
1. True
2. A

Chapter 10

Section 10.1
1. C
2. D
3. A

Section 10.2
1. D
2. C
3. B
4. B

Section 10.3
1. A
2. B

Section 10.4
1. D
2. C
3. B
4. B
5. B

Section 10.5
1. A
2. D
3. C

Section 10.6
1. C
2. A

Section 10.7
1. C
2. B
3. D
4. C

Chapter 11

Section 11.1
1. B
2. Always
3. Sometimes
4. Always
5. False
6. A
7. D

Section 11.2
1. False
2. True
3. B
4. False
5. C
6. True

Section 11.3
1. C
2. False
3. No
4. Yes
5. B

Section 11.4
1. Never
2. Yes
3. No
4. Yes
5. C

Section 11.5
1. Yes
2. Yes
3. B
4. False

Section 11.6
1. False
2. True
3. True
4. A
5. B
6. True

Section 11.7
1. True
2. False
3. True
4. True

Answers

Section 11.8

1. Sometimes
2. True
3. C
4. B
5. C

Section 11.9

1. C
2. C
3. C

Section 11.10

1. C
2. C
3. A
4. True
5. B

Section 11.11

1. B
2. A
3. D

Section 11.12

1. D
2. B
3. B